Predictive HR Analytics, Text Mining & Organizational Network Analysis with Excel

By
Cedric Ng Mong Shen

Copyright Cedric Ng Mong Shen 2020

Praise for "Predictive HR Analytics, Text Mining & Organizational Network Analysis with Excel" book

Yet another gem from an author with a successful track record of publishing easy to read, do-it-yourself HR analytic books. The author's choice of Excel to demonstrate HR analytics is shear brilliance. First, who doesn't already use Excel? Second, it makes the topic of statistical modeling and machine learning far less intimidating. The author facilitates this throughout the entire book with detailed examples, step-by-step explanations and screenshots. A plethora of HR metrics and real-life use cases are described and backed up by recent research publications. These metrics are more than enough to get you going with your own analytic, but if you're an HR expert they will likely also inspire new ideas. Many of the use cases are then followed by step-by-step instructions enabling the reader to create their own HR analytic. The author reviews the entire flow for building any HR analytic; from first defining the business problem you're trying to solve, to in the end using both narratives and Excel visualisations to tell an engaging story. No matter how accurate and useful your analytics are, they won't be a success unless you're able to translate them into a truly engaging story. A review of some basic statistics and machine learning approaches is given, along with the scenarios under which each approach makes sense. The author assumes no prior knowledge and explains each in a clear and easy way to follow that even a complete beginner could understand. Apparently, there are an amazing number of free Excel add-ins relevant to HR analytics (for example, "Analysis ToolPak" for regression and correlation, "Solver" to optimize regression coefficients, "NodeXL" for social network graphs, "Azure" for sentiment analysis). The author shows you how to install each add-in using screenshots and step-by-step instructions. The author also shows you how to add/use a VBA macro in Microsoft Word or a free tool online to parse text into words, and then transform your results into a chart in Excel or a Word Cloud in Microsoft Word. Taking it yet another step, he explains how to apply sentiment analysis and then incorporate that into your Word Cloud. To sum up, the author really brings home the incredible richness of HR data and the amazing number of actionable insights that can be pulled out with HR analytics. And all in Excel! (Who knew?!)

Amy Steier, PhD
Principal Data Scientist, United States

A Gold mine book for HR and Finance practitioner! This book will teach you to be more efficient with excel beyond the vlookup and pivot table; you will learn how to use Excel for Data Visualization and how to build a Decision tree and regression.

Badr Ait Ahmed
Business Intelligence Strategy Advisor (HR), Desjardins, Canada

After searching furiously on Amazon for a book on Predictive HR Analytics, I found this book. I must say that it really has helped me understand Predictive HR Analytics well. The book is not just a high-level source on the topic, but provides a sound introduction to Predictive HR analytics, and most importantly provides the best examples of practical statistical Excel models on sentiment analysis, regression, employee attrition & absenteeism, etc. The one single most objective which is lacking in organizations these days is providing a solid story or narrative when reporting analytics to Executives, and with this book, Cedric makes it so much easier, by providing an entire section on storytelling and the importance of communicating analytics in a structured manner. I am pleased to say that after one week, I have run my first Predictive Employee Flight Risk model, using this book, and it worked out perfectly. A big Thank you to the support of the author, Cedric Ng Mong Shen, and I hope to purchase more of his literature over the next coming months.

Ryan Vorster
Talent Analyst, Nedbank, South Africa

I highly recommend this book "Predictive HR Analytics, Text Mining & Organizational Network Analysis with Excel" by Cedric Ng Mong Shen.
- Firstly, it teaches you the statistical methods used in predictive analytics (probability trees, correlation, multiple & logistic regression).
- Secondly, it shows you how to do Analytics using various free tools such as Excel's Statistical Analysis tools, Azure Machine, NodeXL, and Pro word cloud.
- Thirdly, the book is packed full of case studies that will leave you with many ideas of what you can study in your company to improve business results.
- Lastly, this book is written in a very easy to understand style.

Sharon Miller,
CEO, WorkHuman, Ireland

As someone who has been in the HR analytics business for over 10 years, I like it! The text mining and analytics part is most fascinating to me. It is something that I have in mind for a long time but never really got too deep into it. This book is the first ever book that touches the core of it and it gives me a lot of inspiration in this regard. The case studies are also rich and in-depth. The author certainly has tons of experience in this field and the content is useful for anyone who wants to excel in HR. To be honest, if you were new to either HR or analytics (say less than 3 years of experience), you may find the content a bit deep for your knowledge level. I think one gets the most out of this book with 5 or more years of experience because you'd understand in more depth why some of the analytics are needed and why the case studies are set up the way they are.

Rachel Tan,
CEO, Caripros, Canada

The best part about this book is that you don't have to know anything about people analytics to benefit from it. It reminds me of those fun sized guidebooks that you'll always refer to no matter how much you know about a particular subject. Recommended read for sure!

Chinwendu Ukoha
COO, 31 Thirty, United States

The most comprehensive book on Predictive Analytics, presented in a simple manner and I highly recommend to those who are into using analytics to predict behavior to improve employee experience.

Christina Lu,
Senior Vice President, Human Resources, Volvo Group, Singapore

Cedric has done an excellent job authoring this book. It teaches superb techniques through Excel

Yogesh Dhar
Vice President - Head of Workforce Analytics, APAC, RBS, India

A very easy read, this book is an excellent guide for both beginners to data analysis & to those wanting to sharpen their skills. A great blend of theory & practical examples, this book will help readers quickly incorporate data analysis into their organization's decision making, while also improving the ability to communicate the data and it's results to non-technical personnel. I highly recommend this book for anyone looking for a place to start!

James Hughes
Data Scientist, AGL Institute, Connecticut, United States

The book gives a comprehensive summary of the methodical background and best practices in People Analytics. For those companies still using Excel to analyze their data, the hands-on information and formulas might help a lot. Still, more sophisticated analyses (of bigger data sets) should be done in R, Python or similar tools or languages.

Matthias Hurth
Senior Consultant - Data Analytics/ People Analytics, PwC Deutschland, Germany

Whether you are just learning about HR Analytics or leading a mature and well developed team, Predictive HR Analytics, Text Mining & Organizational Network Analysis: with Excel will provide you with a road map to succeed. This book points out and discusses readily available tools and resources needed to accomplish HR tasks as well as step by step instructions in how to use them. You won't just learn about predictive analytics, text mining, and ONA, you will also learn how to apply them to your work and convert problems into success stories.

Edward Sullivan,
Executive Consultant, People Analytics, Employee Science Solutions,
United States

As a Performance & Reward Analytics professional seeking to develop further understanding in Predictive HR analytics, this book was great in explaining statistical methods and concepts with real life case study examples. Would highly recommend this book to anyone who has a keen interest in this field.

Eleanor Ho
Senior Performance & Reward Analyst, HSBC, United Kingdom

The book covers the essentials of what HR professionals need to know about HR analytics in excel. The format is really easy to understand as there is a helpful step by step guide. I also think the topics covered are very relevant to real life examples.

Catherine Liana Chua
HR Consultant, SAP, Canada

This is an exceptional book. Most HR analytics books have a conversation with you about analytics. This book shows you how to accurately measure people analytics. There was an impressive amount of content that I could use everyday. Great author who wrote a book that I would recommend to anyone.

Blake Cromar
Data Scientist, Human Capital Management Institute, United States

After reading this book I would highly recommend it for everybody who is interested in data analytics. Here is why:
1) The book gives an introduction to a broad range of topics – from regression and correlation analysis, through Organization Network Analytics to text mining and sentiment analysis. Everything is limited to the most important concepts – you don't have to read countless pages of theory, just what you need in the practice.
2) You don't have to be an expert in data science nor in Excel. The book leads you step by step in each topic so you can learn by doing. It explains very well how you can use Excel Data Analysis tool pack for regression and correlation analysis. In fact, I haven't met better demonstration of this highly underestimated tool of Excel.
3) For each topic a large list of additional resources is given so you can delve into details if you wish to. In summary, if you are looking for an easy reading but highly useful book that will teach you how you can analyze your data you can try this one.

Kolyu Minevski,
Head of Performance Measurement and Business Analysis Department at
Postbank,
Bulgaria

Comprehensive & simple, if you're looking to get started with HR Analytics with tools you know, either in pilot mode or at scale, the techniques & methods here will give you a great foundation. Practical examples are relatable with a good elevation towards the advanced stuff towards the end. Great for those getting started & for those brushing up.

Mark Hayton,
HR Analytics, Nokia, Finland

Table of Contents

Table of Contents 6
1) Introduction 10
2) About this book 11
3) How Predictive Analytics Adds Business Value 14
4) Introduction to Machine Learning 16
5) Predictive Analytics Tools (Apache Spark, Excel, Minitab, Python, R, SAS, SPSS, SQL, Stata, Tableau) 20
6) Analytics Maturity Model 23
    6.1) Descriptive Analytics versus Prescriptive Analytics 27
    6.2) HR Metrics (Productivity, Labour Relations, Compensation, HR Efficiency, Recruitment, Learning and Development, Retention, Workforce Demographics) 29
7) ARHAT Predictive HR Analytics Framework 39
    7.1) Step 1 – Asks Questions 40
    7.2) Step 2 – Review Literature 42
    7.3) Step 3 – Hypothesis Formulation 44
    7.4) Step 4 – Analyze Data 45
        (i) Data Gathering 46
        (ii) Data Analysis 49
    7.5) Step 5 – Tell the Story 51
        (i) Know the Audience 53
        (ii) Develop the Narrative 54
        (iii) Develop the Visuals 62
8) Starting Your First HR Analytics Project. 80
    8.1) Stakeholder Analysis 81
        (i) Stakeholders you serve 82
        (ii) Stakeholders you need help from 83
        (iii) Stakeholders affected 85
    8.2) Power Interest Matrix 86
    8.3) Action Priority Matrix 89
    8.4) Court Rulings on Analytics Cases 93
        (i) Case law 1: Griggs v. Duke Power Co. (1971) 94
        (ii) Case law 2: Johnson v. Transportation Agency (1987) 94
9) Basic Statistics (Average, Median, Mode, Percentile) 96
10) Decision & Probability Trees 100

- 10.1) Visualize Probability Trees with Excel 102
- 11) Correlation 108
  - 11.1) Compute correlation coefficient with Excel..................................111
  - 11.2) Correlation Matrix using Excel..112
- 12) Regression..118
  - 12.1) Multiple Regression Example 1: Predict the Output when the Inputs are numerical variables...119
  - 12.2) Multiple Regression Example 2: Predict the Output when the Inputs consist of 1 categorical variable with 2 variables.....................126
  - 12.3) Multiple Regression Example 3: Predict the Output when the Inputs consist of 1 categorical variable with 3 variables.....................135
  - 12.4) Logistic Regression: Predict the value of a categorical output...145
  - 12.5) Logistic Regression: Fixing Solver Error Messages...................161
- 13) Text Mining (Text Analytics)...166
  - 13.1) Count frequently mentioned words with VBA in Microsoft Word 168
  - 13.2) Count frequently mentioned words online 179
  - 13.3) Create a "Word Cloud" to visualize frequently mentioned words using "Pro Word Cloud" Microsoft Word Add-In. 180
- 14) Sentiment Analysis 189
  - 14.1) Real-World Impact of Sentiment Analysis 193
  - 14.2) Glassdoor Company Ratings and Reviews.................................194
  - 14.3) Run Sentiment Analysis in Excel with Azure Machine Learning 196
  - 14.4) Correlation Example: Is there a relationship between "Glassdoor Company Ratings" and "Company Attrition Rate"? 207
  - 14.5) Multiple Regression Example: Predict "Company Attrition Rate" with "Glassdoor Company Ratings"...214
- 15) Employee Engagement Analytics 221
  - 15.1) Employee Net Promoter Score...228
  - 15.2) Case 1: Upwork Global Inc. – Impact Of Flexible Work Arrangements On Employee Morale 230
  - 15.3) Case 2: iNostix - Merging engagement survey outcomes with business data 231
  - 15.4) Correlation Example: Analyze Engagement 233
  - 15.5) Multiple Regression Example: Analyze Engagement 241
- 16) Social Media Analytics 249
- 17) Social Network Analysis (SNA)...254
  - 17.1) Organizational Network Analysis (ONA) 258

17.2) Real world examples of ONA 261
17.3) Applications of ONA in HR 264
17.4) Tools to build an ONA Survey 276
17.5) Tools to visualize your ONA survey results 277
17.6) Visualize ONA with Excel NodeXL 278
17.7) ONA Interventions 290
17.8) Correlations Example: Predict employee churn with ONA graph metrics 292
17.9) Logistics Regression Example: Predict employee churn based on their Organizational Network 299
17.10) Multiple Regression Example: Predict Employee's Sales based on their Organizational Network 315

18) Diversity and Inclusion Analytics..323
  18.1) How to Convert Diversity into an Index 327
  18.2) Multiple Regression Example: Predict Ethnic & Gender Diversity's Impact On EBIT 329
  18.3) Multiple Regression Example: Predict an employee's performance rating based on their "Social Network Diversity Index", "Social Network Size" & "Skillsets" 338

19) Predict Employee Attrition & Absenteeism 348
  19.1) Case 1: AIHR – Turnover Predictors..356
  19.2) Case 2: FlightNetwork "At-Risk" Employees Criteria..................359
  19.3) Case 3: Retention Analytics at Neilson Holdings 360
  19.4) Case 4: iNostix – Flight Risk Scoring 363
  19.5) Case 5: IBM Kenexa Talent Insights 364
  19.6) Probability Tree Example: Predict Employee Resignation 366
  19.7) Correlation Example: Predict Employee Flight Risk....................368
  19.8) Logistic Regression Example: Predict Staff Resignation 376

20) Predict Performance 391
  20.1) Case 1: Deloitte – Characteristics of High-Performing Salesperson In Financial Services 404
  20.2) Case 2: HBR –What Makes Great Salespeople 406
  20.3) Correlation Example: Predict how successful a candidate will be if hired..408
  20.4) Multiple Regression Example: Predict an employee's performance rating based on their "Social Network", "Skillsets", & "Personality Traits". 415

- 21) Compensation & Benefits Analytics 422
  - 21.1) Market-Ratio Analytics 426
  - 21.2) Compa-Ratio Analytics 427
  - 21.3) Flight Risk Formula 428
- 22) Training & Development Analytics 429
  - 22.1) Four Levels Of Training Evaluation - The Kirkpatrick Model 431
  - 22.2) Fifth Level of Training Evaluation - ROI by Jack Philips 433
  - 22.3) How to compute impact of Training on Earnings Per Share (EPS) 436
  - 22.4) Case 1: Hilton Worldwide University Business Impact 438
  - 22.5) Case 2: JetBlue University Training Predictive Analytics 440
  - 22.6) Multiple Regression Example: Predict training's Impact on Customer Service 444
- 23) Health, Safety & Environment (HSE) Analytics 453
  - 23.1) Correlation Example: Predict Workplace Accident 459
  - 23.2) Logistic Regression Example: Predict Workplace Accident 465
- 24) Data Visualization with Excel 477
  - 24.1) Clustered Column Chart 478
  - 24.2) Combination Charts 483
  - 24.3) Pie Charts 488
  - 24.4) Scatter Plot 491
- Annex A) Other Publications by the Author 497
- Index 498

1) Introduction

Predictive HR Analytics, Text Mining & Organizational Network Analysis (ONA) are hot topics and powerful techniques to improve organization effectiveness. Best Buy is able to predict that a 0.1% increase in employee engagement results in an increase of $100,000 in the store's annual income! VoloMetrix found that a salesperson's network size within their company is a more important leading indicator of sales, than the time salespeople spend with customers! You don't need to spend months learning R programming & you don't need to buy expensive SPSS statistical software. This is the **only book that teaches you how to use Microsoft Excel for Predictive HR Analytics, Text Mining & Organizational Network Analysis (ONA)** with step-by-step print-screen instructions:

1) Predictive HR Analytics: Use Excel's Statistical Analysis tools (Decision trees, Correlation, Multiple & Logistic Regression) to run Predictive HR Analytics. You will learn how to predict Ethnic & Gender Diversity's impact on EBIT, predict training's impact on sales revenue, predict employee resignation, predict impact of staff engagement on sales, predict workplace accident, etc.

2) Organizational Network Analysis (ONA): Run ONA using Excel's network analysis tool. Learn how to convert an employee's organizational network into a score & then predict if they will be a high-potential (HiPo). You will also learn how to predict employee performance and resignation with ONA graph metrics. E.g. an employee is predicted to be a HiPo with performance rating of "9", if his "Social Network Score" is "16", "Social Network Diversity Index" is "3" & "Competency Score" is "8".

3) Text Mining, Sentiment Analysis & Word Clouds: Mine text from social network posts, employee engagement surveys & Glassdoor comments, then run Sentiment Analysis using Excel & visualize the insights with "Word Clouds". Learn how to predict a company's average employee attrition rate based on its sentiment. E.g. a company's average employee attrition rate is predicted to be 8%, if unemployment rate is 3%, GDP growth is 2%, Glassdoor public sentiment rating is "5", and engagement score is "7".

2) About this book

Other than this **"Predictive HR Analytics, Text Mining & Organizational Network Analysis with Excel"** book. I have written a few other books covering different areas of People Analytics, and using various techniques, from simple tools such as Excel and Word, to more complicated tools such R programming, and Chi-Square.

"People Analytics & Text Mining with R" book.

https://www.amazon.com/dp/B07PXRLL3Z/

The **"People Analytics & Text Mining with R"** book, teaches People Analytics, Social Media Analytics, Text Mining and Sentiment Analysis - **using R programming**. As R is developed specially for statistical analysis, you can run complicated statistical analyses by simply entering a few commands.

"Predictive HR Analytics" book.

https://www.amazon.com/Predictive-HR-Analytics-Mong-Shen/dp/1790406374/

The "**Predictive HR Analytics**" book - teaches Predictive HR Analytics **using Chi-Square and Excel's statistical** tools (Decision trees, Correlation, Multiple Regression, Logistic Regression). This is the only book that **covers the full HR Analytics scope** (Benefits, Compensation, Culture, Diversity & Inclusion, Engagement, Leadership, Learning & Development, Payroll, Personality Traits, Performance Management, Recruitment, Sales Incentives) with numerous real-world examples.

"IISS: Employee Experience & Engagement with Predictive Analytics" book

https://www.amazon.com/dp/B0859CB567/

Best Buy can predict that a 0.1% increase in employee engagement results in $100,000 increase in income. Though Amazon didn't promise a heart-warming employee experience & didn't promise work-life balance, it is among the top organizations in employee satisfaction. This is the only book that incorporates Employee Experience & Engagement with Predictive Analytics, & offers a new way to cultivate engagement using **"4 Engagement Bags"**:

Bag 1) Inspire Engagement Investment: Learn to inspire Engagement Investment with Predictive Analytics, Stories & Data Visualisation Techniques & **predict the impact of Employee Engagement on, Customer Satisfaction, Workplace Accidents <u>with Excel</u>** using predictive analytics techniques such as Correlation & Regression.

Bag 2) Inspire with Engagement Fertilizers: Making employees happy, doesn't mean they'll work hard for you. Learn to use 5 "Engagement Fertilizers" to build great employee experience & engagement.

Bag 3) Sentiment Gathering: with Company-wide & Pulse Surveys, Focus Groups, Glassdoor Reviews & IISS Engagement Diagnosis Questions.

Bag 4) Sentiment Diagnosis & Prescription: with Engagement Dashboards, Bar & Radar Charts, Correlation, Regression & IISS Engagement Prescriptions, **Sentiment Analysis with "<u>Azure Machine Learning</u>"** & Word Clouds.

3) How Predictive Analytics Adds Business Value

With easy-to-use software becoming more prevalent, predictive analytics is no longer confined to the domain of statisticians. Predictive analytics is becoming prevalent for Sales, Customer Service, and Human Resources as well. More and more companies are turning to predictive analytics to, help solve difficult problems, reduce risk, uncover new opportunities, increase their bottom line and competitive advantage. Common uses of predictive analytics include:

- **Detect fraud.**

    Most companies experienced fraud of some kind. Fraud analytics creates a profile of all the areas where fraud is expected to occur and the possible types of fraud in those areas. For example, standard insurance claims can be automatically paid out. But, if a claim is unusual (outlier), or if the claim exhibits the traits of a fraudulent claim, predictive analytics can highlight the claim for action.

- **Improve operations.**

    Predictive analytics helps companies to function more efficiently as companies can spend less time dealing with low-impact, low-risk operational decisions. Inventories can be forecasted and managed using predictive analytics.

- **Reduce risk.**

    Credit scores are used to assess a person's likelihood of default for purchases. When it comes to loans (e.g. credit cards, car loans, housing loans), your credit is your reputation as a borrower. Your credit score tells lenders how likely you are to repay your loans, which then helps lenders decide whether or not to approve your loan request and how much interest to charge you. Insurers check your credit to determine whether or not to cover you, and at what rates. In some jobs, employers check your credit to ascertain how responsible you are based on your financial history as they want to avoid situations where you may be tempted by bribes.

- **Target marketing efforts on customers who are most likely to buy.**

    Predictive analytics can help businesses to attract, retain and grow their most profitable customers. By examining the data about customers, their demographics, and their purchases, businesses know precisely which customers they should market to, and can target those most likely to buy.

- **Identify customers that are likely to abandon a service or product.**

    Predictive analytics can address customer churn before it happens. Predictive analytics can identify customers that may not renew their membership, and send an alert to the customer service staff to offer the customer an incentive to renew his/her membership.

- **Improve operations.**

    Predictive analytics helps companies to function more efficiently as companies can spend less time dealing with low-impact, low-risk operational decisions. Inventories can be forecasted and managed using predictive analytics.

- **Improve customer service with better planning.**

    Hotels use predictive analytics to predict the number of guests, so they have enough staff and resources to handle demand, and so they can maximize occupancy and increase revenue.

4) Introduction to Machine Learning

Machine learning is becoming increasingly prevalent. Understanding the basics of machine learning will help us to navigate this world, demystifying a lofty concept and allowing us to better understand the technology that we use.

Arthur Samuel is widely deemed as the first person to coin the word "machine learning". In Samuel's journal "Some Studies in Machine Learning Using the Game of Checkers", it studied the use of machine learning in the game of checkers "to verify...that a computer can be programmed so that it will learn to play a better game of checkers than can be played by the person who wrote the program". [1]

Machine learning is the study of teaching a computer program or algorithm how to progressively improve upon a given set task. It studies how to build applications that demonstrate this iterative improvement. There are three main categories of Machine learning: Supervised Learning, Unsupervised Learning, and Reinforcement Learning. [2]

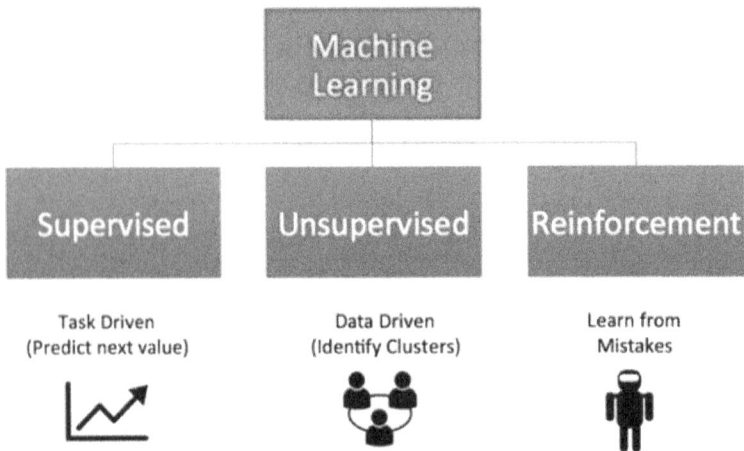

Source: Source: Brent Dykes (2016) Data Storytelling: The Essential Data Science Skill Everyone Needs. https://www.forbes.com/sites/brentdykes/2016/03/31/data-storytelling-the-essential-data-science-skill-everyone-needs/#51c5d24752ad (25 November 2018)

Supervised Learning

Supervised learning works by feeding the machine sample data with various input data, and output data. As both the input and output data are known, the dataset is considered "labeled". The algorithm then deciphers patterns that exist between the input and output data and uses this knowledge to create a model (an algorithmic equation) to make further predictions. Examples of supervised learning algorithms are regression analysis, decision trees, k-nearest neighbours, neural networks, and support vector machine. [3]

Supervised learning common applications include:
- **Predict the market price**: Supervised learning can be used to predict the market price of a used car by studying the relationship between car features (e.g. car brand, year of make, mileage) and the selling price of other cars sold based on historical data. [3]
- **Face Recognition**: In Facebook, your face was probably used in a supervised learning algorithm that is trained to recognize your face. Having a system that takes a photo, finds faces, and guesses who that is in the photo (suggesting a tag) is a supervised process. [2]
- **Email Spam Classification**: Email spam filter is a supervised learning system. Given email examples and labels (spam/not spam), it learns how to filter out malicious emails. [2]

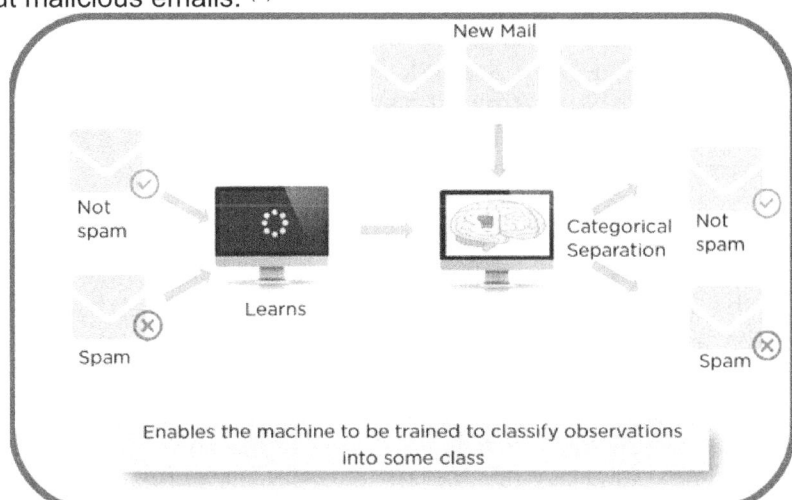

Source: Hunter Heidenreich (2018) What are the types of machine learning?, https://towardsdatascience.com/what-are-the-types-of-machine-learning-e2b9e5d1756f (6 March 2019)

Unsupervised Learning

In unsupervised learning, the output variables are unlabeled (unknown), and combinations of input and output variables are thus unknown. Unsupervised learning focuses on analyzing relationships between input variables and uncovering hidden patterns that can be extracted to create new labels regarding possible outputs. (3)

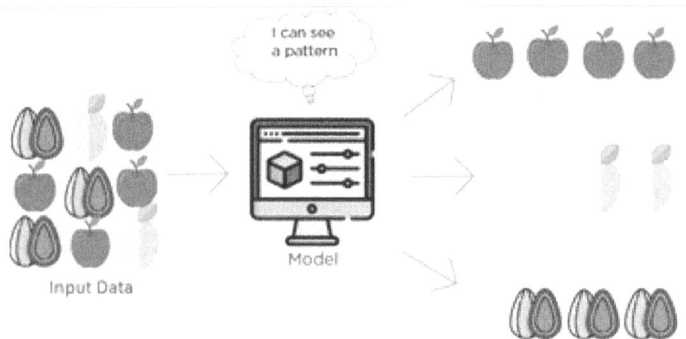

Source: Hunter Heidenreich (2018) What are the types of machine learning?, https://towardsdatascience.com/what-are-the-types-of-machine-learning-e2b9e5d1756f (6 March 2019)

The benefit of unsupervised learning is that it allows you to discover patterns in the data that you were unaware existed. Unsupervised learning is especially useful in fraud detection where some attacks are yet to be classified. Examples of unsupervised learning algorithms are associated analysis, social network analysis, and descending dimension algorithms. (3)

Unsupervised learning common applications include:
- **Recommender Systems**: YouTube probably used a video recommendation system, in the unsupervised domain. Analyzing users that have watched similar videos as you, and then enjoyed other videos that you have yet to see, a recommender system can see this relationship in the data and prompt you with such a suggestion. (2)
- **Buying Habits**: Customer buying habits can be used in unsupervised learning algorithms to group customers into similar purchasing segments. This helps companies market to these grouped segments. (2)

Reinforcement

Reinforcement learning, learns from mistakes. Put a reinforcement learning algorithm into any environment and it will make a lot of mistakes in the beginning. But, over time, its learning algorithm learns to make less mistakes than it used to. As long as the algorithm is given some signals that associates good behaviors with a positive signal and bad behaviors with a negative one, we can reinforce its algorithm to prefer good behaviors over bad ones. (2)

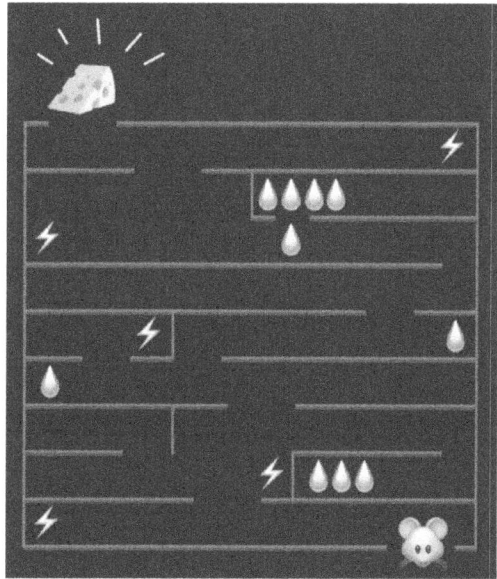

Source: Hunter Heidenreich (2018) What are the types of machine learning?, https://towardsdatascience.com/what-are-the-types-of-machine-learning-e2b9e5d1756f (6 March 2019)

References:
(1) Arthur Samuel, "Some Studies in Machine Learning Using the Game of Checkers," IBM Journal of Research and Development, Vol. 3, Issue. 3. 1959.
(2) Hunter Heidenreich (2018) What are the types of machine learning?, https://towardsdatascience.com/what-are-the-types-of-machine-learning-e2b9e5d1756f (6 March 2019)
(3) Oliver Theorbald (2017), Machine Learning for Absolute Beginners - Second Edition , Independently published

5) Predictive Analytics Tools (Apache Spark, Excel, Minitab, Python, R, SAS, SPSS, SQL, Stata, Tableau)

Adopting HR analytics is a big step for many people and organizations. Here's an overview of some of the most used Predictive HR analytics tools:

Apache Spark

Apache Spark is an open-source distributed general-purpose cluster computing framework with (mostly) in-memory data processing engine that can do ETL, analytics, machine learning and graph processing on large volumes of data at rest (batch processing) or in motion (streaming processing) with rich concise high-level APIs for the programming languages: Scala, Python, Java, R, and SQL. [1]

Excel

Excel has several advantages over specialized statistics software such as SPSS, Minitab, SAS, R, and Stata. The best thing about Excel is that you already have it, so the software is free, and you already know how to use Excel, so it will be the easiest Predictive HR analytics tool for you to learn and use. Furthermore, you don't have to get the IT department involved with your Excel work. If you buy an analytics program you would have to spend a lot of time to learn about it.

Minitab

Minitab is a statistical software, similar to Excel interface. It is quite easy to use because it has drop down menus. Similar to SPSS, Minitab allows users to do most of the analysis without having to understand coding. However, it can be difficult to transfer output from Minitab to other formats to include in reports.

Python

Python is an interpreted, high-level, general-purpose programming language. [2] Python is free and can be download here: https://www.python.org/

R

R is powerful, free software for statistical analysis. With R, you can carry out complicated statistical analyses by simply entering a few commands. R is good for more complicated (machine learning) algorithms. It requires some learning time as it is command-line syntax driven, with limited menu-driven user interface. R enables you to work with much larger datasets compared to Excel. R is free and can be download here: https://cran.r-project.org/

SAS

SAS has a steeper learning curve than SPSS. SAS is basically a command-line-driven software with few menu-driven procedures. It can be difficult to transfer output from SAS to other formats to include in reports.

SPSS

SPSS is another commonly used HR analytics tool. Thanks to its user-friendly graphical interface and point-and-click menus you can use it to analyze data without having extensive statistical knowledge. Once you loaded data into SPSS, the good thing about it, is that it is easy to manipulate that data. SPSS shares many similarities with Excel that makes it easier to work with. Thus, SPSS an easy stepping stone for companies with less mature analytical capabilities.

SQL

Structured Query Language (SQL) is a standard computer language for relational database management and data manipulation. Using SQL, you can query (request information from databases), update and reorganize data, as well as create and modify the schema (structure) of a database system, and control access to the data. SQL is free and can be download from: [3]
https://www.freesqldatabase.com/
https://www.mysql.com/products/workbench/

Stata

Though Stata have some menus, it is basically command line driven and thus takes time to master. It can be difficult to transfer output from Stata to other formats to include in reports.

Tableau

Tableau has an intuitive interface, which helps you to complete sophisticated analytics projects quickly. It has modeling capabilities such as trending and forecasting. You can add a trend line or forecast data with any chart, and see information describing the fit via a simple right-click. [4]

References:
(1) jaceklaskowski (2019) Apache Spark, https://jaceklaskowski.gitbooks.io/mastering-apache-spark/spark-overview.html (6 March 2019)
(2) Wikipedia (2019) Python (programming language), https://en.wikipedia.org/wiki/Python_%28programming_language%29 (6 March 2019)
(3) Rhevanth Mahadev (2018) What is the SQL programming language?, https://www.quora.com/What-is-the-SQL-programming-language (6 March 2019)
(4) Ian Coe (2015) 6 Ways Tableau Can Help with Your Advanced Analytics Projects, https://www.tableau.com/about/blog/2015/8/6-ways-tableau-can-help-advanced-analytics-projects-42296 (6 March 2019)

6) Analytics Maturity Model

Human Resources Professionals have been talking about HR analytics for years, but 2016 saw the biggest leaps toward true HR analytics capabilities. Organizations are progressing from the most fundamental HR data analysis stages of "descriptive and diagnostic analytics" to the more complicated "predictive and prescriptive analytics". According to the "Global Human Capital Trends 2016" report by Deloitte, the percentage of companies that believe they are capable of developing predictive models doubled from four percent in year 2015 to eight percent in year 2016.

Source: Taras Kaduk (2016), 4 Stages Of Data Analytics Maturity: Challenging Gartner's Model. https://medium.com/taras-kaduk/4-stages-of-data-analytics-maturity-challenging-gartners-model-590eb5ebe6d1 (11 November 2018)

In the Analytics Maturity Model, there are four levels of data analytics used across all industries, and the increasing value each step provides to your organization:
- Level 1: Descriptive Analytics
- Level 2: Diagnostic Analytics
- Level 3: Predictive Analytics
- Level 4: Prescriptive Analytics

Fours Levels of Gartner's Analytics Maturity Model

Level 1 Descriptive Analytics	Level 2 Diagnostic Analytics	Level 3 Predictive Analytics	Level 4 Prescriptive Analytics
100 employees resigned	100 employees resigned because they are not given overtime pay	Giving employees overtime pay will improve retention by x%	At least 2 hours of overtime will improve retention

Descriptive and diagnostic analytics are commonly used, while predictive and prescriptive analytics is less common. Progressing from descriptive analytics towards diagnostic, predictive and prescriptive analytics requires much more technical ability (higher difficulty), but also uncovers more insights (value).

Level 1: Descriptive Analytics

The first level of data analytics is descriptive analytics. Descriptive analytics is at the foundation and most commonly used of data analytics. Descriptive analytics answers "what happened" by describing or summarizing raw data usually in the form of dashboards. Most the basic statistics (sums, averages, count) that we use fall into this category. Descriptive statistics are useful to show things like, Key Performance Indicators (KPI's), dashboards, profit per region, how many employees joined/left, total training hours.

HRIS is typically used to obtain descriptive insights about employees as it reports what happened and what is currently happening. HRIS provides information about your employee demographics, characteristics, costs and performance. However, HRIS systems do not have the capability of discovering patterns and the correlations in these patterns.

Level 2: Diagnostic Analysis

The second level of data analytics is diagnostic analysis. After asking "what happened" you may dive deeper and ask "why did it happen"? Diagnostic analysis uses the data that was summarized in descriptive analytics and analyses further to find the cause of that result. It answers the questions raised in Descriptive Analytics. Examples include: why did sales go down, why are shipments slow.

Level 3: Predictive Analysis

The second level of data analytics is Predictive analytics. Predictive analytics tries to answer the question "what is likely to happen". It uses HRIS to gain descriptive insights about employees, and then use Predictive HR Analytics to build predictive models. Predictive analysis uses the insight from Diagnostic analytics to predict of the results of various variables with statistical modeling. Uses of predictive analytics include: predicting employee churn, predicting sales, etc. Predictive analytics answers questions such as: Which employees are most likely to leave in the next six months? Which employees are most likely to perform?

Level 4: Prescriptive Analysis

The fourth level of data analytics is Prescriptive analytics. Prescriptive analytics prescribe what action to take to remove a future problem or take advantage of a promising trend. It combines the insight from all previous analyses to ascertain the course of action to take. Whilst predictive analytics forecasts for what might happen, prescriptive analytics recommends one or more courses of action along with its predicted future scenarios, allowing companies to assess several possible outcomes based upon their actions. Examples of prescriptive analytics include: which customer segment shall we target next year to improve profitability.

6.1) Descriptive Analytics versus Prescriptive Analytics

The easiest way to understand the differences between descriptive and predictive HR analytics is to look at the answers they generate. Descriptive analytics answer questions about the demographics and characteristics of your employees. Whereas, predictive HR analytics go beyond the descriptions generated by descriptive HR analytics by providing predictive answers and prescribing specific actions or recommendations.

For most descriptive HR analytics, predictive HR analytics questions can be developed as you can see in the list below.

Descriptive vs Predictive HR Analytics

Descriptive Analytics	Predictive Analytics
How long has John been with the company	Hong long will john stay with the company
What is John's marital status, salary market-ratio, performance rating, traveling time to office?	What variables affect John's tenure with the company?
Who are our best performing call center employee?	Which of our call center employees are likely to be best performers?
How many hours of training did our Sales employees clock last year	What is the impact of the training programme on Sales?
How many Sales employees leave last quarter?	Which Sales employee is likely to leave next quarter?
What is the engagement score?	What impact does engagement scores have on accident rates?

According to Dr. Jac Fitz-Enc, these predictive Human Capital Measures are valid, based on his experience working with different organizations [1]:

- **Professional and Managerial Ratio**: The higher the number of Professional and Managerial Ratio as a percentage of the total full-time equivalent employees, the higher chance for future growth and profitability, subject to the nature of the business and/or industry. For example, a company with 2000 employees, and 1000 Professionals and Managers would have a Professional and Managerial Ratio of 50% (1000/2000= 0.50).
- **Readiness Ratio**: Readiness Ratio is the percentage of key jobs with at least one qualified person ready to take over, as it means the company is resilient against disruption. The closer is Readiness Ratio is to "1", the better. If a company with 100 key jobs has 30 qualified employees ready to take over if those key jobs are vacated, their Readiness Ratio would be 0.30 (30/100 = 0.3). In this example, the company has a big gap in their succession plan for key positions. This may result in predicted higher cost and slower growth while these positions are vacant during the recruitment period.
- **Accession Ratio**: Accession Ratio is the ratio of new and replacement hires as the percentage of total employment. A company with 2000 employees, and 1000 new and replacement hires, would have an Accession Ratio of 0.5 (1000/2000= 0.50). A high Accession Ratio usually means delays which will lower engagement, increase cost and reduce productivity.
- **Climate Culture Rating**: This refers to the percentage of employees giving top ratings in engagement surveys, for "the organization is a good place to work". This rating is predictive of the employee retention and turnover rates as employees often quit their jobs due to poor working culture. SuveyGizmo (https://www.surveygizmo.com/) or SurveyMonkey (https://www.surveymonkey.com/) can be used to create and administer free employee surveys.
- **Depletion Ratio**: Depletion Ratio is the annual percentage of lost top talents. An company with 1000 employees, that lost 100 of their top talents in the prior year would have a depletion ratio of 0.10 (100/1000=0.10). The higher the depletion ratio, the worse your predicted company profitability as the company loses their best leaders and innovators.

References:
(1) Jac FITZ-ENZ (2010), The New HR Analytics: Predicting the Economic Value of Your Company's Human Capital Investments, AMACOM

6.2) HR Metrics (Productivity, Labour Relations, Compensation, HR Efficiency, Recruitment, Learning and Development, Retention, Workforce Demographics)

There are numerous types and perspectives of metrics that can be used to measure impact of HR strategies. The HR Metrics Service, Canada's leading HR benchmarking service, has produced a HR Metrics Standards and Glossary document to allow HR practitioners to easily review a common set of HR metrics and choose those that best fit their organization (refer to Section on Resources). The metrics have been classified into eight categories for ease of reference.: [1]

- Productivity
- Labour Relations
- Compensation
- HR Efficiency
- Recruitment
- Learning and Development
- Retention
- Workforce Demographics

(i) Productivity Metrics

Metric Name	Formula	Metric Description
Revenue per FTE	Revenue / Permanent FTE	The number of dollars of revenue generated per permanent FTE. This metric can be used to ascertain if a growing organization should hire. Or warn when staff is growing at a faster rate than revenue.
Profit per FTE	(Revenue- Operating Cost) / Permanent FTE	The number of dollars of profit generated per permanent FTE. It indicates if an organization has the right people, in the right place, at the right price.
Absenteeism Rate	Sick Days / Permanent FTE	The number of work days missed due to illness per permanent FTE. An increase in absenteeism may also indicate a disengaged workforce and can be a leading indicator of future employee turnover.
Overtime per Individual Contributor Headcount	Overtime Hours / Permanent Individual Contributor Headcount	The average number of overtime hours worked by each Permanent Individual Contributor. When overtime is excessive, absence rates and turnover may increase. Consistently high levels of overtime can be fixed easily by hiring additional employees.

(ii) Compensation Metrics

Metric Name	Formula	Metric Description
Labour Cost per FTE	Labour Cost / Permanent FTE	The average labour cost to the organization for each permanent FTE. If your Labour Cost per FTE is too low, it may a cause of your turnover.
Labour Cost Expense Percent	Labour Cost / Operating Cost	The total labour costs as a percentage of total expenses. Labour cost can be a significant operating expense, in an organization that relies on human capital to generate value. Review this with Revenue per FTE to ensure that revenues are not being sacrificed through Labour Costs reduction.
Benefits as Percentage of Labour Costs	Benefits / Labour Costs	The total cost of benefits as a percentage of the total Labour Costs. If your benefits percentage is low compared to the benchmark, your organization may experience employee higher absenteeism and sick leave.

(iii) Recruitment Metrics

Talent Acquisition is an area where predictive analytics can be used to help recruiters become more effective by focusing on the factors that matter.

Metric Name	Formula	Metric Description
Vacancy Rate	Permanent Vacant Positions / Permanent Headcount	Permanent positions being actively recruited for at the end of the reporting period as a percentage of permanent headcount. A high percentage of open positions can indicate high attrition. The organization may have a problem attracting good candidates; review the competitiveness of the compensation package or organization reputation.
1st Year Resignation Rate	Permanent Resignations within 0-1 Year of Service / Permanent Headcount within 0-1 Year of Service	Percentage of employees with less than one year of service who resigned. This metric is one indicator of quality of hire. If this resignation rate is higher for the first year of employment than for employees with a greater length of service, it indicates a larger issue. This metric indicates cultural misfit, mismatch between job description and actual job, overselling of the job by the recruiter, poor on-boarding.
Diversity Hire Ratio	New Hires who are Aborigine, Disabled, or Minority / New Hires	The percentage of new hires who are Aborigine, Disabled, or Minority / New Hires.
Female Hire Ratio	The percentage of new hires who are female.	New Hires who are Female / New Hires
Quality-of-source	Quality-of-source = number of candidates hired / number of candidates from source	See how many qualified applicants you receive from each source-of-hire (e.g. Monster.com, Jobstreet.com, Jobsdb.com), to find out if you should drop some of your sources.
Success Ratio	Success Ratio = number of hires rated at least "meets-expectations"/ total number of candidates hired	Success Ratio measures the quality of candidates.

Success ratio for each source	Success ratio for each source = number of hires rated at least "meets-expectations" from a source / total number of candidates hired from a source	Success ratio for each source measures the quality of candidates from each source.
Time-to-Fill	Time-to-Fill = (Days between the publishing a job vacancy and getting an offer accepted) / Permanent External Recruits	Time-to-fill is the number of days between the publishing a job vacancy and getting an offer accepted. Underestimating Time-to-fill can throw off a company's growth plans. A short time-to-fill a job means less overtime and better morale. The average time-to-fill worldwide is 47 days. [2] If your time-to-fill too high compared to industry benchmarks, breakdown the time-to-fill analytic by department, and find out which are jobs takes the longest to fill.
Time-to-start	Time-to-start = (Days between the initial approval or posting of a job vacancy and the actual day when the new hire starts work) / Permanent External Recruits	Time-to-start measures the days between the initial approval or posting of a job vacancy and the actual day when the new hire starts work. In some companies, it is important to have both time-to-fill and time-to-start analytics, because new employees for certain jobs need to undergo several weeks of training before they can start performing the job. Due to the costs of leaving certain jobs vacant, some companies intentionally pre-hired, pre-trained, and over-staff these jobs to reduce time-to-start. However, this strategy only makes sense if the cost of carrying extra headcount is less than the value gained.

Revenue loss for a vacant Sales role	Revenue loss for a vacant Sales role = (Annual Sales Quota/ 365 days) x 47 average time-to-fill	Having an open vacancy in your sales department for even one-day has a negative impact on your organization. Note that the negative impacts of a vacant Sales headcount go beyond top-line revenue and includes drop in morale and increased employee burnout due to the understaffed sales-force. If you're need to fill a vacant Sales role with an annual quota of $1,000,000, you will be losing $128,767 over the 47 time-to-fill average.
Offer acceptance rate	Offer acceptance rate = Number of candidates offers accepted / Number of candidates offers	The offer acceptance rate compares the number of candidates offers accepted, with number of candidates offers. A low offer acceptance rates may indicate compensation problems. To improve offer acceptance rates, ask your candidates for their expected package at the beginning of the recruitment process (i.e. when their fill-up the job application form). The formula for offer acceptance rate is,
Employee referral rates	Employee referral rates = Number of your new hires that are recommended by your current employees/ total number of new hires	Employee referral rates metrics refers to the number of your hires that are recommended by your current employees. Referral rates lets you know how many of your new hires were referred by your employees. Example, your employee referral rates might make up 30 percent of all hires. Employee referrals is an important source-of-hire because it often leads to better quality candidates. Your employees are unlikely to refer someone they know, if they feel that the person would not fit the job and culture, as it would affect their reputation and not reflect well on their judgment. You may need to review your Employee incentive program if the employee referral rate is too low. Maybe your employees prefer additional off-days instead of cash rewards?

Candidate experience rating.	To analyze the candidate experience survey is to use the Net Promotor Score (NPS) method: [3] • "Promoters", are candidates who rated 9 or 10 in the candidate experience survey. You can ask these candidates: What did you like most during applying to the company X? • "Passives", are candidates who rated 7 or 8 in the candidate experience survey. You can ask these candidates: In your opinion, what should we change first in our recruitment process? • "Detractors", are candidates who rated answer between 0 and 6 in the candidate experience survey. You can ask these candidates: What are the areas in our recruitment process that in your opinion need to be changed?	Candidate experience measures job applicant's experience of a company's recruitment and onboarding process, and is measured using a candidate experience survey. Candidate experience is important, as increasingly, candidates share their positive and negative experience with others through word-of-mouth, and through websites like Glassdoor. According to Career Arc, about 60 percent of candidates have had a poor recruitment and onboarding experience, and 72 percent of them shared the experience either online or with someone directly. [2]

(iv) Retention Metrics

Metric Name	Formula	Metric Description
Voluntary Turnover Rate	Permanent Employee Resignations / Permanent Headcount	Permanent employees who left the organization voluntarily as a percentage of permanent headcount. An increase in voluntary turnover can point to a lack of competitiveness in salary, a drop in leadership credibility, poor retention practices or an improved job market.
Total Top Quartile Performer Resignation Rate	Permanent Resignations in the Top Quartile of Performance / Permanent Headcount	Permanent employees who resigned and were within the top quartile of performers, as a percentage of permanent headcount.
Succession Planning Rate	Executive Candidates / Permanent Executive Level Headcount	The percentage of permanent executive roles for which there is a succession candidate.

(v) Labour Relations Metrics

Metric Name	Formula	Metric Description
% of Grievances Closed	Number of Permanent Employee Grievances Closed / Number of Open Permanent Employee Grievances	Permanent employee grievances closed as a percentage of open permanent employee grievances.
Arbitrated Grievances as a % of Grievances Open	Permanent employee grievances gone to arbitration as a percentage of permanent employee grievances open.	Number of Permanent Employee Grievances Gone to Arbitration / Number of Open Permanent Employee Grievances

(vi) Workforce Efficiency Metrics

Metric Name	Formula	Metric Description
HR Headcount Ratio	Permanent Headcount / Permanent HR Headcount	The number of permanent employees per individual permanent HR employee.
HR Costs per Employee	HR Costs / Permanent Headcount	The HR cost for each permanent employee.

(vii) Learning & Development Metrics

Metric Name	Formula	Metric Description
Learning & Development Investment per FTE	Learning & Development Cost / Permanent FTE	The number of dollars invested in learning and development per permanent FTE.
Learning & Development Cost Payroll Percentage	Learning & Development Cost / Labour Cost	The learning and development cost as a percentage of labour cost.
Learning & Development Hours per FTE	Learning & Development Hours / Permanent FTE	The numbers of hours spent on learning per permanent FTE.

(viii) Workforce Demographics Metrics

Metric Name	Formula	Metric Description
Promotion Rate	Permanent Promotions / Permanent Headcount	Permanent employees promoted as a percentage of permanent headcount.
Churnover	Permanent (Promotions + Demotions + Transfers) / Permanent Headcount	Internal movement by permanent employees as percentage of permanent headcount.
Female Percent	Permanent Female Employees / Permanent Headcount	Permanent employees who are female as a percentage of permanent headcount.
Diversity Percentage (Aborigine, Disabled, or Minority)	Permanent Employees who are Aborigine, Disabled, or Minority / Permanent Headcount	Permanent Employees who are Aborigine, Disabled, or Minority as a percentage of permanent headcount.
Management Span of Control	Permanent Headcount / Permanent (Management + Executive Level Headcount)	Average number of permanent employees per permanent manager level employee / executive level employee.
Average Length of Service	Permanent Employee Total Length of Service / Permanent Headcount	Average length of permanent employee service.

References:
(1) HR Metrics Service (2019) HR Metrics Service - Standards & Glossary. https://www.hrmetricsservice.org/wp-content/uploads/2013/07/HR-Metrics-Service-Standards-and-Glossary-v9.6.pdf (19 June 2019)
(2) Daniel Howden (2018) What is time to fill? KPIs for recruiters. Workable. https://resources.workable.com/blog/recruiting-kpis (18 November 2018)
(3) CareerArc (2016) 23 Surprising Stats on Candidate Experience – Infographic https://www.careerarc.com/blog/2016/06/14/candidate-experience-study-infographic/ (17 November 2018)
(4) Erik van Vulpen (2016) 17 Recruiting Metrics You Should Know About. AIHR. https://www.analyticsinhr.com/blog/recruiting-metrics/ (17 November 2018)

7) ARHAT Predictive HR Analytics Framework

"If you can't explain it simply, you don't understand it well enough" – Albert Einstein.

Similarly, until you can put a structure to complex and ambiguous problems and explain it simply, you don't understand predictive HR analytics well enough.

Many people struggle with analytics because they do not have a structured framework for unstructured problems.

This section explains how to use the structured five-steps ARHAT Predictive HR Analytics Framework:

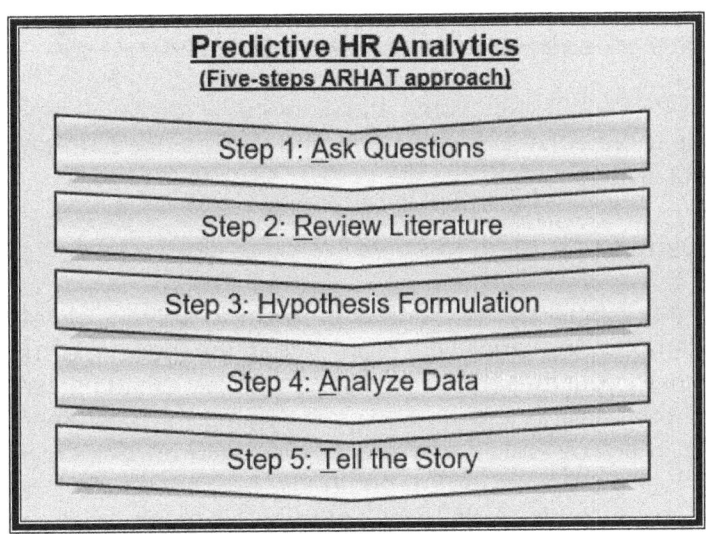

7.1) Step 1 – Asks Questions

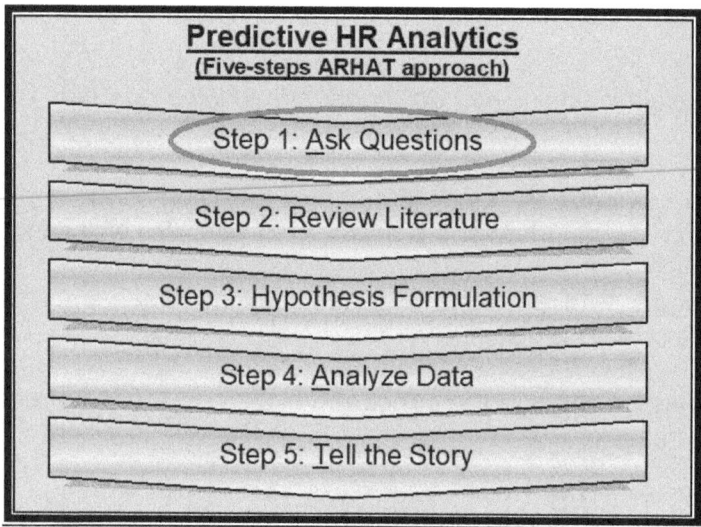

Network around with various functions and find out if there is any business problem that needs to be solved. A well-defined business question ensures that the analytics project is successful.

- **What is the business problem to solve?**

For every project, there must be a business problem to solve. This step comes first to avoid undertaking an analytics project that the business heads don't support. Stay close to your business problems such as customer satisfaction, quality, sales and profitability. If your business strategy is customer intimacy, focus on customer analytics first. If you are dealing with an employee turnover problem, ask how is it a problem for the company, and how does it prevent the business unit from achieving its KPIs? Don't start investigating attrition if it is not a real problem for your business.

What are your project sponsor's expectations?

Project sponsors are people who have a vested interest in the HR analytics project. Set expectations with your project sponsors before taking on the analytics project. Setting expectations includes defining the project scope, timeframe, and budget with the project sponsor. Project sponsors are responsible for approving the project, securing the resources needed, rallying support, removing roadblocks, getting you access to the necessary data for analysis, and leveraging their relationships and position to help the team.

Considerable resources (time, budget, etc.) might be required to run the project, so it is essential to get stakeholder agreement and clarity on the problem being investigated, before proceeding.

7.2) Step 2 – Review Literature

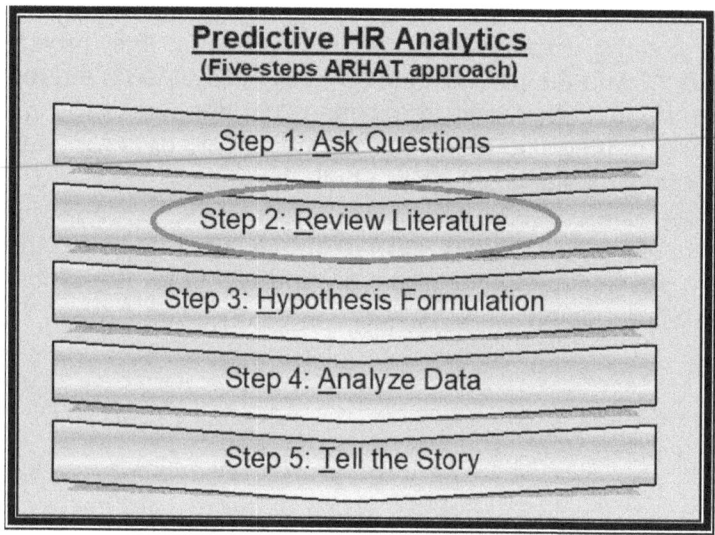

Don't reinvent the wheel. Literature review is about researching what knowledge other people have already discovered on a subject. i.e. it is a prelude to further research or hypothesis. Read books or search the internet for analytics literature to find out how other companies solve the problem, what is their hypothesis, and what statistical techniques they used. There should already be a wealth of findings on the causes of various problems. A good literature review will cover a wide range of material with as many different perspectives as possible.

- **Books & Bibliography**

A good place to start your literature review is by checking out the bibliography. Pick a few good Predictive HR Analytics books or articles and then look through the book's bibliographies to get more information.

- **Internet**

Another way is to search for articles via internet search engines. You may need to keep trying different keywords in search engines before you can find the relevant information that you need. Keywords that you can try for Predictive HR Analytics include: People Analytics, Human Capital Analytics, Human Resources Analytics, etc.

- **Academic journals**

Get ideas from the latest academic journals on predictive HR analytics to develop your hypothesis.

Remember, this is not a one size fits all situation, and all answers will vary depending on the business. What works in other companies, may not work in your company. You need to test other company's hypothesis in your company.

7.3) Step 3 – Hypothesis Formulation

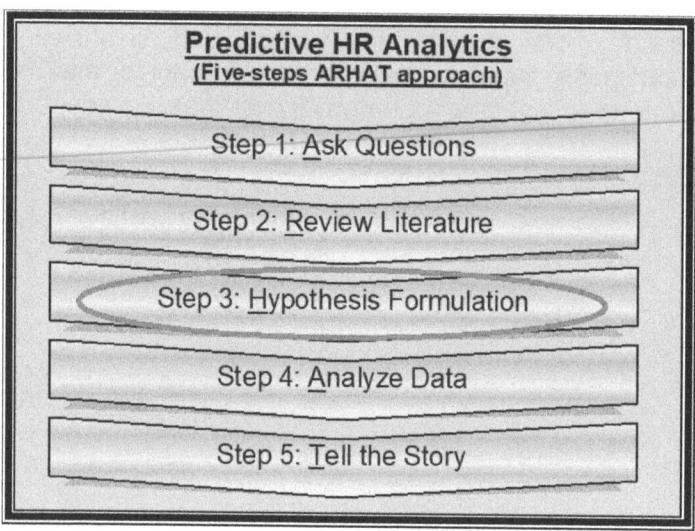

Formulate your hypothesis on the variables that you thinking might affect the outcome. You can formulate your hypothesis based on feedback from the business about the observed relationships. Discuss and share your hypothesis with your sponsor to get their agreement and support.

A good hypothesis is clear and simple. Your hypothesis should testable, and should look something like this:

If {X is done}
➔ {Y} will happen
Or
If {A and B is done}
➔ {Y} will happen

If your company has a higher than industry employee attrition rate, you can formulate your hypothesis as:

If we increase salaries by x%
➔ employee attrition will decrease by y%

7.4) Step 4 – Analyze Data

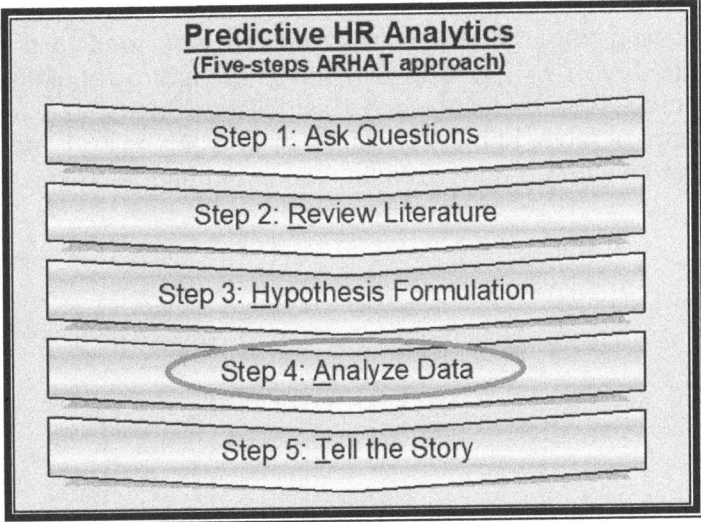

The Analyze Data step involves where and how to get your data, assessing if the data quality is good to use, and testing your hypothesis. Analyze Data has two sub-steps:

(i) Data Gathering

(ii) Data Analysis

(i) Data Gathering

Before you can do data analysis, you need to decide what data to collect, whether to use existing data, or collect new data. If you need to collect data currently not captured, you will need to explore data-capture requirements and the systems and processes that should be established to capture such data. Data refers to records of facts, numerical values or text belonging to an event, or observation that can be stored and processed into information and knowledge.

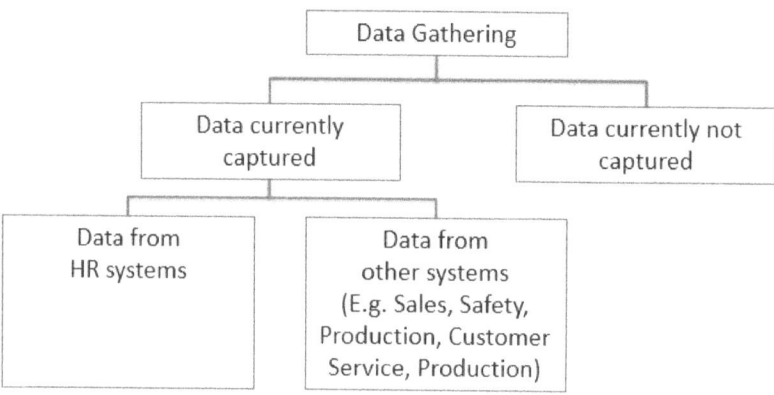

Data currently captured

For data that are already captured in business systems during its usual business operations, there is no further need to create data entry requirements. Instead, we just need to determine the method to extract the data from different sources, gather them onto a common platform with all other data for analysis purposes.

Always start with the data you already have access to, and evaluate whether it is good enough for your hypothesis testing, before you collect new data.

Sometimes, data can be fragmented, outdated, missing, incorrect, kept in various disconnected systems, or even non-existent.

- **Missing data**

    To deal with missing data, either remove them from the analysis, or use regression or average to estimate the missing values.

- **Missing promotion history**:

    To deal with missing promotion history, check your HR system (E.g. SAP, Peoplesoft, Workday) for the incumbents' title changes.

- **Outliers:**

    Outliers are data records that differ dramatically from other observations, and can negatively bias the result of an analysis. To decide what to do with Outliers, talk to the domain experts to get their insights. Generally, it is better to conduct two analyses: one model that includes outliers, and another model that exclude outliers. How to deal with data outliers depends on the circumstances:

    - Outliers may be due to input error, where it would be advisable to remove the outliers, cap the outliers' values, or convert the outliers to average. Find the behaviour of the observations similar to the outlier and make inferences of what would be the best approximate value. You can replace the missing data with the average of other data, or by deriving the value through regression.

    - On the other hand, outliers can lead to a discovery. Outliers can highlight fraud, quality issues, or a customer about to churn.

Data Currently Not Captured

For data which is currently not captured, the following should be fulfilled:

- Create a new avenue to capture new data type through data entry.
    - Creation of templates (such as spreadsheet templates)
    - Creation of a new data fields or new platforms within existing business systems for data entry purposes

- Establish a process for accurate data entry through
    - Identifying the source of data entry (example – from additional field to be inserted at the point-of-sale system, or from a created field in a spreadsheet template).
    - Efficient design of data fields and templates (example – user drop down buttons that will allow mouse-click selection instead of typing data into data field).
    - Detailed instructions to guide users on data entry.
    - Training and briefing for data entry users.
    - Setting limits on field types (example - data field to only accept text string and not values).
    - Automating fields (example – an entry in one field will lead to automation of one other field).
    - Automated validation checks on unacceptable data input, such as missing data, field length, data range, comparison against stored data, consistency across data fields.
    - Limit access using passwords and log-in identification.
    - Review and approval of data entry workflow and system transactional logs.

- In circumstances where data capture is too costly, or when data cannot be sourced, proxies must be identified before applying the process mentioned above.

(ii) Data Analysis

There are two ways to do analysis:
- Qualitative analysis.
- Quantitative analysis.

Quantitative Analysis

Quantitative analysis is used to analyze numerical data using statistical means. Quantitative Statistical techniques such as Decision trees, Correlation, and Multiple Regression will need to be applied to the data to test the hypotheses to discover data patterns.

- **Decision Trees** is a simple method of creating a predictive model is to use the decision tree. A decision tree is a tree-like model consisting of decisions and their possible consequences.

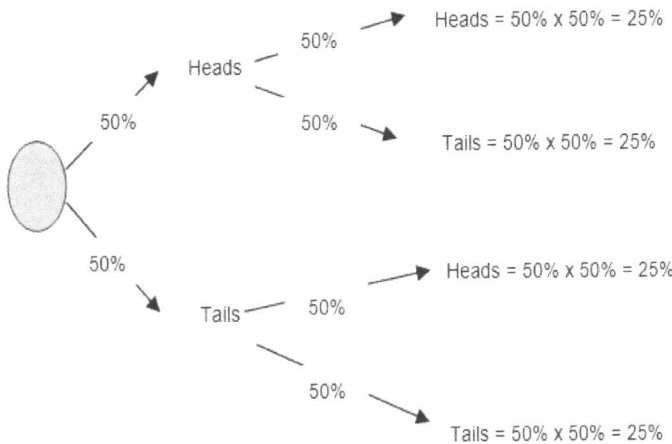

- **Correlation** is a measure of the extent to which two variables are related. In other words, if variable A changes does variable Y change? If an increase in one variable leads an increase in another variable, it is called positive correlation. If an increase in one variable leads to a decrease in another variable, it is called negative correlation. When there is no relationship between two variables, it is called zero correlation.

- **Regression** analysis is a statistical technique that helps you predicts the level of one variable (dependent variable) based on the level of another variable (independent variable). Correlation tells you if a relationship exists, whereas regression tells you which variables have the most impact on the dependent variable.

Qualitative Analysis

Qualitative analysis is used to analyze descriptive data such as: opinions, attitudes, and life experiences. Qualitative research is useful for interpreting or contextualizing quantitative results. Contextualizing helps to explain technical quantitative analysis by adding color and depth. For instance, quotes from focus groups can bring to life quantitative findings, turning it into a story.

The sections in the later part of this book will show you step-by-step how to do Predictive analytics using statistical techniques such as Decision trees, Correlation, and Multiple Regression using Excel.

7.5) Step 5 – Tell the Story

According to Brent Dykes (2016), Data storytelling is a structured approach for sharing data insights, and it uses three key elements: **data**, **visuals**, and **narrative**.

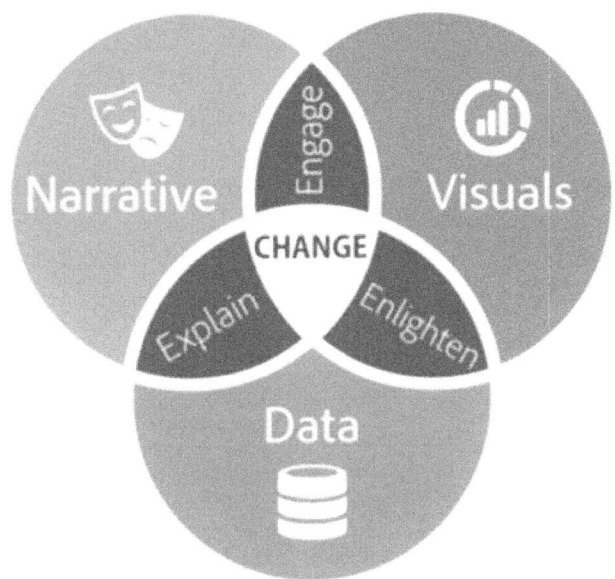

Source: Brent Dykes (2016) Data Storytelling: The Essential Data Science Skill Everyone Needs. https://www.forbes.com/sites/brentdykes/2016/03/31/data-storytelling-the-essential-data-science-skill-everyone-needs/#51c5d24752ad (25 November 2018)

- **Data** and **Narratives** explain: When narrative is combined with data, it helps to **explain** your data's context, source, relevance, what you did with it, what it is and why it is important.

- **Data** and **Visuals** enlighten. When visuals are combined with data, it can **enlighten** the audience to insights that wouldn't be visible without graphs or charts. Without data visualizations, interesting trends and data outliers would be hidden in the data tables rows and columns.

- **Narratives** and **Visuals** engage. When narrative and visuals are combined together, they **engage** the audience.

Data and **Narratives** and **Visuals** drive change. When you combine the right visuals and narrative with the right data, you have a data story that influences and drives **change**.

Data isn't as memorable as stories. After analyzing your data, you need to communicate the project insights and recommendations to the stakeholders (i.e. Tell the Story). Tell the Story has three sub-steps:
(i) Know the Audience
(ii) Develop the Narrative
(iii) Develop the Visuals

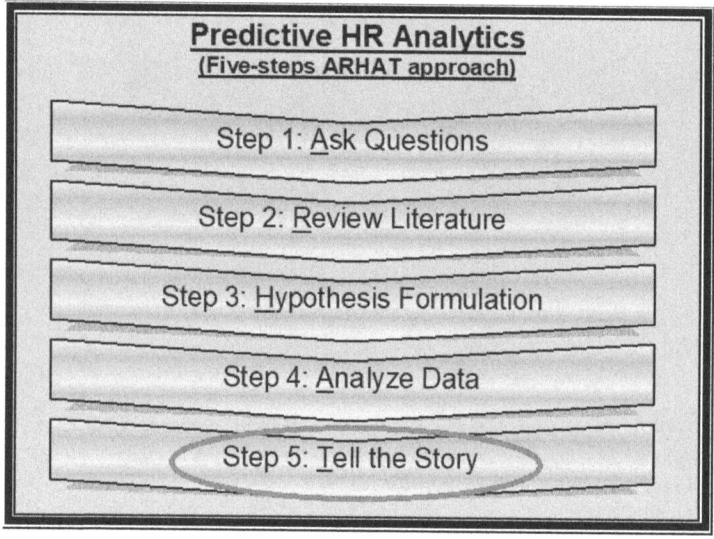

(i) Know the Audience

Customize your story to your audience. What and how you present your story depend on the audience that you are presenting to:
- Who are your audience?
- What are their needs and challenges?
- What do you want your audience to know or do?

There are different types of audiences that you will encounter during your presentation:
- **Neutral Audiences** sit on the fence about your proposal.
- **Critics Audiences** strongly disagree with your proposal.
  - think of reasons why they don't agree with your proposal and come up with a common ground to relate to them.
- **Uninformed Audiences** are unfamiliar with your topic of discussion - decide how much they know about this topic, and how much information to provide them.
- **Subject-Matter-Expert Audiences** already know what you are telling them
  - give them something new that they are uninformed about.
- **Top Management Audiences** don't have patience
  - keep it concise.

(ii) Develop the Narrative

Different people can interpret the same data differently because data don't provide contextual information. Thus, you can't just show your data, you need to tell a story with it. Data storytelling connects the data, analysis, recommendations and visuals with the audience.

After you have analyzed the data, you need to convert them into a story. Steve Jobs used a three-act structure, with heroes, villains, and victims for his narrative:
- Act 1) The Setup (<u>Why</u> should I care?)
- Act 2) The Confrontation (<u>How</u> will your idea make my life better?)
- Act 3) The Resolution (<u>What</u> action do I need to take?)

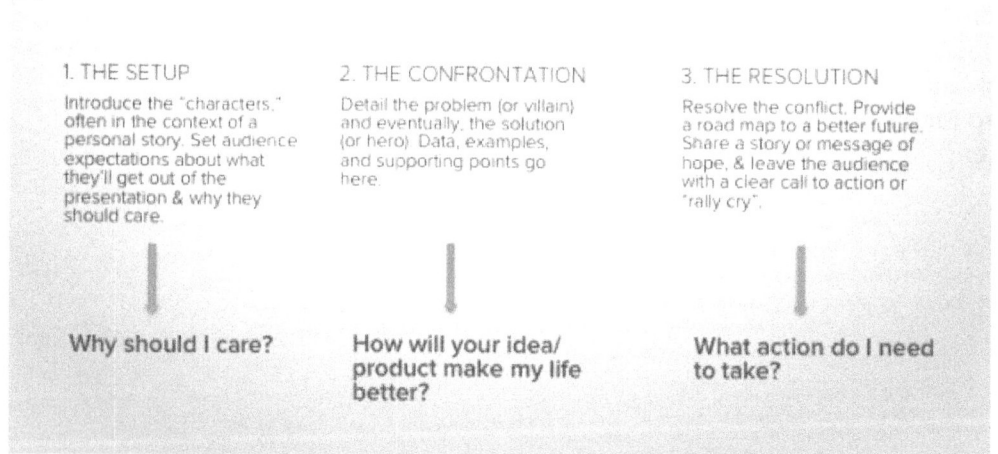

Source: Marta Kagan (2017) https://blog.hubspot.com/blog/tabid/6307/bid/34274/7-lessons-from-the-world-s-most-captivating-presenters-slideshare.aspx (25 November 2018)

Act 1) The Setup - <u>Why</u> should I care?

The beginning set audience expectations about what they'll get out from the presentation and why they should care. At the beginning of your story, you need to hook your audience to get them interested. Start by providing some background and context to ensure your audience understands the issue to addressed, or opportunity to tap.

The setup introduces the characters (heros, villains, victims) in the context of a personal story.
- **Villains** are the **problems** that you are trying to overcome. For example, high staff turnover, absenteeism, performance issues, etc.
- **Heroes** are the **solutions** to the problems. For example, introducing incentive programmes, running training programmes, etc. You get to cast your organization, your team, yourself, or your project as the hero of your story.
- **Victims** are those harmed by the villains.

Source: Sunday Mancini (2016) https://www.ethos3.com/2016/04/storytelling-tactics-for-presentations-creating-the-right-villain/ (25 November 2018)

In the chart below, the problem (Villain) is that, for the first time in the company's history, sales decreased in February.

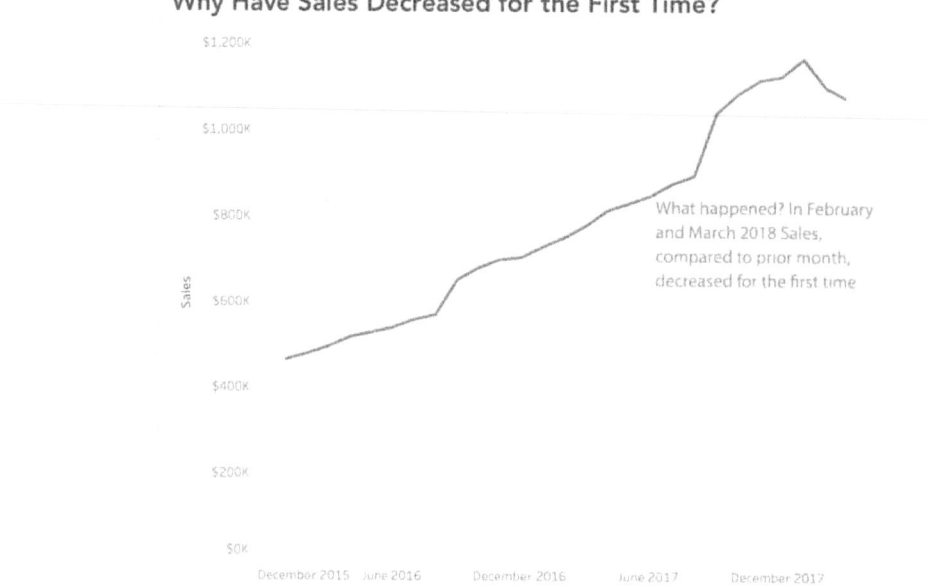

Source: Nick Mannon (2018) Persuasive Storytelling with Data Visualization. https://www.blastam.com/blog/persuasive-storytelling-with-data-visualization (25 November 2018)

Data is your story's supporting character, not main character. Start your data storytelling with people, then data. Positioning your data story with the people you're addressing helps to make it more relatable to your audience.:

- Rather than saying, "Our company's average engagement score is seventy percent", say "two out of three of the people here are engaged."

- Rather than just saying, "Attrition rate is high", say "One-third of our new hires in Sales leave the company within 6 months, making it tough for us to achieve our sales targets.

- Rather than saying, "We want to achieve 100% completed goal-setting," say "We want all our people here to be successful, thus we want all our manager here to tell our people what is expected of them at work through goal setting.

Act 2) The Confrontation - How will your idea make my life better?

The middle details the problem (villain) and eventually, the solution (hero) and supporting points go here. Here is where your show your analysis, findings, and the supporting facts, and answer how your solution will make the audience's life better.

Where possible, communicate your insights in a visual format, instead of complicated correlation tables and regression tables. Data visualization is the presentation of data in a graphical or pictorial format. Visualization helps audiences to can grasp difficult analytics concepts easily and identify new patterns. Avoid showing raw data or analysis, without sharing with the audiences your insights or interpretation. If you present your data and analysis without your interpretation of it, your audiences may draw their own conclusions to reinforce their preconceptions.

Select the most critical insights to share with your audience. Don't overwhelm your audiences with the large number of insights that you generated. Examples of Apple's' memorable headlines are:

>"iPhone - Apple reinvents the phone"
>"MacBook Air - The world's thinnest notebook".
>"iPOD - 1,000 songs in your pocket"

In the chart below, it was found that total sales decreased in February. Then after breakdown the total sales by channel, the team found that it was the Internet Sales channel that dropped significantly because there was no stock. Customers are not able to buy as the company do not have stocks of the product. (1)

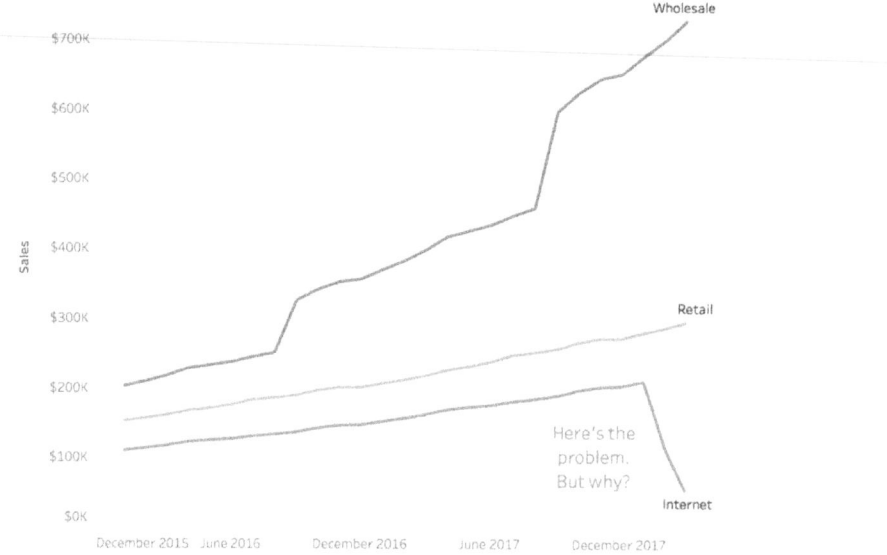

Source: Nick Mannon (2018) Persuasive Storytelling with Data Visualization. https://www.blastam.com/blog/persuasive-storytelling-with-data-visualization (25 November 2018)

Act 3) The Resolution - <u>What</u> action do I need to take?

The ending is where you provide a road map to a better future, and leave the audience with a call to action. The Resolution has three steps:
- Tell the Recommendations
- Encourage Action
- Get approval and implement it.

It is important to make an impact on your closing statement because when your audience is given a series of information, they usually remember the first and last items most. Here are a few ways on how to end your presentation to create a memorable moment, and help your audience retain your message:

Tell the Recommendations

Write each recommendation clearly as a statement: *Our analytics insight demonstrated that if X is changed ... Y will change... thus we recommend...*

- If your analytics project affects only a few employees in a department, your recommendation can be a one-off training course.

- If the issue is high turnover, and your analytics insights uncovers that low salary is a cause, your recommendation can be to allocate a higher salary increase budget.

- If the issue is high turnover, and your analytics insights uncovers that lack of career growth is a cause, your recommendation can be to job rotate people.

Encourage Action

Tell your audience what you'd like them to do. When you want your audience to commit to an action, don't be vague., communicate confidently and clearly. Make it easy for your audience to act by breaking a big project into smaller milestones.

In the diagram below, the decrease in sales was because the company does not stock the products our customers want to buy. As such, we recommend the following... if we did these recommendations, our sales through March 2018 would have continued our stellar month over month increase, which would look like the following... [1]

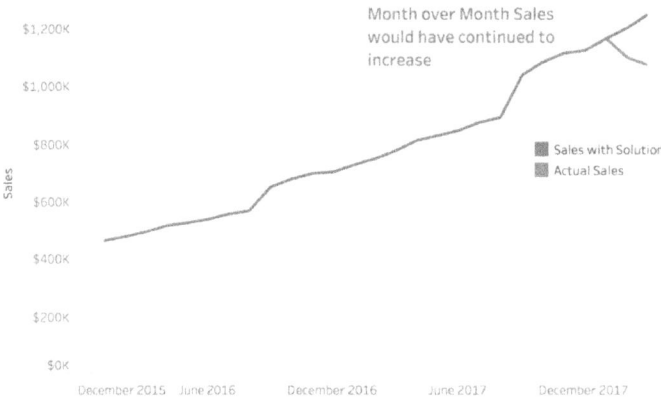

Source: Nick Mannon (2018) Persuasive Storytelling with Data Visualization. https://www.blastam.com/blog/persuasive-storytelling-with-data-visualization (25 November 2018)

Get approval and implement it.

After telling the story, get approval from the sponsors and implement it. Evaluate your project along the way, analyzing the effectiveness of the implemented recommendations. In Analytics, we measure ourselves on the impact we made, not on the number of reports we produce, or the number of projects we completed.

After you have made your recommendations, ask your sponsor whether they agree with your recommendations, and who will be responsible for implementing the recommendations?

References:

(1) Nick Mannon (2018) Persuasive Storytelling with Data Visualization. https://www.blastam.com/blog/persuasive-storytelling-with-data-visualization (25 November 2018)

(iii) Develop the Visuals

Data visualizations help to transforming complex information into something easier to understand. Use interesting data visualizations that capture attention. Data visualization is the presentation of data in graphical or pictorial format, so that it is easier for audiences to see patterns or understand complex concepts. Because of how our brain processes information, it is easier to tell a story using pictures or graphs to visualize complex data than poring over columns and rows of Excel data.

Color Groups

Colors combinations are important for effective and professional-looking Powerpoint presentations. Colors are classified into two broad groups: [1]
- **Warm colors (reds, orange, and yellow):** Warm colors pop out and attract attention, especially bright red.
- **Cool colors (greens, blue, and purple):** Cool colors recede in the background and don't draw attention, especially darker shades.

White and light colors also catch the eye, while black and dark colors are less noticeable.

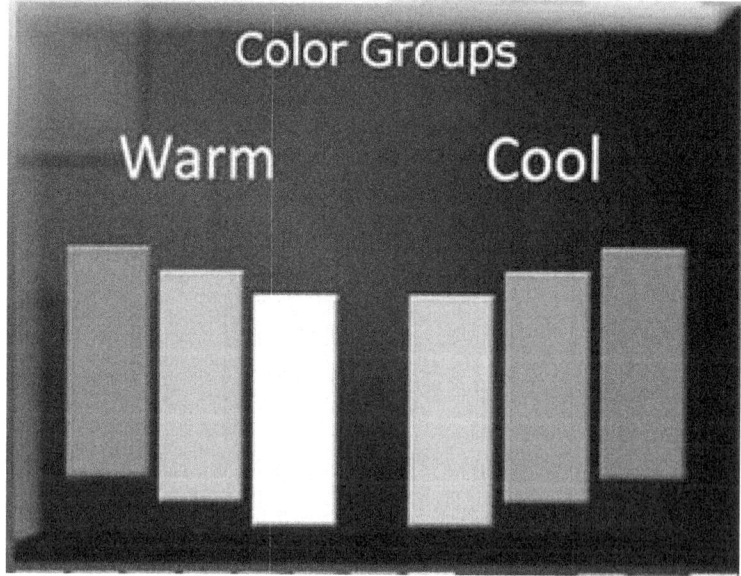

Source: Robert Lane (2018) Combining colors in PowerPoint – Mistakes to avoid. https://support.office.com/en-us/article/combining-colors-in-powerpoint---mistakes-to-avoid-555e1689-85a7-4b2e-aa89-db5270528852 (23 November 2018)

Color Groups and Contrast

It is important to consider color groups and contrast, when combining colors for your PowerPoint slides. White, black, and beige are neutral colors and goes well with all colors. [1] It is ok to combine warm colors with each other and shades of brown.

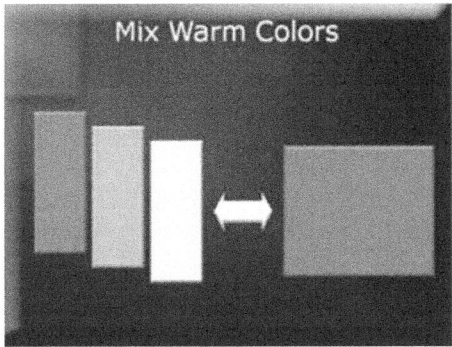

Source: Robert Lane (2018) Combining colors in PowerPoint – Mistakes to avoid. https://support.office.com/en-us/article/combining-colors-in-powerpoint---mistakes-to-avoid-555e1689-85a7-4b2e-aa89-db5270528852 (23 November 2018)

It is ok to combine cool colors with each other and shades of gray.

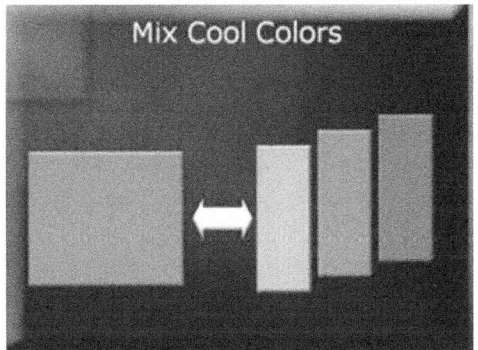

Source: Robert Lane (2018) Combining colors in PowerPoint – Mistakes to avoid. https://support.office.com/en-us/article/combining-colors-in-powerpoint---mistakes-to-avoid-555e1689-85a7-4b2e-aa89-db5270528852 (23 November 2018)

As our brain is better at discerning differences in color rather than shape, data visualizations use shade differences to draw attention to key information.

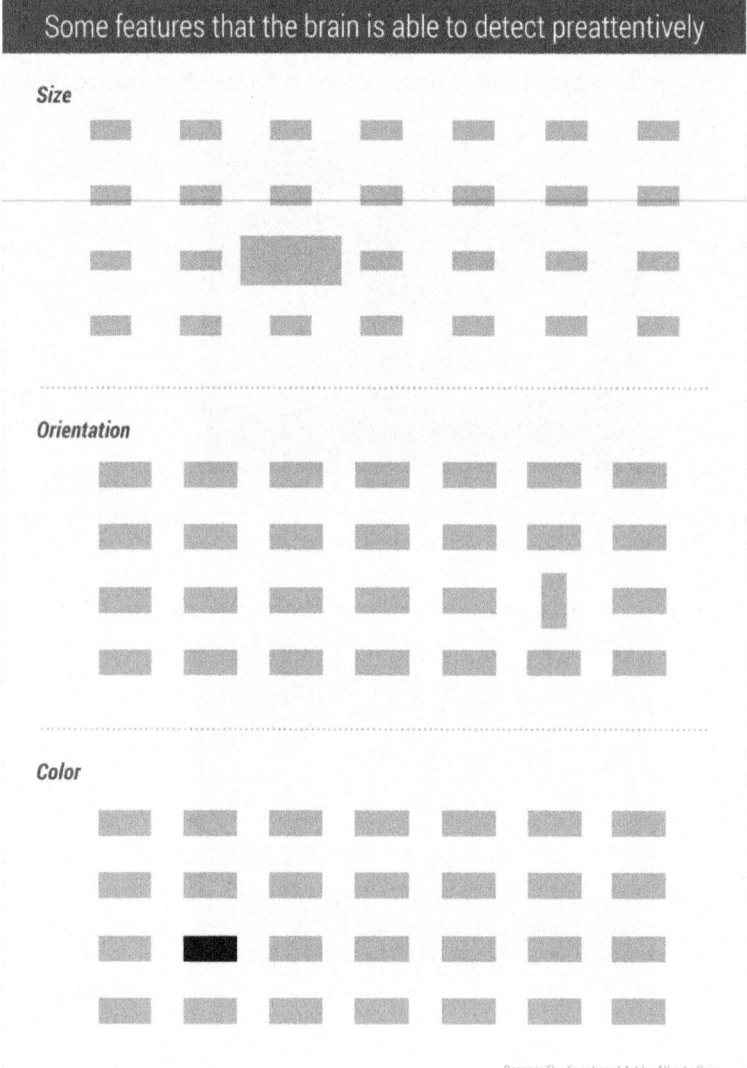

Source: Nayomi Chibana (2017) https://www.crazyegg.com/blog/data-storytelling-5-steps-charts/ (23 November 2018)

Avoid combining colors across the warm/cool boundary as they cause eye strain, especially for bright blue + red combination, and red + green combination. If the text color darkness has little contrast with the background color, some audience might not be able to read that text. An issue with the red + green combination is that 7 percent of men and 1 percent of women have cannot differentiate red and green colors because it is the most common type of color blindness. [1]

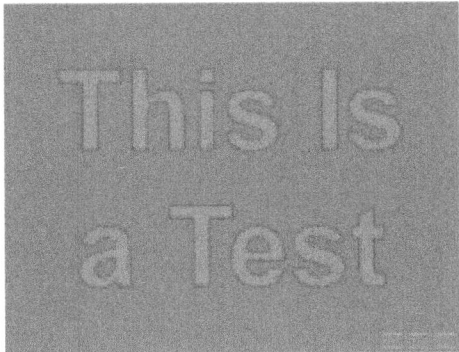
Red + Blue color combinations cause eye strain.

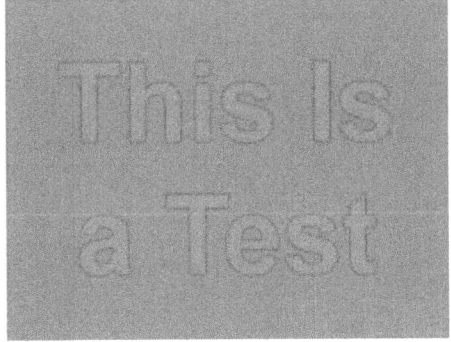
Red + Green color combinations cause eye strain.

Color and Text Considerations

Using a simple background that contrasts with the text color helps to pop out your message. For professional looking slides, use white or light beige on a dark background. Alternatively, use black or very dark color on a light background. Avoid using brightly colored text and red text as it washes out when projected on screen. [1]

Text Color should Contrast Sharply with a Background

Source: Robert Lane (2018) Combining colors in PowerPoint – Mistakes to avoid. https://support.office.com/en-us/article/combining-colors-in-powerpoint—-mistakes-to-avoid-555e1689-85a7-4b2e-aa89-db5270528852 (23 November 2018)

Fewer Bullet Points, More Narrative

Our hearts and minds crave stories. Data without a story cannot emotionally connect and engage your data with your audience. By combining the power of your data with the art of storytelling, your audience will connect emotionally to the why behind the numbers. And when your people are engaged, they will contribute to the success of your project. [2]

Steve Jobs never used a single bullet point. His presentations were always remarkable spare, relying on a few powerful images and carefully selected words or phrases. [3]

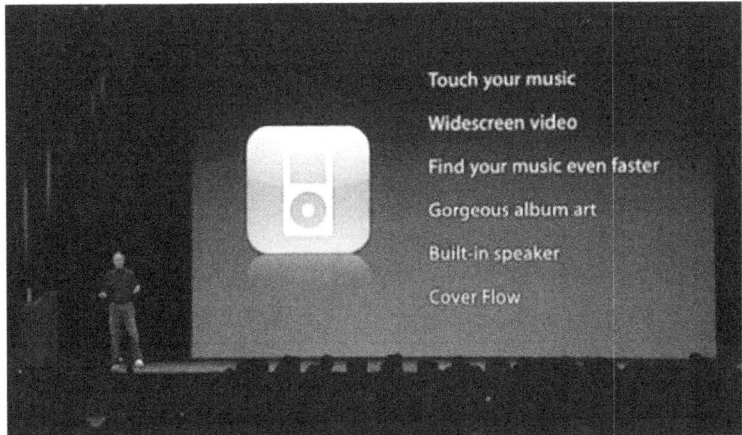

Source: Marta Kagan (2017) https://blog.hubspot.com/blog/tabid/6307/bid/34274/7-lessons-from-the-world-s-most-captivating-presenters-slideshare.aspx (26 November 2018)

Avoid this,

Dogs Have an Incredible Sense of Smell

- A dog's sense of smell is much keener and stronger than a human's.
- Their noses are about a million times more sensitive than ours.
- Their noses are so powerful that they can even smell disease.
- Research has found that dogs have an incredible ability to detect the smell of a range of organic compounds that indicate the human body isn't working properly.
- There are now diabetic alert dogs who signal the human when they pick up the distinct scent that is released when a human's insulin levels drop.

Source: Rebekah Pearson (2018) https://patronmanager.com/blog-power-up-your-powerpoint-with-storytelling/ (23 November 2018)

Use this,

Lifesaving Power of Smell

Marie has Type 1 Diabetes. Her service pup, Fitz, uses his keen sense of smell to detect and signal Marie when her insulin levels drop.

Research has found that dogs have an incredible ability to detect the smell of a range of organic compounds that indicate when the human body isn't working properly.

Source: Rebekah Pearson (2018) https://patronmanager.com/blog-power-up-your-powerpoint-with-storytelling/ (23 November 2018)

Use color to draw the audience's attention to the important aspects, Reduce the amount of text, and Quantifying the data if it provides important context. (4)

Avoid this,

> When responding to the question "What could have been improved?", customers' top 3 responses were complaints with A/V (poor video quality), the presenter's visuals and an unclear agenda.
>
> Complaints with A/V quality:
> - "The main challenge was logging into the webinar."
> - "The AV quality is was difficult."
> - "Sometimes the presenter's voice was very garbled for me. Very distracting to the flow and when asking a question, I didn't know if I was heard/understood."
>
> The presenter's visuals were lacking:
> - "Technical issues impacted my ability to see the presenter's visuals."
> - "Many times the PPT froze or didn't work."
>
> The agenda was unclear:
> - "The webinar started late with a lot of time wasted."
> - "Bad pre-sharing of information, no time keeper, no clear goal."
> - "The agenda seemed very fluid and I was confused around the main point."

Source: Elizabeth Ricks (2018) http://www.storytellingwithdata.com/blog/2018/10/10/three-tips-for-storytelling-with-qualitative-data (26 November 2018)

Use this,

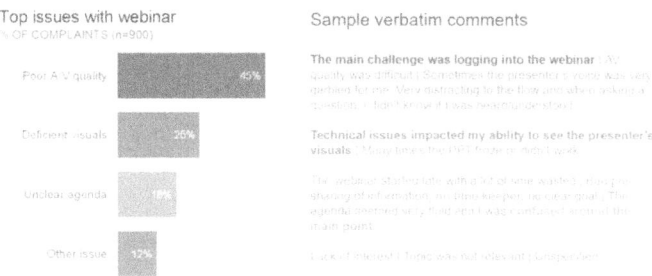

Source: Elizabeth Ricks (2018) http://www.storytellingwithdata.com/blog/2018/10/10/three-tips-for-storytelling-with-qualitative-data (26 November 2018)

Less Text, More Visuals

Avoid putting blocks of text on your PowerPoint presentation. Use pictures or diagrams with keywords and short phrases. Let your speech to attract your audience.

Avoid this,

Source: Rebekah Pearson (2018) https://patronmanager.com/blog-power-up-your-powerpoint-with-storytelling/ (23 November 2018)

Use this,

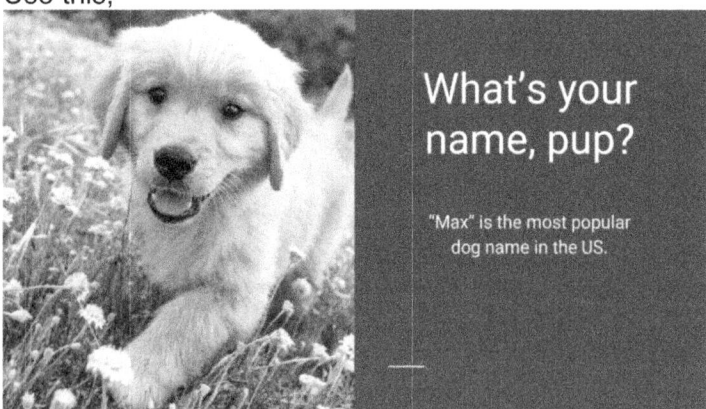

Source: Rebekah Pearson (2018) https://patronmanager.com/blog-power-up-your-powerpoint-with-storytelling/ (23 November 2018)

Minimize Stand Alone Data

Don't just present a chart on its own. Present the figures with a story — they create connection and action.

Avoid this,

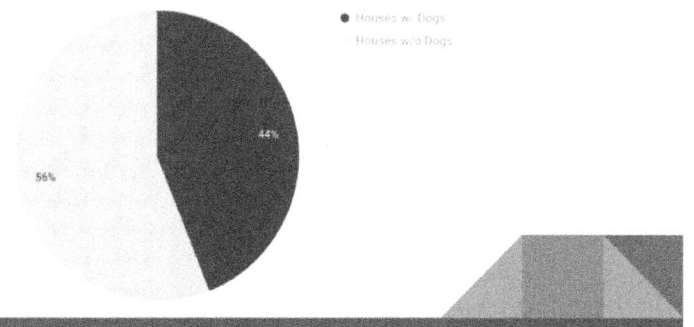

Source: Rebekah Pearson (2018) https://patronmanager.com/blog-power-up-your-powerpoint-with-storytelling/ (23 November 2018)

Use this,

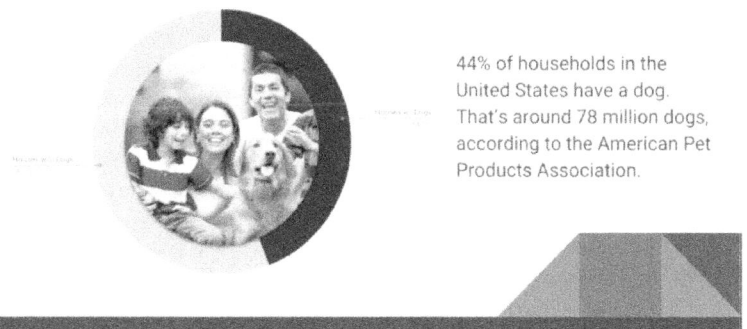

Source: Rebekah Pearson (2018) https://patronmanager.com/blog-power-up-your-powerpoint-with-storytelling/ (23 November 2018)

Declutter Your Chart

Declutter your visualizations by removing unnecessary and repetitive information. Summarize your information if your audience don't need to know the details. Ask yourself:

- Do I need these details to convey my message across?

- Can I summarize this information?"

Here are tips to declutter your visualizations:

- Avoid using more than 3 colors.

- Use grey color for items in the background that are not so important.

- Label columns, lines or segments directly instead of using a legend, so that your audience eyes do not need to move all over your slide.

- Remove chart borders and gridlines.

Avoid this,

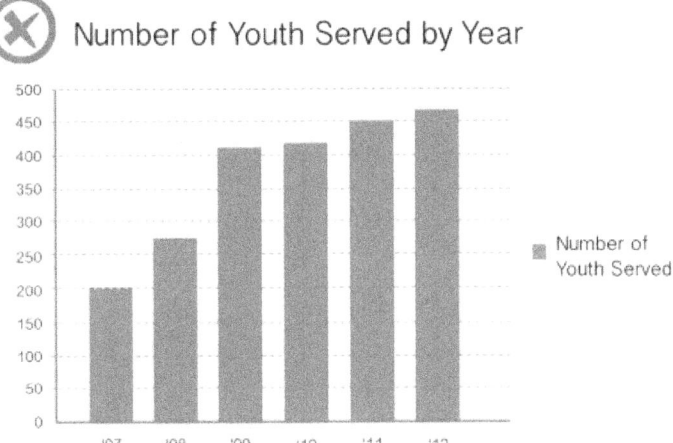

Source: Nayomi Chibana (2017) https://www.crazyegg.com/blog/data-storytelling-5-steps-charts/ (23 November 2018)

Use this,

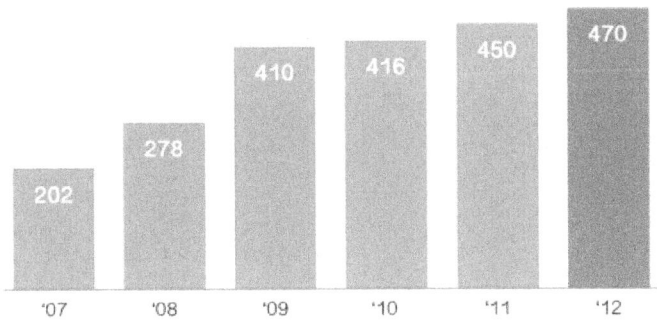

Source: Nayomi Chibana (2017) https://www.crazyegg.com/blog/data-storytelling-5-steps-charts/ (23 November 2018)

Push Secondary Information to the Back

To focus your audience's attention, highlight the most important items by using a different color and thicker line. Push all the secondary information to the background by using light gray color. [5]

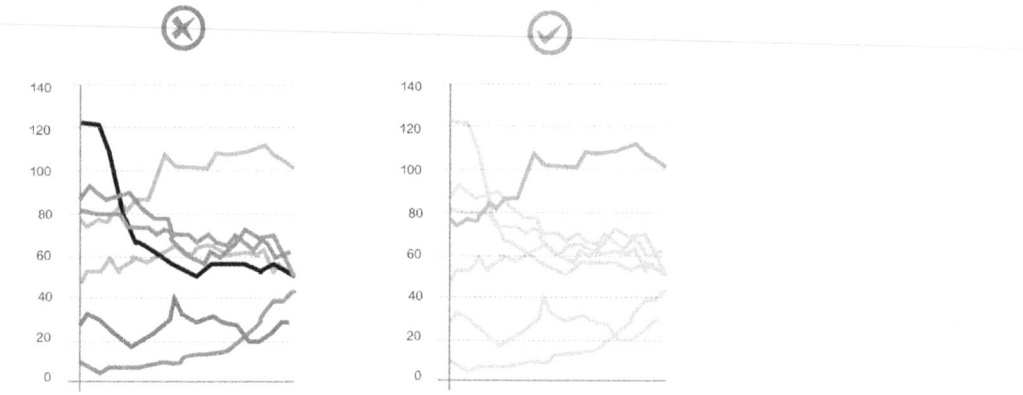

Source: Nayomi Chibana (2017) https://www.crazyegg.com/blog/data-storytelling-5-steps-charts/ (23 November 2018)

Push Secondary Information to the Back

In the graph below, the message is: (5)

> Since mid-2014, the supply of oil has exceeded demand, resulting in excess (blue bars).

However, the main message was distracted because too many colors was used, and secondary information were in bold.

Avoid this,

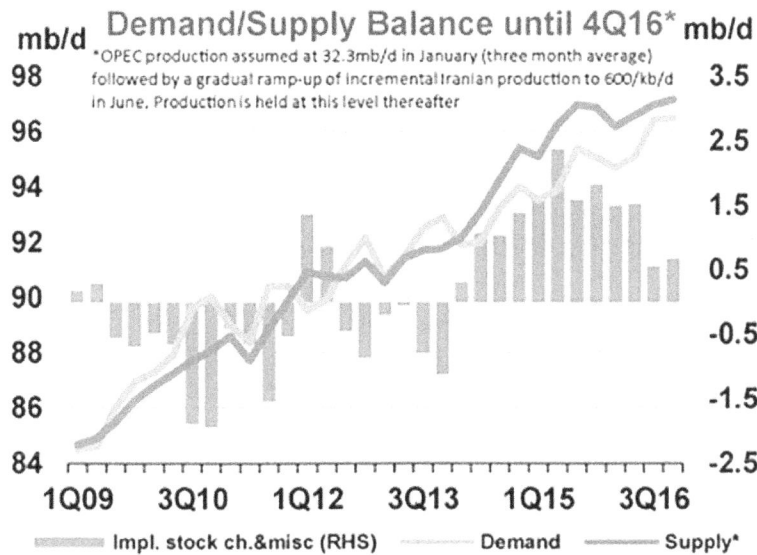

Source: Nayomi Chibana (2017) https://www.crazyegg.com/blog/data-storytelling-5-steps-charts/ (23 November 2018)

Below is an improved version of the above graph. To focus your audience's attention, the most important items (change in supply and demand starting in mid-2014) were highlighted by using a different color. All the secondary information was push to the background by using light gray color, and the axis values, title and subtitle are deemphasized. [5]

Use this,

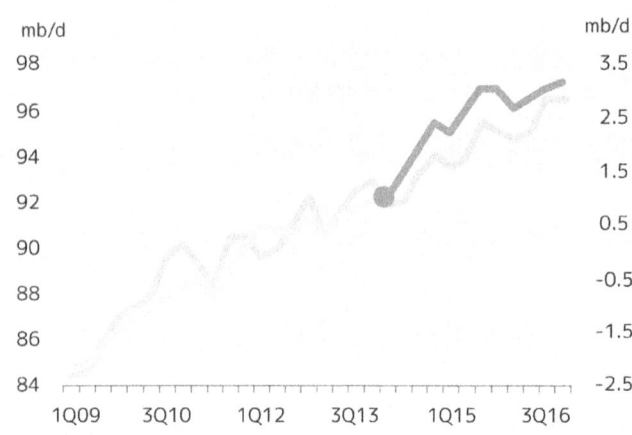

Source: Nayomi Chibana (2017) https://www.crazyegg.com/blog/data-storytelling-5-steps-charts/ (23 November 2018)

Use Slope-graphs to Compare Rate of Change

When you want to compare the rate of change between two points in time, it is easier to visualize through the slope of the line, than through bar charts. [5]

Avoid this,

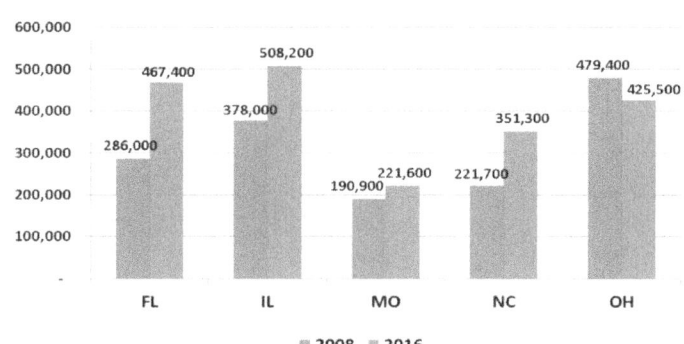

Source: Nayomi Chibana (2017) https://www.crazyegg.com/blog/data-storytelling-5-steps-charts/ (23 November 2018)

Use this,

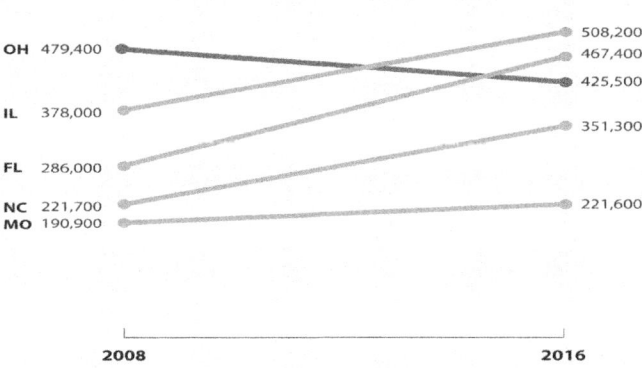

Source: Nayomi Chibana (2017) https://www.crazyegg.com/blog/data-storytelling-5-steps-charts/ (23 November 2018)

Avoid using legends

It is tedious to read the chart below because the bubbles are not proportional to each other. The reader needs to look back and forth between the bubbles and the corresponding legend. (5)

Avoid this,

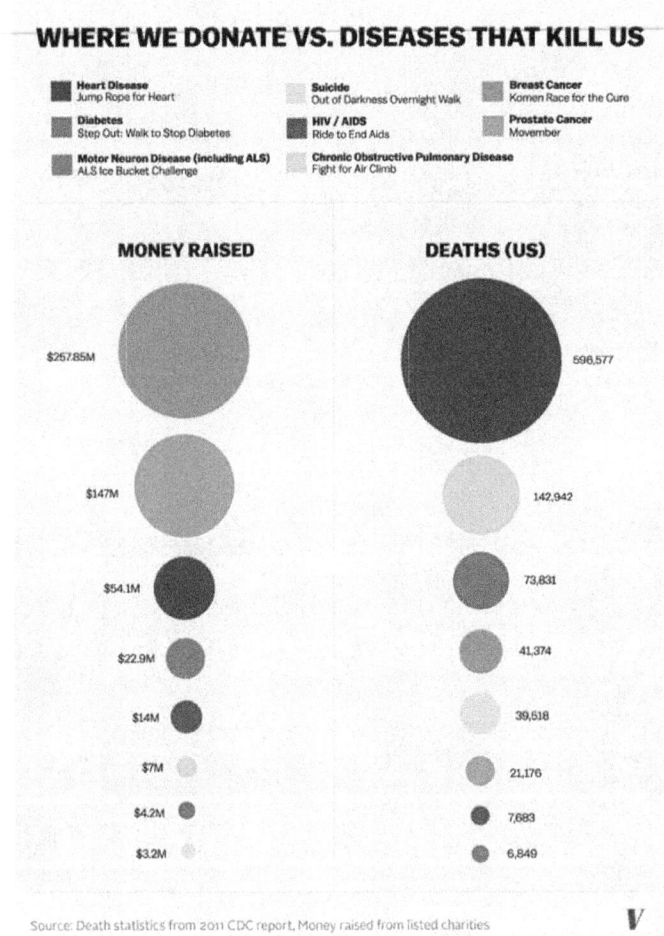

Source: Nayomi Chibana (2017) https://www.crazyegg.com/blog/data-storytelling-5-steps-charts/ (23 November 2018)

Side-by-side bar charts is a better way to present this information because they all start at the same baseline. The difference between values can be easily compared by just looking at the length of the bars, instead of a legend. [5]

Use this,

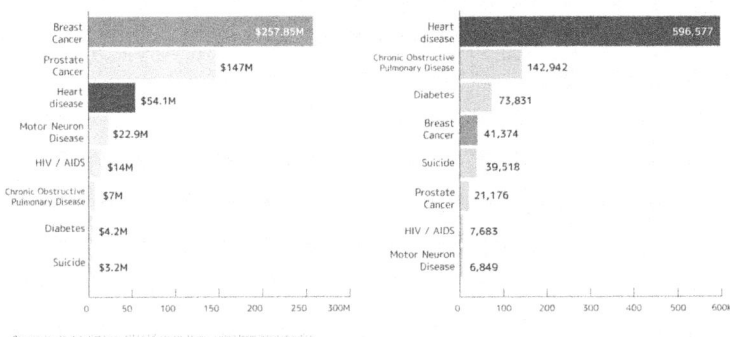

Source: Nayomi Chibana (2017) https://www.crazyegg.com/blog/data-storytelling-5-steps-charts/ (23 November 2018)

References:
(1) Robert Lane (2018) Combining colors in PowerPoint – Mistakes to avoid. https://support.office.com/en-us/article/combining-colors-in-powerpoint---mistakes-to-avoid-555e1689-85a7-4b2e-aa89-db5270528852 (23 November 2018)
(2) Rebekah Pearson (2018) https://patronmanager.com/blog-power-up-your-powerpoint-with-storytelling/ (23 November 2018)
(3) Marta Kagan (2017) https://blog.hubspot.com/blog/tabid/6307/bid/34274/7-lessons-from-the-world-s-most-captivating-presenters-slideshare.aspx (26 November 2018)
(4) Elizabeth Ricks (2018) http://www.storytellingwithdata.com/blog/2018/10/10/three-tips-for-storytelling-with-qualitative-data (26 November 2018)
(5) Nayomi Chibana (2017) https://www.crazyegg.com/blog/data-storytelling-5-steps-charts/ (23 November 2018)

8) Starting Your First HR Analytics Project.

The role of the HR Analytics lead is to provide analytically driven HR recommendations that improve performance. As the HR Analytics lead has to "sell" their projects to stakeholder, they need to understand the business, possess data storytelling skills, and be skilled in stakeholder management. Awareness of the hurdles allows you to minimize them and enhance your success chances.

To deliver successful HR analytics projects, stakeholder management, and the selection of HR analytics projects is critical. This section covers:
- Stakeholders Analysis
- Power Interest Matrix
- Action Priority Matrix
- Court Rulings on Analytics Cases

8.1) Stakeholder Analysis

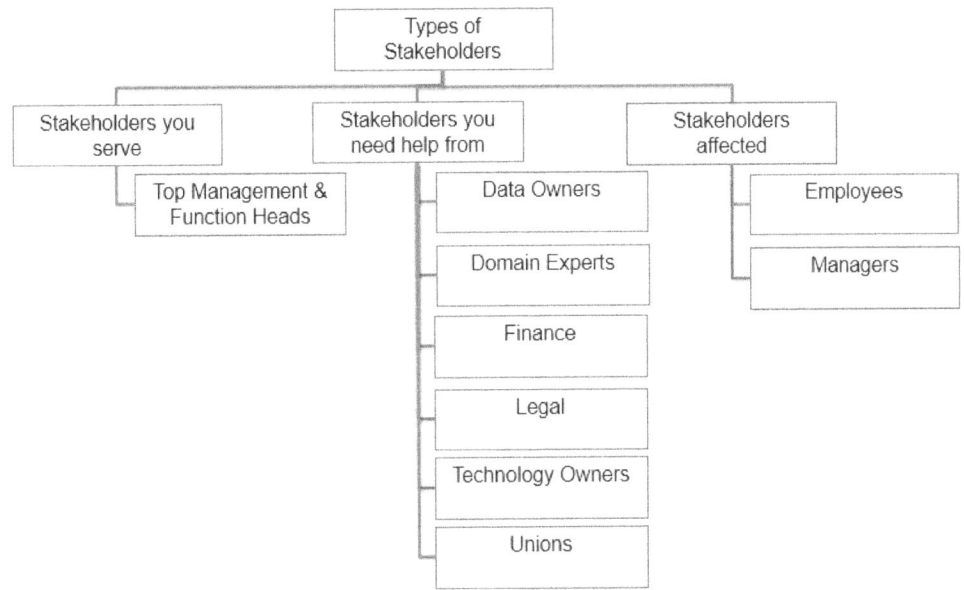

Stakeholder management is critical for your Analytic project to be successful. You need to build relationship with the stakeholders, manage all the stakeholders and political landscape to ensure that the projects can be delivered successfully.

Stakeholders reside at different levels within the company, and even outside the company. Other than those higher up in the hierarchy, it is just as important to establish relationships with people to whom you depend for data, expertise, and implementation.

Awareness that stakeholders have different degrees of influence is essential in planning who to cultivate relationships with. Analytics projects that lack a strong sponsor are likely to fail. Strong project sponsors are senior management who are well connected, able to rally support, remove road blocks, secure the resources, get you access to the necessary data for analysis, and leverage their relationships and position to help the team. Analytics projects can fail, even when you have a well-defined analytics hypothesis, if you cannot get access to the data that you need to test your hypothesis.

(i) Stakeholders you serve

Stakeholders you serve, are typically the leaders of the organisation.

- **Top Management & Function Heads**

To identify potential HR analytics projects, talk to the top management and function heads to understand their top priorities and concerns.

Share with the function heads, examples of successful HR analytics projects from within and outside the company that produced measurable results, and how their teams benefited.

Get the function heads inputs on their beliefs (hypotheses) regarding the possible causes of the issues and challenges.

Identify the data needed to address the concerns, identify the data owners, request help to facilitate your access to the data, and agree on realistic timelines.

Finally, prepare your function heads for all the possible results in advanced so as to manage the stakeholder's expectations. Sometimes, unexpected findings will emerge, or the findings will be inconclusive. Sometimes, the analytics will just confirm what was already suspected – but your effort is not wasted as you have found evidence to support the beliefs.

(ii) Stakeholders you need help from

- **Data Owners**

Finding out who can grant you access to the various data that you require for your analytics can be challenging. After you identified the data owners, getting their cooperation to give you what you need, in a usable format, when you need it, might be another challenge.

Knowledge-hiding (i.e. deliberately withholding or hiding information that is requested by another colleague) in the workplace is common and takes different forms. Sometimes, the data owners might be genuinely too busy with their daily workload to provide you with the data that you need. Sometimes, the data owners genuinely felt that they were justified in withholding the information as they felt that it is confidential. Sometimes, a colleague might try to withhold information that another colleague legitimately needs it, by claiming that they will provide the information later and never following through, or by giving incorrect or incomplete information. These people act contrary to their employer's interests probably because they fear that you outshine them. Thus, sometimes, you may need to get help from influential stakeholders if you have difficulty getting access to the data, but try to maintain good relationships with the data owners where possible. After receiving the data from the data owners, be sure to express your appreciation and publicly recognize the data owners' contributions.

- **Domain experts**

Other than data owners, you may also need help from colleagues who understand the context behind the data. For example, after generating the customer turnover rate, you will need to engage the domain experts to understand the context behind the high customer turnover rate for certain months. Sometimes high customer turnover was deliberate as certain products was removed from the market – inputs from domain experts can provide these insights to help you avoid wrong analysis and recommendations.

Share preliminary analysis with the domain experts to get their inputs on the patterns and context of the analysis. Then brainstorm with the domain experts on the implications and recommendations. Don't forget to publicly acknowledge the help provided by the domain experts.

- **Finance**

Align your data definitions with Finance to increase your credibility. Understand how Finance interprets manpower related cost from a financial perspective.

- **Legal**

HR data generally contain sensitive employee personal information such as employees' career and salary history, performance ratings, date of birth, ethnic origin, and religious beliefs. Seek your legal department advice to ensure that the types of information that you need adhere to all the countries laws regarding employee data.

Analytics brings new privacy concerns. Privacy issues arises in a few ways. Firstly, employees may not be aware that their data was collected. Secondly, employees may be aware that their data was collected, but does not agree with how the data is being used. Analytical HR models should be built in a regulatory compliant way whereby the employees are informed about what data is collected and how it is used.

Recruitment Analytics can subject companies to lawsuits. Some companies screen candidates for variables such as attitudes toward alcohol use, or the distance a candidate lives from the job. Even unintentionally filtering out older or minority applicants can be illegal under equal opportunity laws. In the event that a company is sued in court for discriminatory recruitment practice, a company must show that the recruitment criteria it is using are proven to predict success in the job.

- **Technology Owners**

For data which is currently not captured in the HR systems (e.g. SAP, Peoplesoft), you may need to work with IT or HRIS to create a new data fields or new platforms to capture new data type through data entry.

- **Unions**

Initial success is important for any projects. To achieve early success, start your analytics projects in countries where the laws allow and in non-unionized companies.

Unions would be concerned when employees' careers are affected by analytics recommendations. When Unions are engaged in the early stages of the analytics initiatives, they can help you to anticipate various implications and ensure that the initiatives are managed effectively.

Be open and engage the Unions early on, listen to their concerns and suggestions, find common ground. Demonstrate how employees will benefit from the analytics initiatives with success stories examples, such as: more equitable HR practices, or career growth for employees regardless of ethnicity or gender. In some unionized companies, companies may have to provide clear justifications of how they intend to utilize employee data & analytics. To address this, it is recommended to introduce analytics as an opportunity rather than a looming threat, and illustrate how working conditions and employee satisfaction can be improved as a result of analytics.

(iii) Stakeholders affected

- **Employees**

HR analytic affects people, thus we need too be aware of the impact on people. Employees are likely to question what HR analytics can do for them, and how does it benefit them. It is recommended to be open, communicate, and educate employees that the organization's objective is to manage HR in a systematic fact-based way.

- **Managers**

As Managers are often the ones implementing the HR analytics projects, they should be educated on the rationale, value, and limitations of it. As Managers deal with people, they should not totally rely on the statistical models for decision making, as statistical models are based on probability.

8.2) Power Interest Matrix

A common stakeholder analysis technique is a power interest matrix. Two primary variables define stakeholders and how they influence the project. The stakeholder's interest level is on the x-axis, and the power level is on the y-axis.:
- **Power** is the ability of the stakeholder to change or stop the project.
- **Interest** is the amount of involvement the stakeholder has in the project. It is the size of the overlap between the stakeholder's and the project's needs.

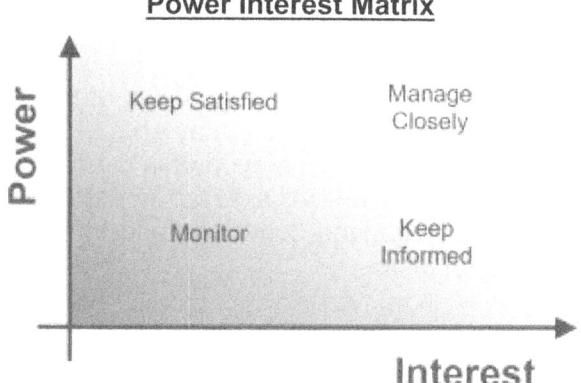

Source: Project Engineer (2019) How to Perform a Stakeholder Analysis https://www.projectengineer.net/how-to-perform-a-stakeholder-analysis/ *(17 May 2019)*

A dot on the power-interest matrix shows where the stakeholder sits on the Power Interest Matrix, and their type of support (Resistant, Neutral, or Supportive).

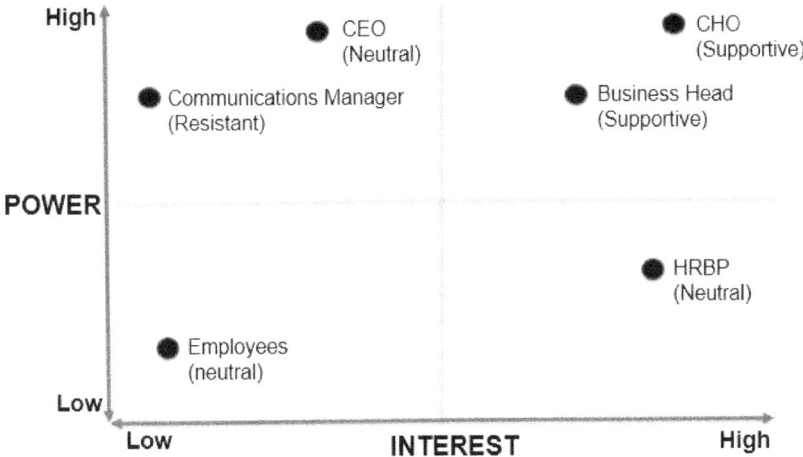

The power interest matrix provides lots of information about a stakeholder and how to manage them. Stakeholders can move around the matrix. Some stakeholders interest may increase as the project progresses. Some stakeholders power increase because of promotion or transfer.:

- **Low power and low interest stakeholders:** must by monitored to ensure that they are not able to stop or change the project.
- **High power and low interest stakeholders:** must be kept satisfied so they don't derail the project.
- **Low power and high interest stakeholders:** must be kept informed to ensure they are on your side.
- **High power and high interest stakeholders:** must be actively managed as they have a major influence on the project. Discuss all recommendations and actions with them.

Stakeholder engagement strategy

A stakeholder engagement strategy template can be used to analyze the power, interest, type of support, and engagement strategy of each stakeholder.

Stakeholder engagement strategy template

Stakeholder Name	Power (Ability to stop or change the project)	Interest (High or low)	Support (Resistant, Neutral, or Supportive)	Engagement Strategy (Type and frequency of communication)
Stakeholder 1				
Stakeholder 2				
Stakeholder 3				
Stakeholder 4				
Stakeholder 5				
Stakeholder 6				

8.3) Action Priority Matrix

Delivering successfully HR analytics projects is critical especially when establishing your credibility. A Priority Matrix is useful to help you decide which projects to prioritize and which projects to avoid. You need a few HR analytics projects that aligns with your company's KPIs before you can prioritize them.

(i) The Action Priority Matrix two axis: "Impact" and "Effort": [1]

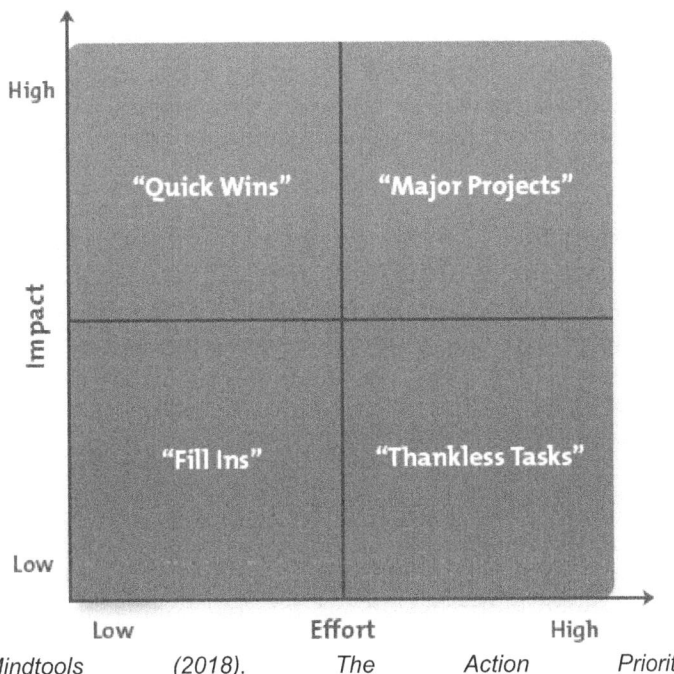

Source: Mindtools (2018), The Action Priority Matrix https://www.mindtools.com/pages/article/newHTE_95.htm (7 November 2018)

- **Horizontal Axis - Effort**

Effort refers to the challenges that you might encounter for the project. Factors that affect effort include political landscape, analytics capability, data access and quality, and level of change.

> **Political landscape** - You will encounter people who support your goals, as well those who will block or sabotage it. To succeed in HR analytics projects, you need to know where you have backing for your projects, where you don't, and which of your projects are worth pursuing when you face roadblocks.

> **Analytics capability** – Does your team have the analytics capability to deliver the project? Effort needed for the project is higher when your team lacks the technical statistical skills required for the project. If your team do not have the capability, you may need to recruit, train, or hire external consultants.

> **Data access and quality** – Effort needed for the project is higher when the data quality is not suitable, or when you have difficult accessing the data. For example, you may face difficulty linking your SAP employee salary data and Payroll system employee overtime data, if the employee number used in SAP is different from the employee number used in the Payroll system. You may also encounter difficulty getting access to the required data because of colleagues who deliberately withholding information.

> **Level of change** – The more people affected by the change, and the more change needed, the greater the project effort, because more communication, training and stakeholder management effort is required to successfully bring about change. However, if change affects too few people, the impact will be lower – thus a balance is needed. The best approach to deal with resistance to change is to get the people who will be affected by the change to participate in deciding how to make the change.

- **Vertical Axis - Impact**

"Impact" is on the vertical axis of The Action Priority Matrix. When deciding whether your project has high or low impact, bear in mind that the benefits should give a higher net return compared to other HR analytics projects. Usually cost reductions analytics projects are easier and faster achieve, compared to revenue gains analytics projects. Hence, for your first projects, it is better for you to focus on cost reduction analytics projects, as they can give you the initial quick wins to establish your credibility.

(ii) Four Quadrants of The Action Priority Matrix

The Action Priority Matrix has four quadrants (Quick Wins, Major Projects, Fill Ins, Thankless Tasks) [1]:

- **Quick Wins (High Impact, Low Effort)**

Projects that produce significant business results, and are easy to accomplish are called "Quick Wins". Quick wins are the best projects to focus on when you are new on a job, because they give good return with little effort, thereby helping you to establish credibility early.

- **Major Projects (High Impact, High Effort)**

Major projects are complex time-consuming long-term projects with high returns. Major projects are complicated because you need to have good relationship with stakeholders to deliver it successfully. You may have difficulty getting access to the required data because or the country's privacy laws, or knowledge-hiding by colleagues (i.e. deliberately withholding or hiding information that is requested by another colleague), or because the data is kept in the vendor's IT systems in other formats, or because you need help from people that you do not have formal authority over.

- **Fill-Ins (Low Impact, Low Effort)**

Fill-Ins have low impact and low effort. If you have spare time, you can do some Fills-Ins projects, but drop them or delegate them if a better project comes along.

- **Thankless Tasks (Low Impact, High Effort)**

Thankless Tasks have high effort, but low impact. These projects are done because of personal interests rather than what the business really needs. Avoid projects of personal interest, as they might attract adverse attention to yourself.

8.4) Court Rulings on Analytics Cases

HR data generally contain sensitive employee personal information such as employees' career and salary history, performance ratings, date of birth, ethnic origin, and religious beliefs. Considering the potential risk of handling HR information, it is essential to take precautions. Seek your legal department advice to ensure that you are adhere to all the countries laws regarding employee data. Share with your legal department, your HR analytics objectives and types of information that you need.

Analytics brings new privacy concerns. Privacy issues arises in a few ways. Firstly, employees may not be aware that their data was collected. Secondly, employees may be aware that their data was collected, but does not agree with how the data is being used. Analytical HR models should be built in a regulatory compliant way whereby the employees are informed about what data is collected and how it is used.

Recruitment Analytics can subject companies to lawsuits. Some companies screen candidates for variables such as attitudes toward alcohol use, or the distance a candidate lives from the job. Even unintentionally filtering out older or minority applicants can be illegal under equal opportunity laws. In the event that a company is sued in court for discriminatory recruitment practice, a company must show that the recruitment criteria it is using are proven to predict success in the job.

Employment discrimination is taken seriously in U.S. Below are some of the U.S. Supreme Court cases involving employees' rights and employment discrimination:

(i) Case law 1: Griggs v. Duke Power Co. (1971)

In this court case, the U.S. Supreme Court ruled that certain education requirements and intelligence tests used as a condition of employment in or transfer to jobs at the plant served to exclude African American job applicants, did not relate to job performance, and were prohibited. These requirements were not directed at or intended to measure ability to learn to perform a particular job or category of jobs. While 703 (a) of the Act makes it an unlawful employment practice for an employer to limit, segregate, or classify employees to deprive them of employment opportunities or adversely to affect their status because of race, color, religion, sex, or national origin, 703 (h) authorizes the use of any professionally developed ability test, provided that it is not designed, intended, or used to discriminate [2].

(ii) Case law 2: Johnson v. Transportation Agency (1987)

The Court ruled that a county transportation agency appropriately considered an employee's sex as a factor in deciding whether to promote her. In 1978, an Affirmative Action Plan (Plan) for hiring and promoting minorities and women was voluntarily adopted by Santa Clara County transportation Agency (Agency). The Plan provides, that when making promotions to positions which women have been significantly underrepresented, the Agency is authorized to consider the sex of a qualified applicant as one of the factors. The plan's objective is to achieve measurable annual improvement in the recruitment and promotion of women and minorities in positions where they are underrepresented, and in the long-term, achieve a work force composition that reflects the proportion of women and minorities in the area's labor force. When the Agency announced a vacant promotional position of road dispatcher, none of the dispatcher positions was held by a woman. The Agency, passed over a male employee, and promoted a female, Diane Joyce, both of whom were rated as well qualified for the job, and the male employee sued the Agency. Held: The Court held that the Agency had considered Joyce's sex as one factor in deciding that she should be promoted, and it is a moderate, flexible, case-by-case approach to achieve a gradual improvement in the representation of minorities and women in the Agency's work force [3].

References:
(1) Mindtools (2018), The Action Priority Matrix https://www.mindtools.com/pages/article/newHTE_95.htm (7 November 2018)
(2) Findlaw for legal professionals, Griggs v. Duke Power Co. (1971). https://caselaw.findlaw.com/us-supreme-court/401/424.html (7 November 2018)
(3) Findlaw for legal professionals, Johnson v. Transportation Agency (1987). https://caselaw.findlaw.com/us-supreme-court/480/616.html (7 November 2018)

9) Basic Statistics (Average, Median, Mode, Percentile)

(i) Average

$$\text{average} = \frac{\text{sum}}{\text{number}}$$

Pros:
- Easy to calculate: just add and divide.

Cons:
- Can be skewed by outliers

How to annualize turnover rate:

	Q1		Q2
Resignations	9		11
Headcount	100		103
Quarterly Turnover	9.0%		10.7%
Annualized Turnover		((9.0% + 10.7%)/2) x 4 = 39.4%	

(ii) Median

The median takes the **number in the middle of a sorted list**. If there's even number of items, take their average.

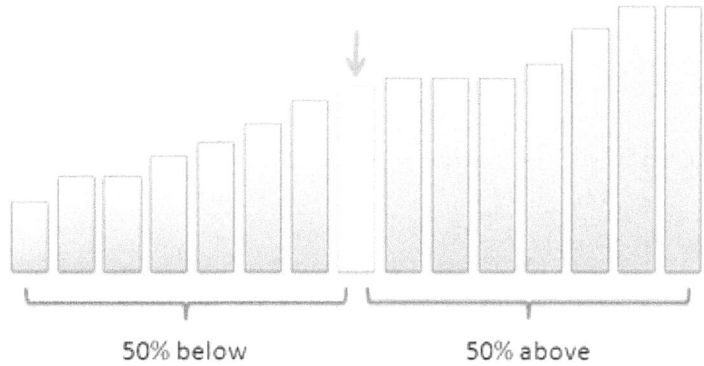

Source: https://betterexplained.com/articles/how-to-analyze-data-using-the-average/

Pros:
- Handles outliers well.
- Splits data into two groups, each with the same number of items

Cons:
- More steps to calculate: you need to sort the list first.

(iii) Mode

The mode is the value that occurs most frequently in a given set of data.

Sometimes mode, not a calculation, is the best way to get a representative sample of what people want. If you need to select a day next week for a party (Monday, Tuesday…., Saturday, Sunday). The best day would be the option that satisfies the most people (average may not make sense). Similarly, it is better to use mode for colors, and movie preferences.

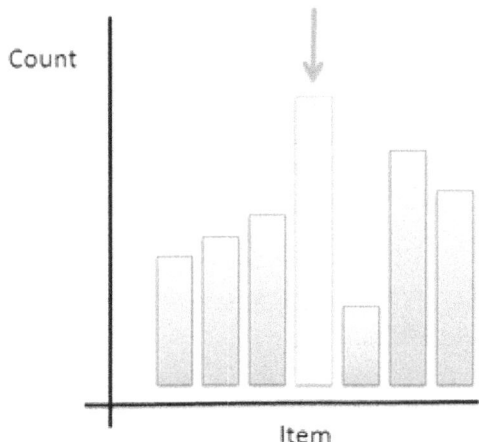

Source: https://betterexplained.com/articles/how-to-analyze-data-using-the-average/

Pros:
- For exclusive voting situations (this or that - no compromise).
- Gives a choice that the most people wanted (whereas the average can give a choice that nobody wanted).
- Simple to understand.

Cons:
- Requires more effort to compute (have to tally up the votes).
- Winner takes all (no middle path).

(iv) Percentile

A Percentile is a measure denoting the particular value corresponding to the particular percentage. If you are the fourth tallest person in a group of 20. 80% of people are shorter than you. That means you are at the 80th percentile. If your height is 1.85m then "1.85m" is the 80th percentile height in that group.

Source: https://www.mathsisfun.com/data/percentiles.html

Steps to calculate percentile

Step 1: Assigning Values
Find the 90th percentile of a data set with 8 values as given below.
Data Set = 8, 9, 10, 12, 1, 2, 5, 7
N = 8

Step 2: Ordering Data Set
Arrange the numbers in the data set in ascending order.
Data Set (Ordered) = 1, 2, 5, 7, 8, 9, 10, 12

Step 3: Finding Index Number
Index Number (I) = 90% x 8
= 0.90 x 8
= 7.2
= 7 (Rounded Off)

Step 4: Percentile Calculation
90th Percentile
= Value corresponding to the Index Number (7)
= Value at 7th place in data set
= 10

10) Decision & Probability Trees

A simple method of creating a predictive model is to use the decision tree. A decision tree is a tree-like model consisting of decisions and their possible consequences. In the decision tree, every node represents a test on a specific attribute and each branch represents the possible outcomes of this test. When you use your decision tree with a probability model, you can use it to calculate the conditional probability of an event happening, given that another event happens. Begin with the initial event, then follow the path from that event to the target event, multiplying the probability of each of those events together. In this way, a decision tree can be used like a tree diagram, which maps out the probabilities of certain events, such as flipping a coin twice.

Figure below shows the probability of getting flu. To work out the chance (probability) for a single branch we multiply each branch that makes up the path.

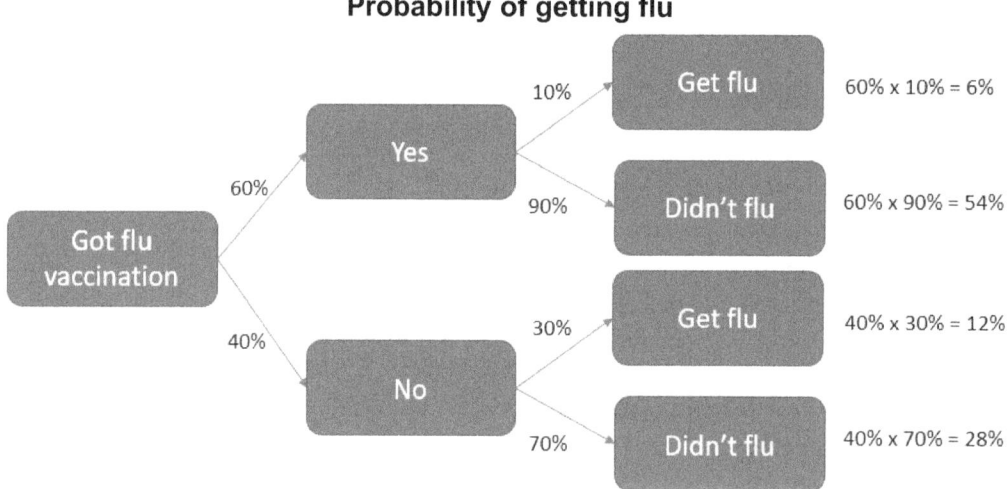

Probability of getting flu

Figure below shows 5 job candidates. To calculate the probability for a single branch we multiply the path of each branch. The probability of selecting a male candidate (2/5) followed by a female candidate (3/4) is calculated by multiplying 2/5 by 3/4.

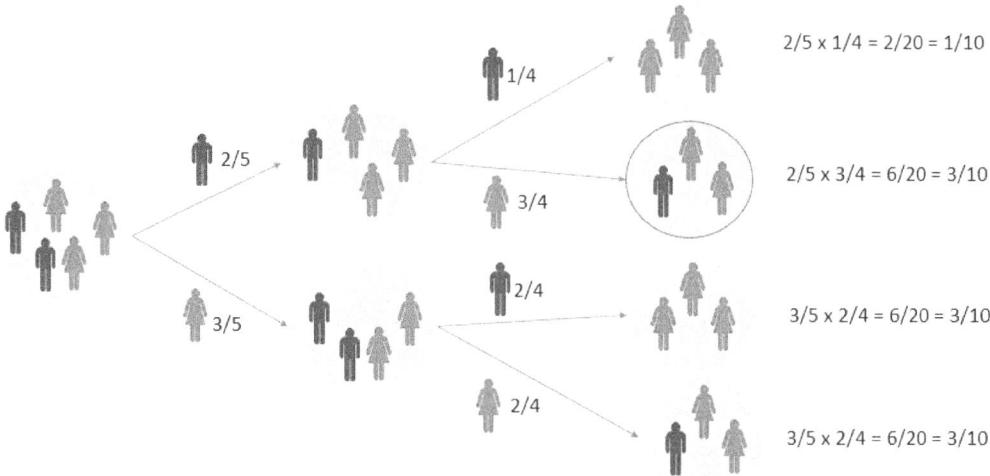

Probability of selecting a Male and Female candidate

10.1) Visualize Probability Trees with Excel

1) At cell B7, type "Male or female", then click "All Borders".

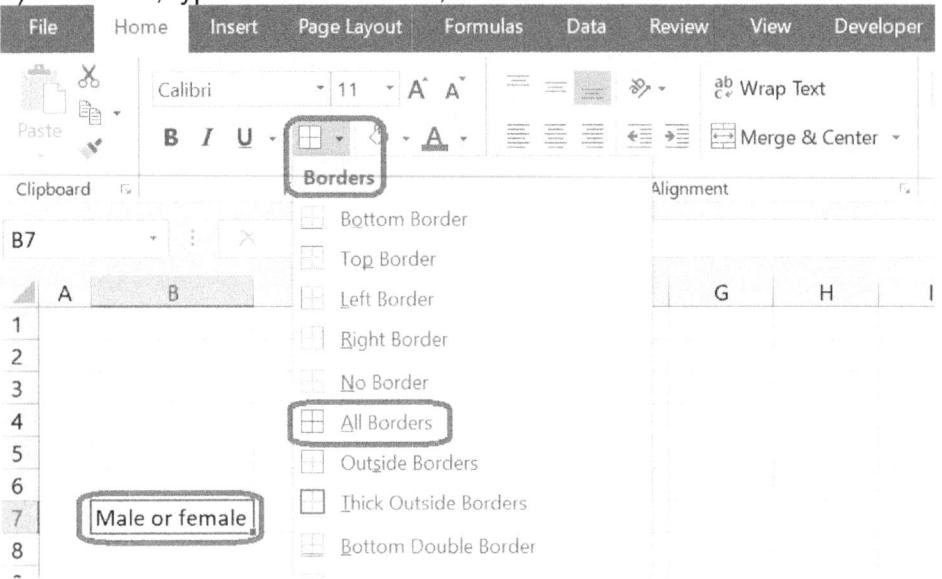

2) Select cells C5 and C6, click "Merge & Center", then input "0.4".

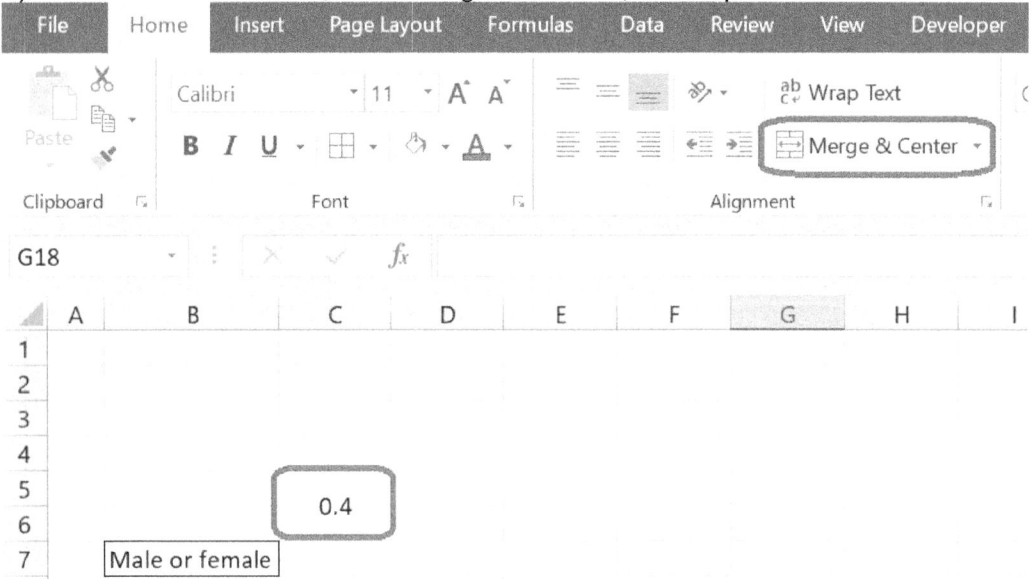

3) Right click, cells C5 and C6, and click "Format Cells".

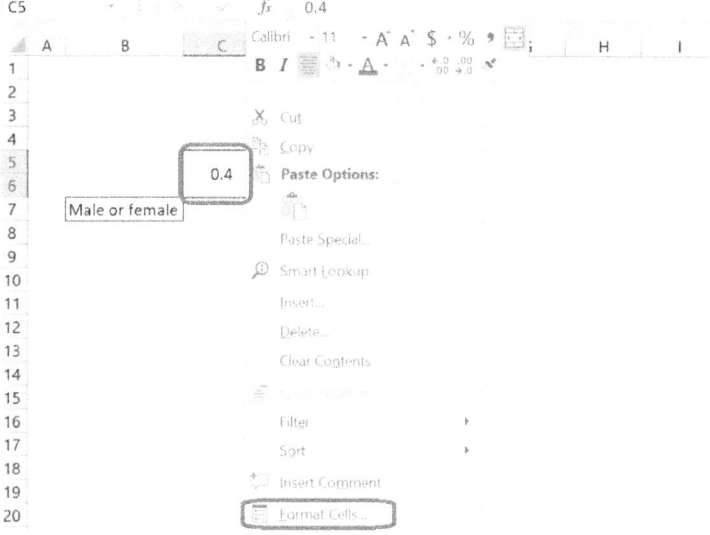

4) At the "Format Cells" window, click the diagonal icon, then click "OK".

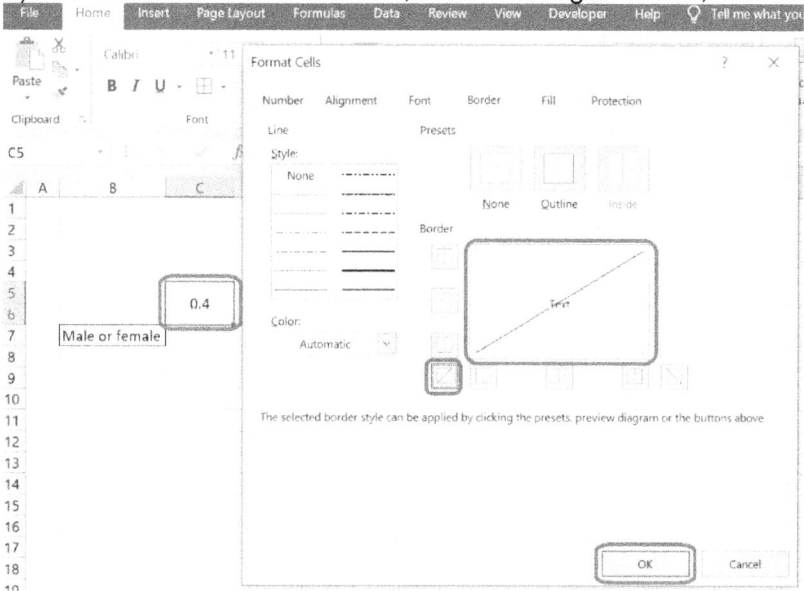

5) After you click "OK", you will see a diagonal line drawn in cells C5 and C6. Repeat the process for Cells C7 and C8.

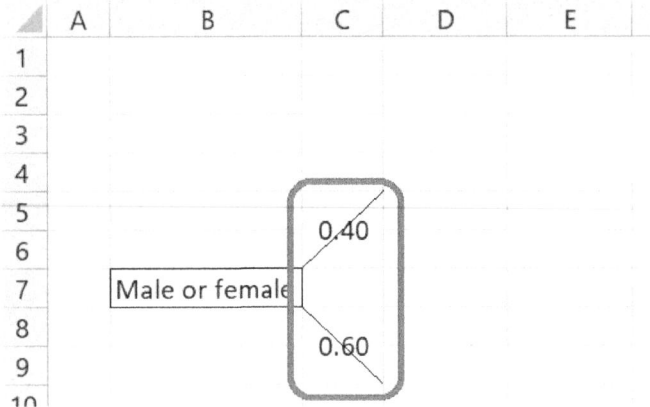

6) At cell D4, type "Male", then click "All Borders". Repeat the process for cell D9.

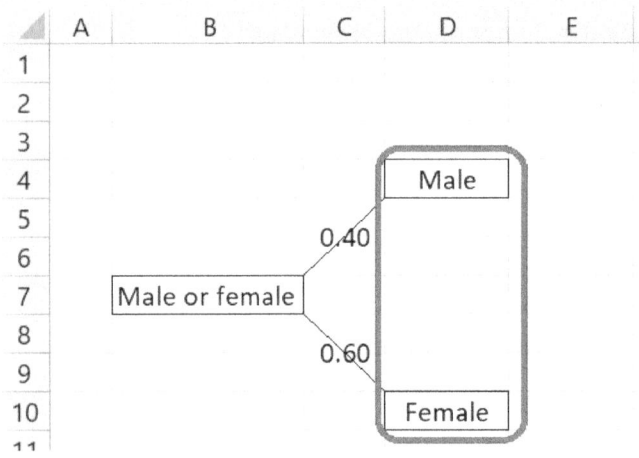

7) Input "0.25" in cell E3, then "right click" and select "Format Cells".

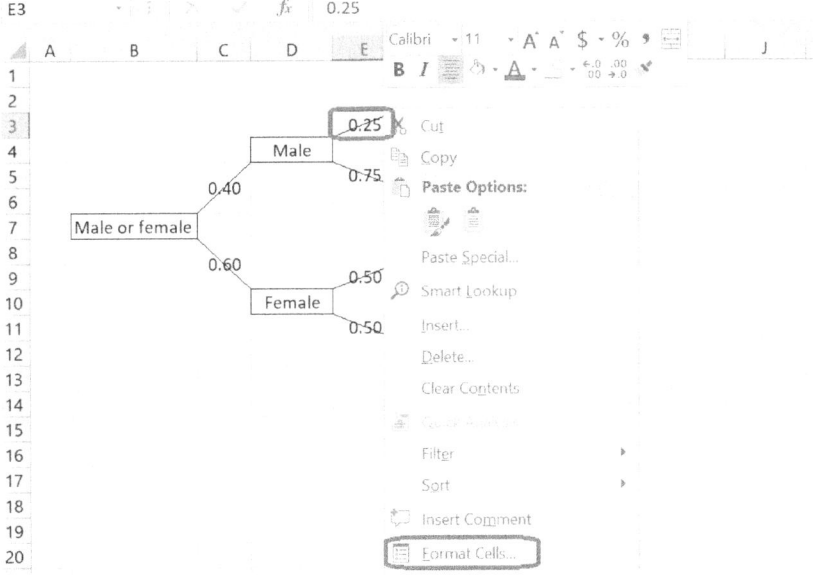

8) At the "Format Cells" window, click the diagonal icon, then click "OK". After you click "OK", you will see a diagonal line drawn in cell E3. Repeat the process for Cells E5, E9 and E11.

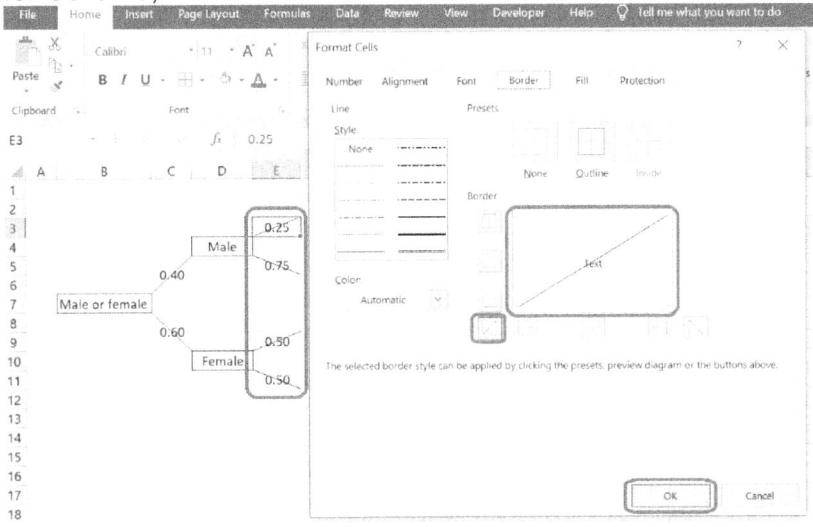

9) At cell F2, input "Male =". Repeat the process for Cells F6, F8 and F12.

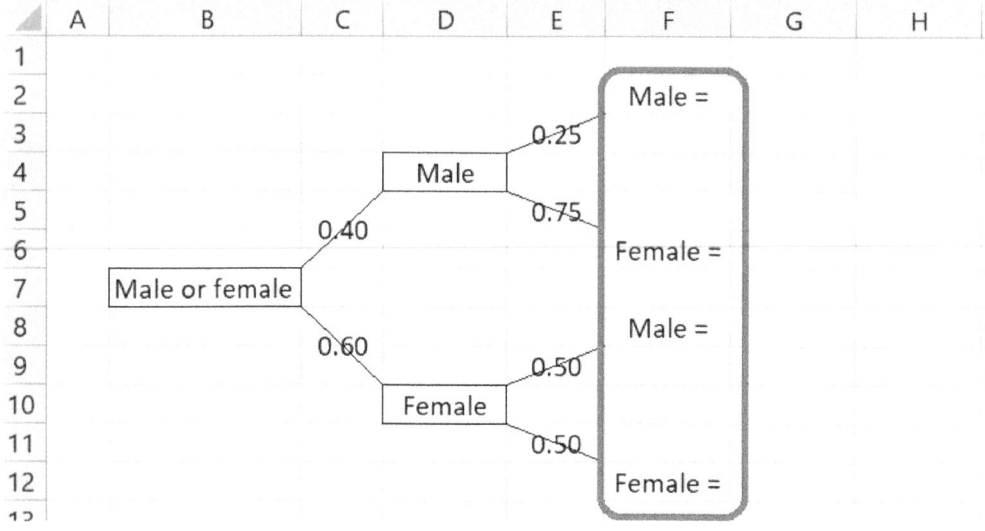

10) Input the following formulas
- For cell G2, input "=C5*E3"
- For cell G6, input "=C5*E5"
- For cell G8, input "=C8*E9"
- For cell G12, input "=C8*E11"

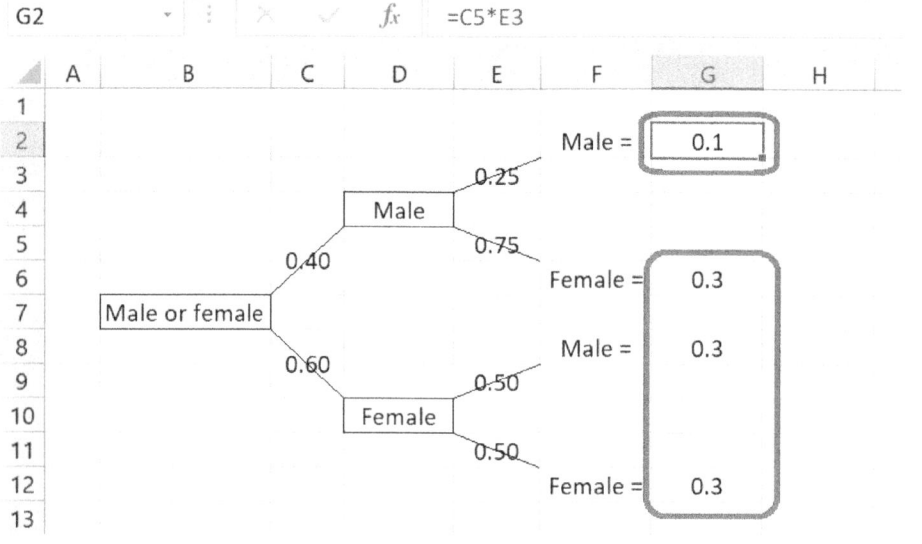

11) Select cells F2 to G2, and click "Outside Borders". Repeat the process for Cells G6, G8 and G12.

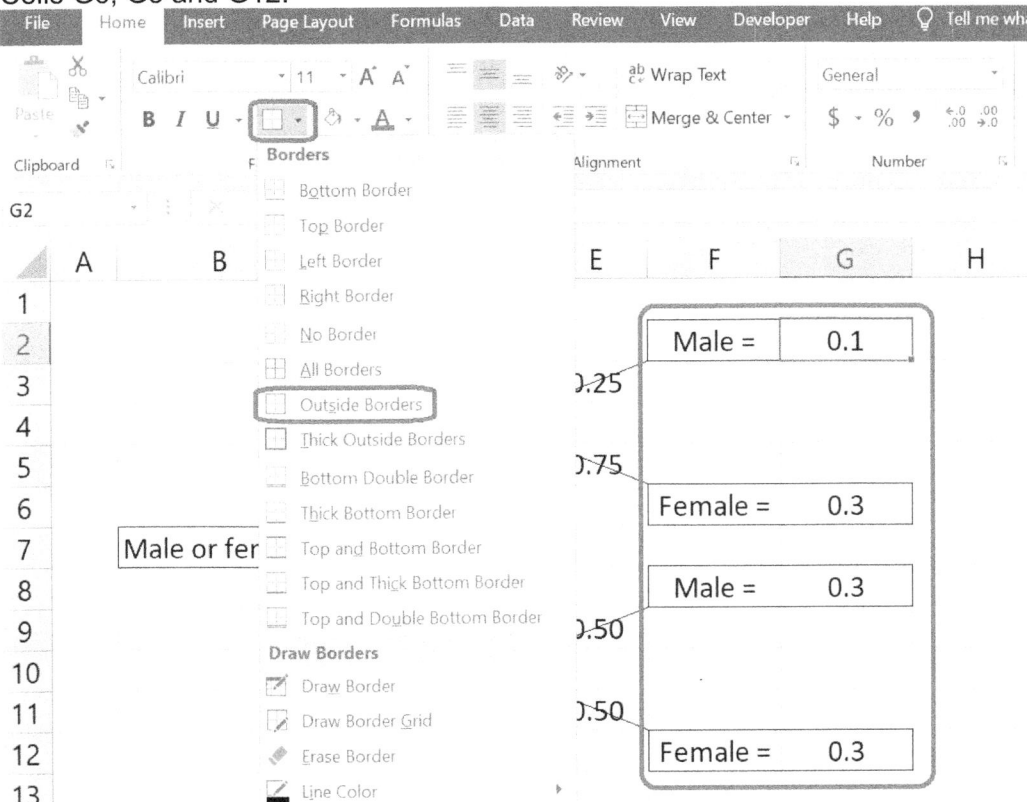

11) Correlation

Correlation means association - it is a measure of the extent to which two variables are related:

- **Positive Correlation:** If an increase in one variable leads an increase in another variable, it is called positive correlation. E.g. The number of calories you eat and your weight.

- **Negative Correlation:** If an increase in one variable leads to a decrease in another variable, it is called negative correlation. E.g. The temperature outside and your heating bills.

- **Zero Correlation:** When there is no relationship between two variables, it is called zero correlation. E.g. The colour of your eyes and your height

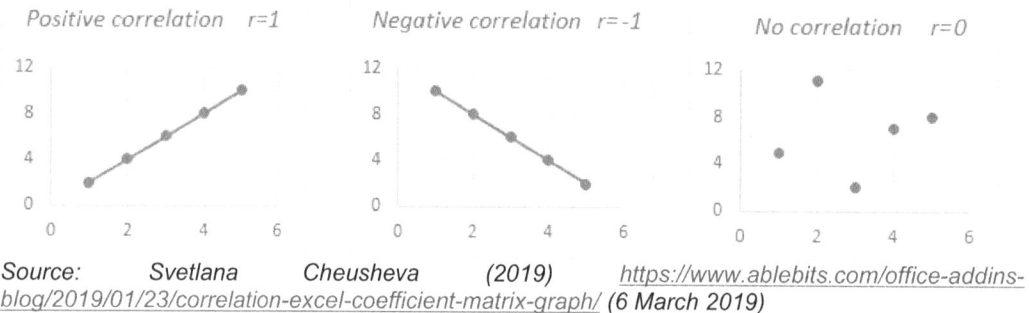

Source: Svetlana Cheusheva (2019) https://www.ablebits.com/office-addins-blog/2019/01/23/correlation-excel-coefficient-matrix-graph/ *(6 March 2019)*

Correlation Coefficients

The coefficient sign (plus or minus) indicates the direction of the relationship. The closer the number is to 1, the stronger the correlation is. For the correlation to be considered significant, the correlation must be 0.5 or above in either direction.:

- **A coefficient of 1**: denotes direct correlation. i.e. as one variable increases, the other increases proportionally.

- **A coefficient of -1:** denotes inverse correlation. i.e. as one variable increases, the other decreases proportionally.

- **A coefficient of 0:** denotes no correlation. i.e. the data points are scattered all over the graph.

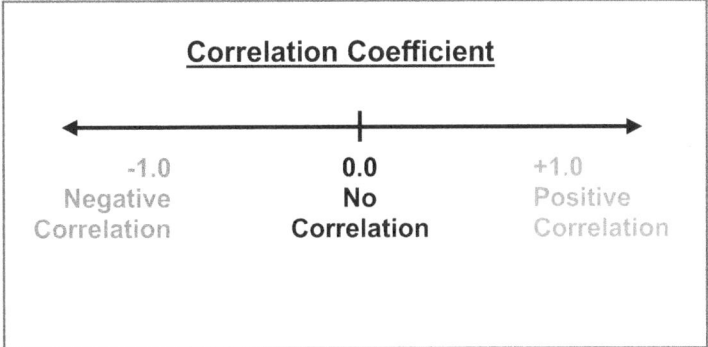

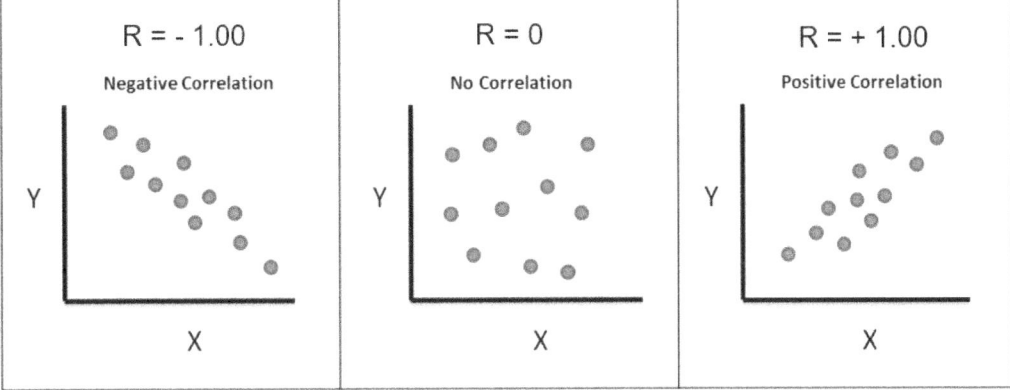

Limitations of Correlation

- **Correlation may not imply causation.**

It is important to note that a strong correlation relationship between two variables (variable X and Y) does not mean that there is a "causal" relationship. There might be a third, variable Z that is actually causing the change. It is impossible to determine what causes what, just from a correlation analysis. More analysis is needed to define the cause and effect relationship. However, correlation analysis does highlight implied relationships that we can combine with other knowledge to make better choices. The fact that the variables have a correlation is enough to take relevant action.

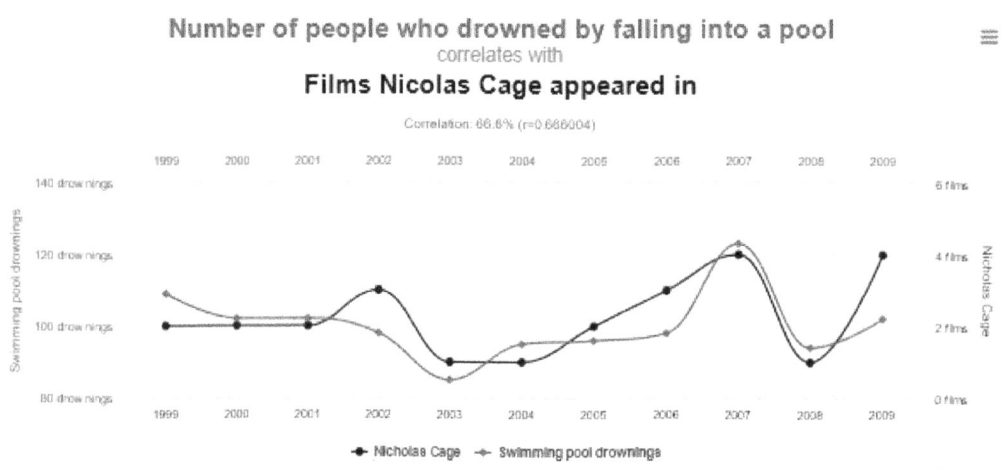

Source: tylervigen.com (2019) Spurious correlations. tylervigen.com/spurious-correlations (9 November 2018)

- **Correlation does not allow us to go beyond the data that is given.**

Even if there is a correlation between time spend studying and a student's grades, we cannot mathematically infer that spending 6 hours studying, will generate grade A or grade B.

11.1) Compute correlation coefficient with Excel

Correlations can be calculated very simply in Excel. Inside the correl() function, just identify both data sets with a comma between them. Where:
- Array1 is the first range of values.
- Array2 is the second range of values.

The two arrays should have equal length.

CORREL(array1, array2)

A8   fx   =CORREL(A2:A6,B2:B6)

	A	B
1	Variable X	Variable Y
2	30	100
3	20	70
4	40	90
5	50	160
6	60	130
7	Correlation	Description
8	0.80	Correlation coefficient of variable X and Y in columns A and B.

To use the CORREL function in Excel successfully, take note of these:
- If the cells in the array contains text or blanks, these cells are ignored.
- Cells with zero values are calculated.
- If the arrays are of different lengths, the #N/A error is returned.
- If either of the arrays is empty, the #DIV/0! error is returned.

11.2) Correlation Matrix using Excel

When you need to test interrelations between more than two variables, you will need construct a correlation matrix (also called multiple correlation coefficient). The correlation matrix is a table that shows the correlation coefficients between the variables at the intersection of the corresponding rows and columns.

The correlation matrix in Excel is built using the Correlation tool from the Analysis ToolPak add-in. This add-in is not enabled by default. "Analysis ToolPak" is an add-in for Microsoft Excel that comes with Microsoft Excel. To be able to run regression using Excel, you need to first install "Analysis ToolPak", an Excel add-in program that provides data analysis tools. To load the Analysis ToolPak add-in, follow these steps:

1) Install "Analysis ToolPak", an Excel add-in

- On the File tab, click Options.

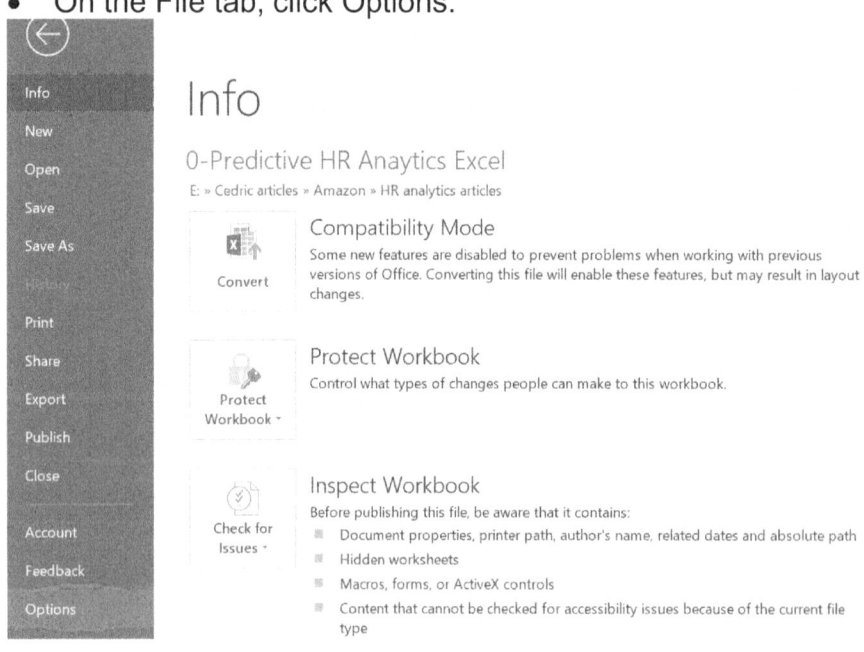

- Under Add-ins, click Analysis ToolPak and click the "Go" button.

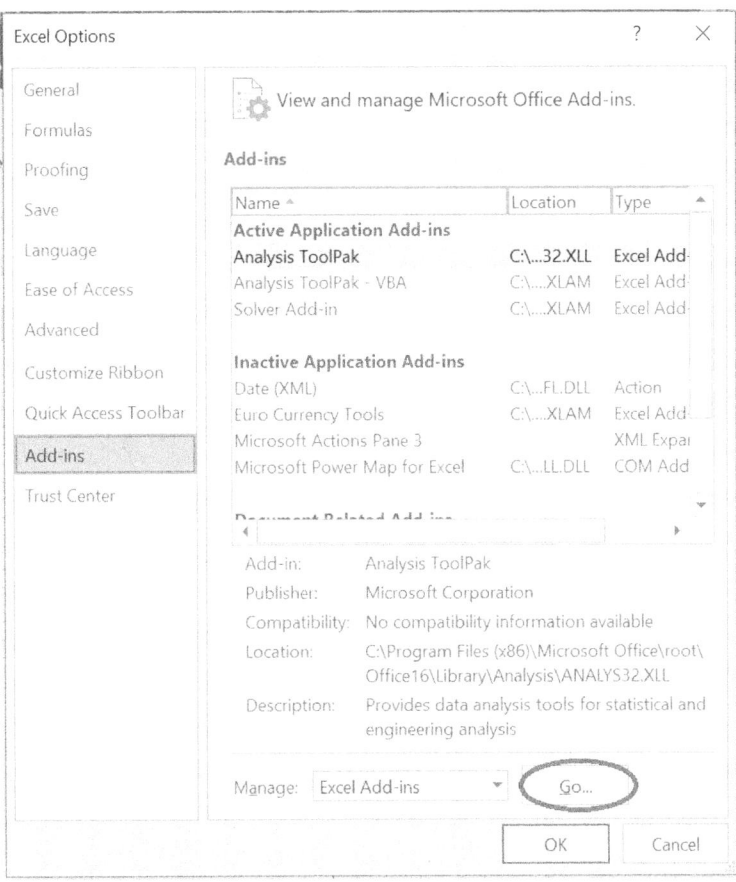

- Click "Analysis ToolPak" and click on OK.

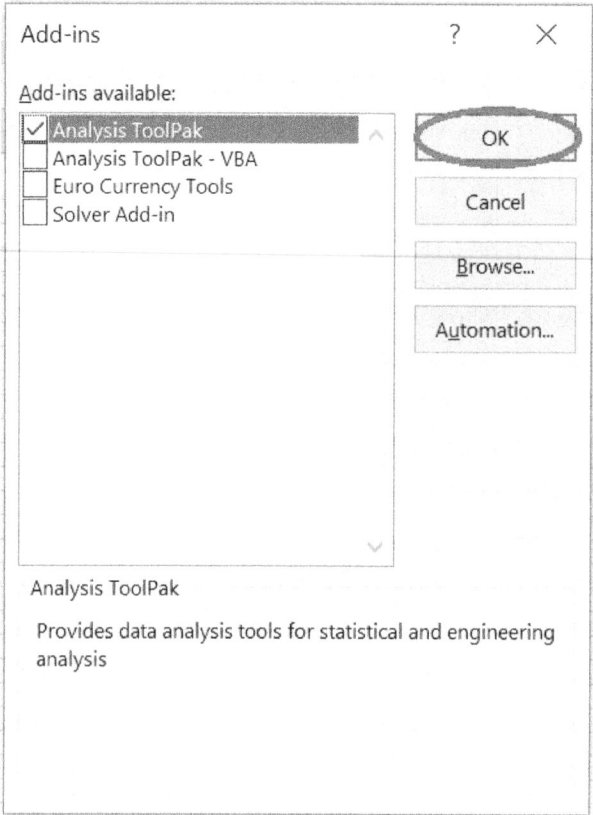

- On the Data tab, in the Analysis group, you are now able to click on "Data Analysis".

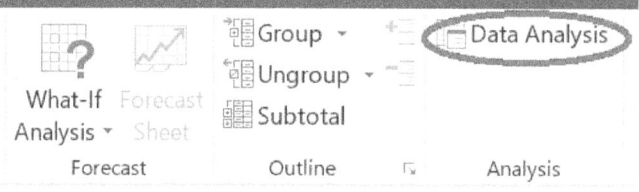

2) Copy the example data in the following table, and paste it in cell A1 of a new Excel worksheet.

	A	B	C	D
1	Month	Temperature	Number of Salesperson	Drinks Sold
2	Jan	-15.00	100	900
3	Feb	-10.00	110	990
4	Mar	5.00	90	850
5	Apr	20.00	70	750
6	May	25.00	45	300
7	Jun	20.00	40	250
8	Jul	35.00	50	250
9	Aug	25.00	30	300
10	Sep	20.00	70	400
11	Oct	10.00	90	550
12	Nov	5.00	80	900
13	Dec	-5.00	100	850

3) Select "Correlation" and click "OK".

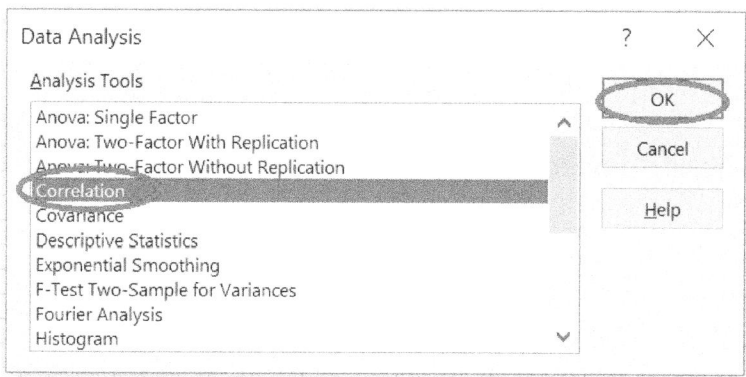

4) After you click OK in the "Data Analysis" dialog box, you will see a "Correlation" dialog box.

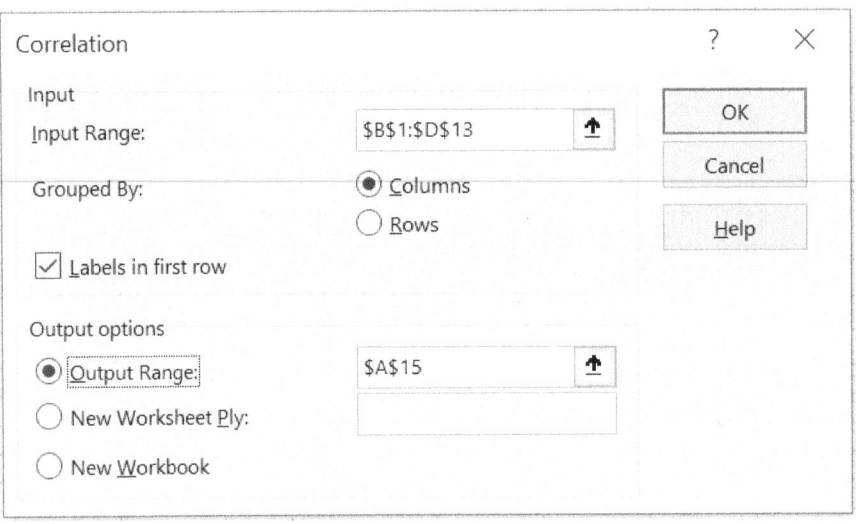

5) For "Input Range", select cells (B1:D13).

6) Check "Labels in first row".

7) For "Output Range", select cells (A15).

8) Click "OK".

Interpreting correlation analysis results

After you click "OK", Excel generates the below Correlation analysis. In your Excel correlation matrix, the coefficients are at the rows and columns intersection. If the column and row coordinates are the same, the value is 1. As we want to know the correlation between the dependent variable (Drinks Sold) and two independent variables (Temperature and Number of Salesperson), we need the highlighted numbers at the intersection of these rows and columns.
- The negative coefficient of -0.87 shows a strong inverse correlation between the "Temperature" and "Drinks Sold".
- The positive coefficient of 0.89 indicates a strong direct connection between "Number of Salesperson" and "Drinks Sold".

	A Month	B Temperature	C Number of Salesperson	D Drinks Sold
2	Jan	-15.00	100	900
3	Feb	-10.00	110	990
4	Mar	5.00	90	850
5	Apr	20.00	70	750
6	May	25.00	45	300
7	Jun	20.00	40	250
8	Jul	35.00	50	250
9	Aug	25.00	30	300
10	Sep	20.00	70	400
11	Oct	10.00	90	550
12	Nov	5.00	80	900
13	Dec	-5.00	100	850
14				
15		Temperature	Number of Salesperson	Drinks Sold
16	Temperature	1		
17	Number of Salesperson	-0.88	1	
18	Drinks Sold	-0.87	0.89	1

12) Regression

Correlation tells you if a relationship exists, whereas Regression can be used to infer causal relationships between the independent (input) and dependent (output) variables.

Regression analysis is a statistical technique that helps you predicts the level of one variable (dependent variable) based on the level of another variable (independent variable).
- Dependent Variable is the factor that you are trying to predict.
- Independent Variables are the factors that you hypothesize have an impact on your dependent variable.

The Dependent and Independent variables can be "numerical" or "categorical".
- **Numerical variables** (also called quantitative or continuous) are: age, income, temperature.
- **Categorical variables** (also called discrete, qualitative, nominal) are: gender, occupation, eye color.

12.1) Multiple Regression Example 1: Predict the Output when the Inputs are numerical variables

In this section, we will derive the multiple regression equation for continuous variables, using Excel and use this equation to test for significance in order to evaluate a hypothesis. The difference between simple linear regression and multiple linear regression is that, multiple linear regression has (>1) independent variables, whereas simple linear regression has only 1 independent variable. This example shows you step-by-step how to run an Excel multiple linear regression analysis, and how to read the Summary Output. The objective is to predict the number of Umbrellas sold (Output), from the historical Selling Price and Advertising Cost (Input).

1) Install "Analysis ToolPak", an Excel add-in.

"Analysis ToolPak" is an add-in for Microsoft Excel that comes with Microsoft Excel. To be able to run regression using Excel, you need to first install "Analysis ToolPak", an Excel add-in program that provides data analysis tools. To load the Analysis ToolPak add-in, follow these steps:

On the File tab, click Options.

Under Add-ins, click Analysis ToolPak and click the "Go" button.

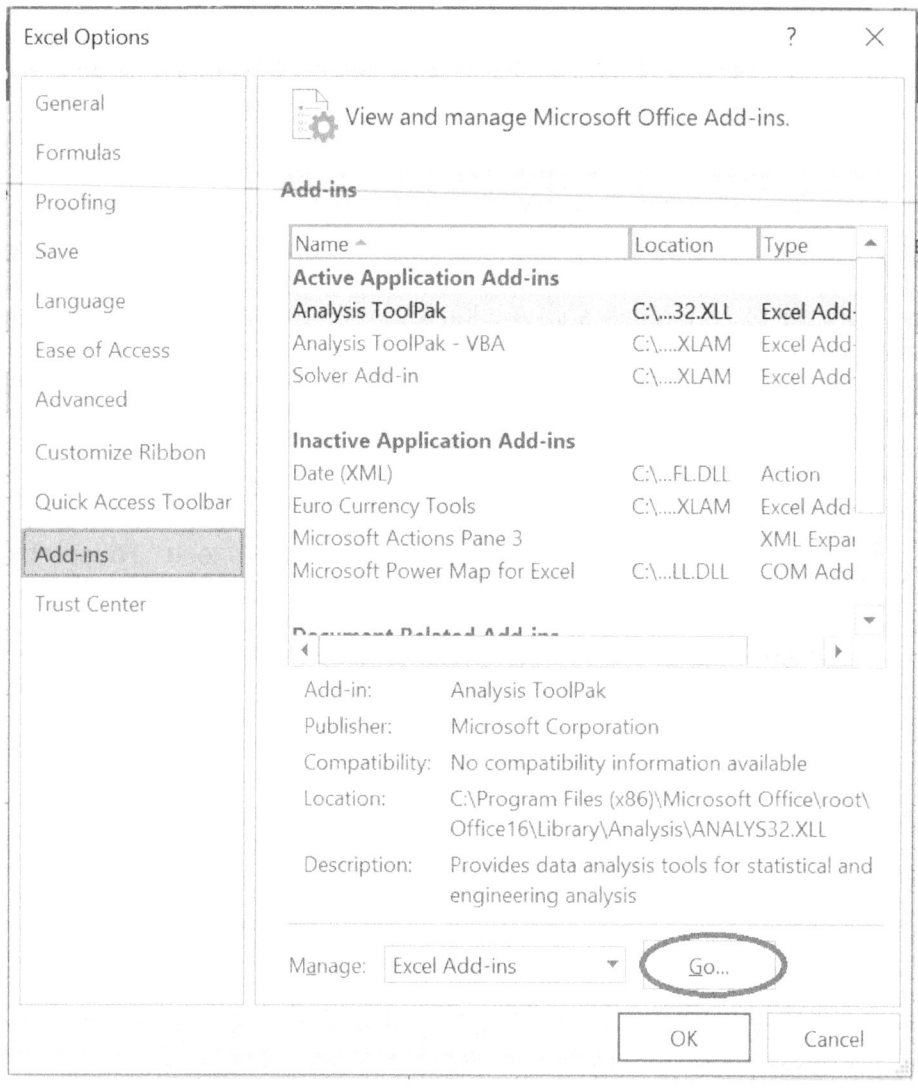

Click "Analysis ToolPak" and click on OK.

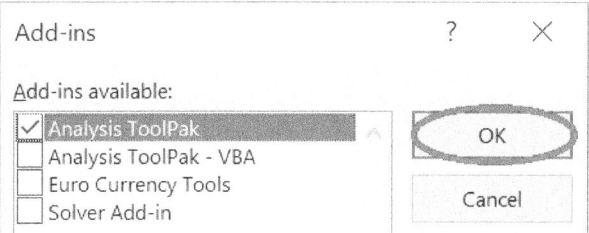

On the Data tab, in the Analysis group, you are now able to click on "Data Analysis".

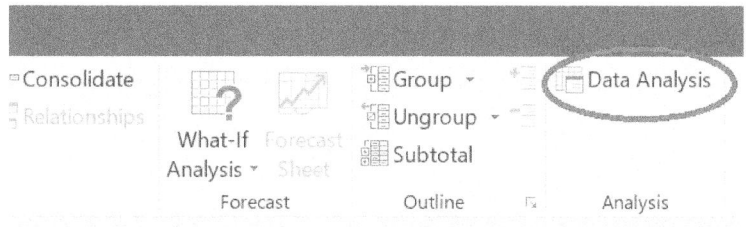

2) Copy the example data in the following table, and paste it in cell A1 of a new Excel worksheet.

	A	B	C
1	No. of Umbrellas sold	Selling Price	Advertising cost
2	850	$3	$280
3	470	$6	$20
4	580	$4	$40
5	740	$3	$50
6	620	$6	$320
7	730	$4	$180
8	560	$5	$90

3) On the Data tab, in the Analysis group, click on "Data Analysis".

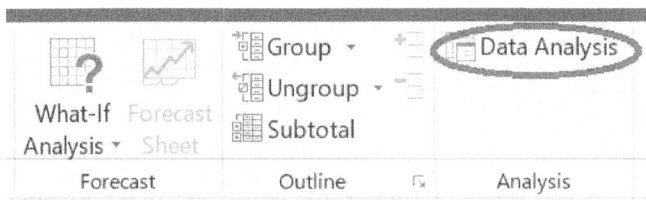

4) Select "Regression" and click "OK".

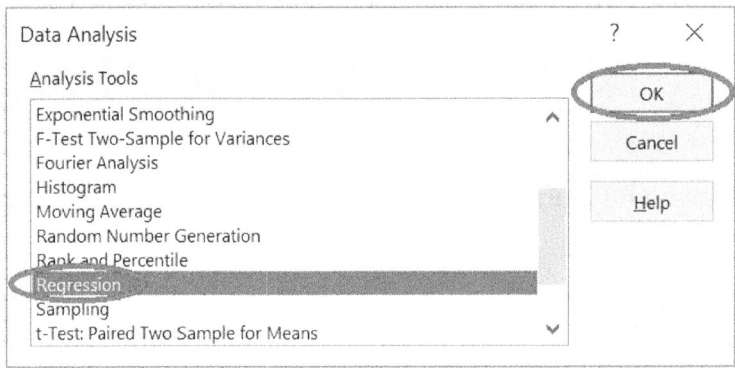

5) After you click OK in the "Data Analysis" dialog box, you will see a "Regression" dialog box.
6) For "Input Y Range", select cells (A1:A8). This is the predictor variable or dependent variable.
7) For "Input X Range", select cells (B1:C8). These are the explanatory variables or independent variables.
8) Check "Labels" box.
9) Click the "Output Range" box, and select cell A12.
10) Click "OK".

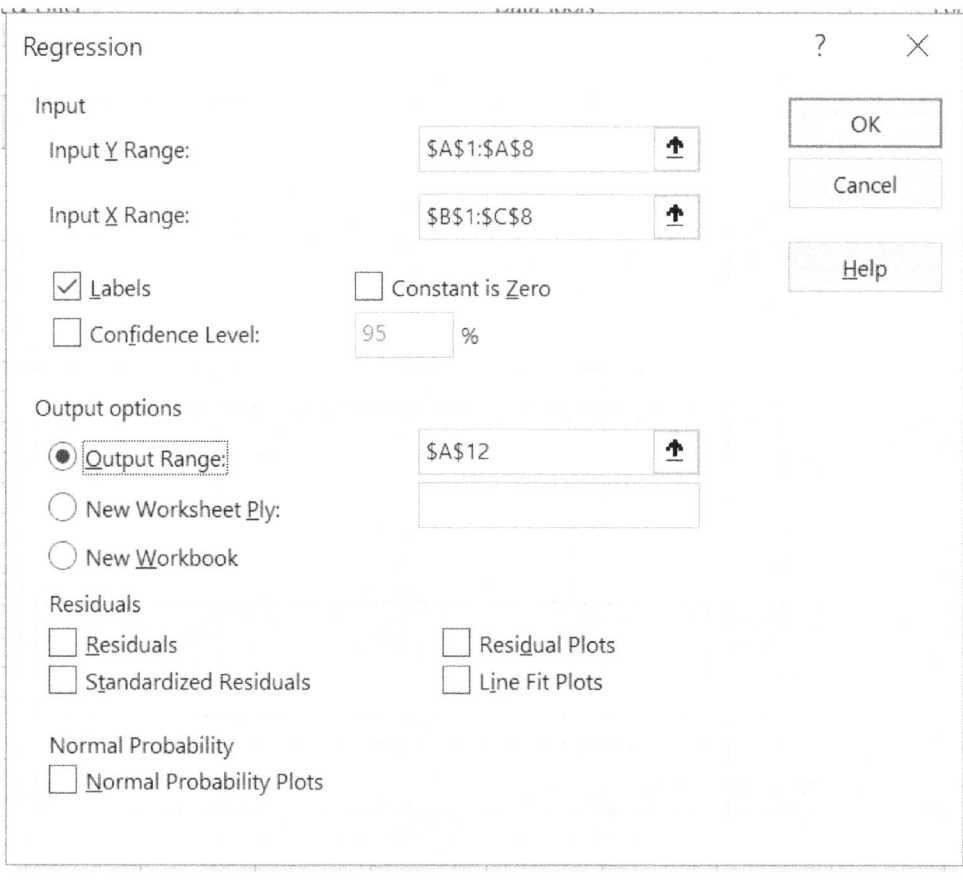

After you click "OK", Excel generates the following Summary Output. Round the numbers to 3 decimal places.

SUMMARY OUTPUT

Regression Statistics	
Multiple R	0.981
R Square	0.962
Adjusted R Square	0.943
Standard Error	31.052
Observations	7

ANOVA

	df	SS	MS	F	Significance
Regression	2	96942.996	48471.498	50.269	0.001
Residual	4	3857.004	964.251		
Total	6	100800			

	Coefficients	Standard Error	t Stat	P-value	Lower 95%	Upper 95%	Lower 95.0%	Upper 95.0%
Intercept	937.194	47.655	19.666	0.000	804.882	1069.505	804.882	1069.505
Selling Price	-83.572	9.965	-8.386	0.001	-111.240	-55.904	-111.240	-55.904
Advertising cost	0.592	0.104	5.676	0.005	0.303	0.882	0.303	0.882

R Square: In the output, R Square is 0.962, which means it is a very good fit. 96% of the variation in No. of Umbrellas sold (Output) is explained by the independent variables (Input), Selling Price and Advertising Cost. The closer R Square is to "1", the better the regression line fits the data.

Significance F and P-values: To determine if your results are statistically significant (i.e. reliable), check "Significance F" (0.001). If the value of "Significance F" is less than 0.05, it is statistically significant (i.e. reliable). If "Significance F" is bigger than 0.05, don't use this set of independent variables. Delete those variables with "P-value" that is bigger than 0.05 and run the regression again until "Significance F" drops below 0.05. Most or all of your P-values should be lower than 0.05. In our example below "P-value" is 0.000, 0.001 and 0.005, for Intercept, Selling Price, and Advertising Cost, respectively.

ANOVA	df	SS	MS	F	Significance
Regression	2	96942.996	48471.498	50.269	0.001
Residual	4	3857.004	964.251		
Total	6	100800			

	Coefficients	Standard Error	t Stat	P-value	Lower 95%	Upper 95%	Lower 95.0%	Upper 95.0%
Intercept	937.194	47.655	19.666	0.000	804.882	1069.505	804.882	1069.505
Selling Price	-83.572	9.965	-8.386	0.001	-111.240	-55.904	-111.240	-55.904
Advertising cost	0.592	0.104	5.676	0.005	0.303	0.882	0.303	0.882

Coefficients

From the Summary Output, the regression line is:
No. of Umbrellas sold (Output)
= 937.194 – (83.572 * Selling Price) + (0.592 * Advertising cost)

Based on the above regression formula,
- For each unit increase in Selling Price, No. of Umbrellas sold decreases with 83.572 units.
- For each unit increase in Advertising cost, No. of Umbrellas sold increases with 0.592 units.

Coefficients can also be used for forecasting. For example, if "Selling Price" is $5 and "Advertising Cost" is $300, **predicted "No. of Umbrellas sold"**
= 937.194 – (83.572 * Selling Price) + (0.592 * Advertising cost)
= 937.194 - 83.572 * 5 + 0.592 * 300
= 697.

12.2) Multiple Regression Example 2: Predict the Output when the Inputs consist of 1 categorical variable with 2 variables

Variables can be: numerical or categorical.
- **Numerical variables** (also called quantitative or continuous) are: age, income, temperature.
- **Categorical variables** (also called discrete, qualitative, nominal) are: gender, occupation, eye color.

The Regression data analysis tool cannot analyze non-numeric data. You need to convert the alphanumeric data to numeric form.

In this section, we show you how to analyze regression equations when one of the independent variables is categorical. The key to the analysis is to express categorical variables as dummy variables. A dummy variable is a numeric variable to represents categorical data, such as gender, ethnic group, etc. Dummy variables can take on only two quantitative values. Typically, 1 represents the presence of a qualitative attribute, and 0 represents the absence.

Consider the table below. Three variables are used to describe 10 students. "Gender" variable is categorical. "Entry test score" and "IQ" variables are quantitative. In this example, we want to test if IQ and Gender are predictors of "Entry test score":

	A	B	C	D
1	Candidate	Entry test score	IQ score	Gender
2	1	96	131	Male
3	2	89	126	Female
4	3	99	121	Male
5	4	83	116	Female
6	5	95	110	Male
7	6	77	105	Female
8	7	87	100	Male
9	8	79	95	Female
10	9	75	89	Male
11	10	76	84	Female

Dummy Variable

The first step is to express the "gender" variable as one or more dummy variables. To determine how many dummy variables, look at the number of values (k) "Gender" variable has. You need k - 1 dummy variables to represent a categorical variable. As the "Gender" categorical variable has two values (male or female), only one dummy variable (X1) is needed to represent "Gender":
- $X_1 = 1$ for male candidates.
- $X_1 = 0$ for non-male candidates.

Now, replace the "Gender" variable in the table with X_1:

	A	B	C	D
1	Candidate	Entry test score	IQ score	X_1
2	1	96	131	1
3	2	89	126	0
4	3	99	121	1
5	4	83	116	0
6	5	95	110	1
7	6	77	105	0
8	7	87	100	1
9	8	79	95	0
10	9	75	89	1
11	10	76	84	0

Regression Equation

For this example, your regression equation is:
$$y = a_0 + a_1 IQ + a_2 X_1$$

- y => represents "Entry test score" predicted value
- IQ => represents "IQ score"
- X1 => represents the "Gender" dummy variable
- a_0, a_1, a_2 => represents regression coefficients

IQ and X_1 are known inputs where we can get the values from the data table. Unknow values are the regression coefficients (a_0, a_1, a_2), which we will estimate through regression. The next task in our analysis is to assign values to coefficients in our regression equation using Excel.

1) Install "Analysis ToolPak", an Excel add-in.

"Analysis ToolPak" is an add-in for Microsoft Excel that comes with Microsoft Excel. To be able to run regression using Excel, you need to first install "Analysis ToolPak", an Excel add-in program that provides data analysis tools. To load the Analysis ToolPak add-in, follow these steps:

On the File tab, click Options.

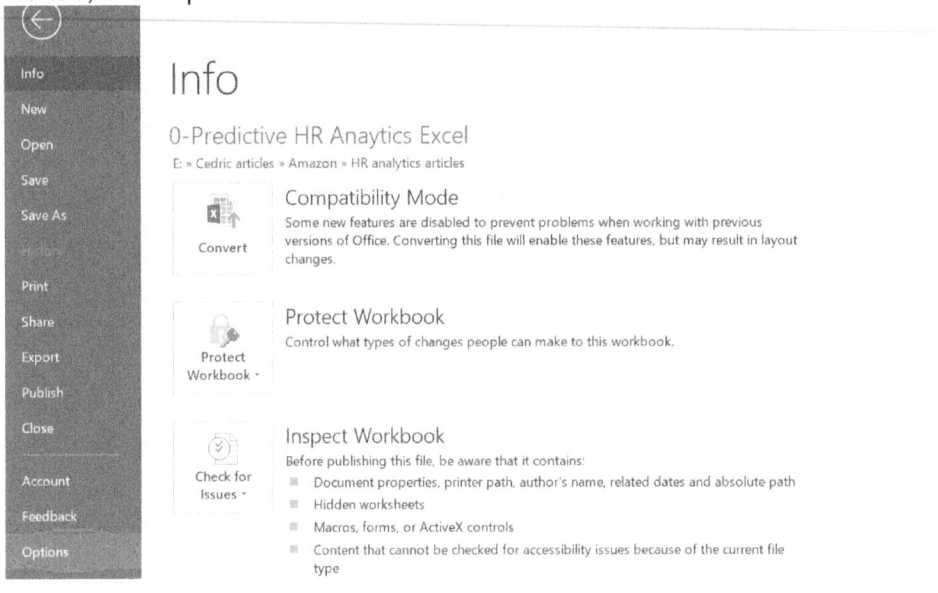

Under Add-ins, click Analysis ToolPak and click the "Go" button.

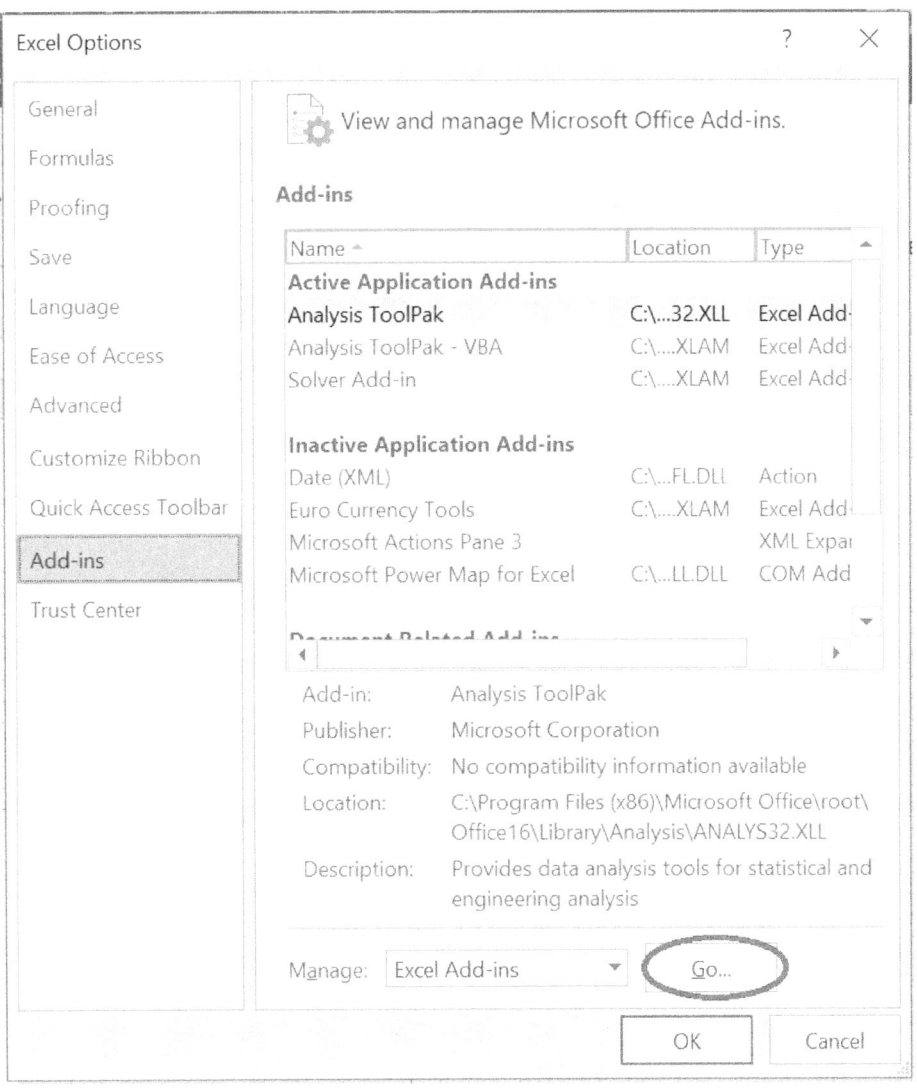

Click "Analysis ToolPak" and click on OK.

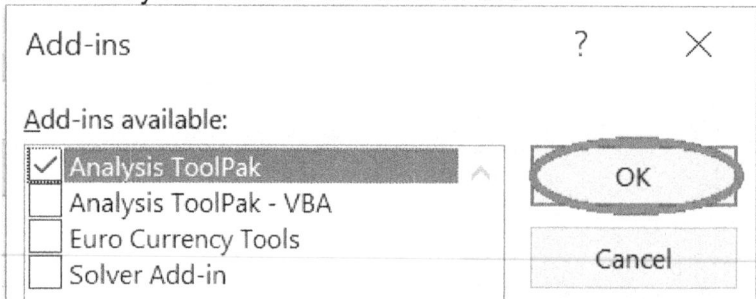

On the Data tab, in the Analysis group, you are now able to click on "Data Analysis".

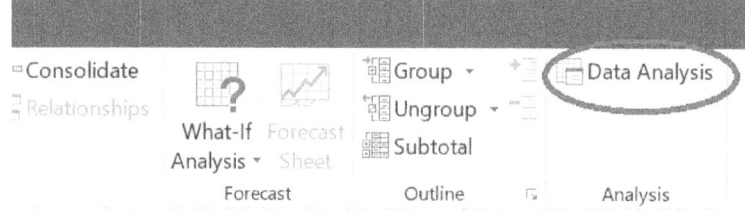

2) Select "Regression" and click "OK".

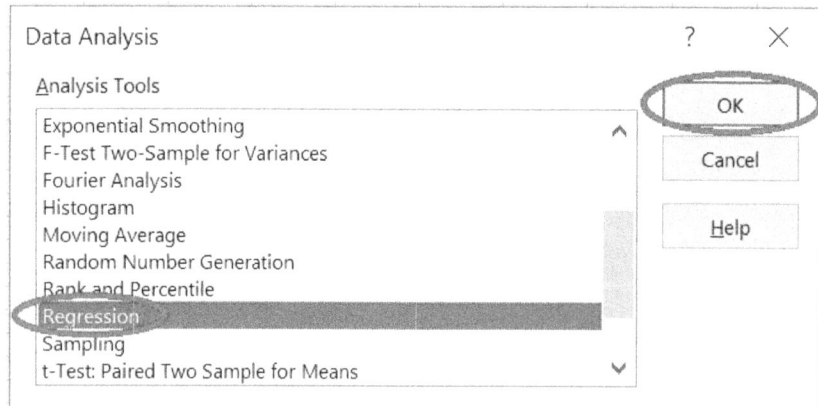

3) After you click OK in the "Data Analysis" dialog box, you will see a "Regression" dialog box.
4) For "Input Y Range", select cells (B1:B11). This is the predictor variable or dependent variable.
5) For "Input X Range", select cells (C1:D11). These are the explanatory variables or independent variables.
6) Check "Labels" box.
7) Click the "Output Range" box, and select cell A14.
8) Click "OK".

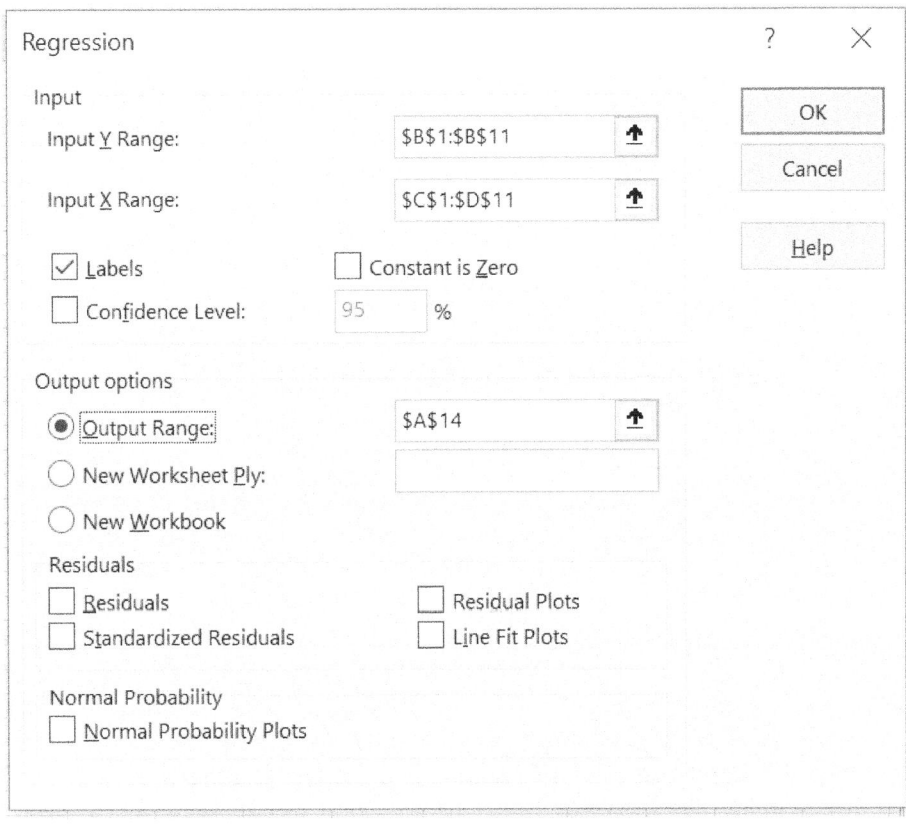

After you click "OK", Excel generates the following Summary Output. Round the numbers to 3 decimal places.

SUMMARY OUTPUT

Regression Statistics	
Multiple R	0.903
R Square	0.815
Adjusted R Square	0.762
Standard Error	4.360
Observations	10

ANOVA

	df	SS	MS	F	Significance F
Regression	2	585.347	292.673	15.398	0.003
Residual	7	133.053	19.008		
Total	9	718.400			

	Coefficients	Standard Error	t Stat	P-value	Lower 95%	Upper 95%	Lower 95.0%	Upper 95.0%
Intercept	38.598	9.959	3.876	0.006	15.049	62.146	15.049	62.146
IQ score	0.401	0.093	4.321	0.003	0.182	0.621	0.182	0.621
X1	7.594	2.796	2.716	0.030	0.982	14.206	0.982	14.206

R Square: In the output, R Square is 0.815, which means it is a very good fit. 81% of the variation in Annual Salary (Output) is explained by the independent variables (Input), "IQ Score" and X1 (Gender). The closer R Square is to "1", the better the regression line fits the data.

Significance F and P-values: With multiple regression, there is more than one independent variable - thus we need to determine whether each independent variable contributes significantly to the regression after effects of other variables are considered. To determine if your results are statistically significant (i.e. reliable), check "Significance F" (0.001). If the value of "Significance F" is less than 0.05, it is statistically significant (i.e. reliable). If "Significance F" is bigger than 0.05, don't use this set of independent variables. Delete those variables with "P-value" that is bigger than 0.05 and run the regression again until "Significance F" drops below 0.05. Most or all of your P-values should be lower than 0.05.

ANOVA	df	SS	MS	F	Significance F
Regression	2	585.347	292.673	15.398	0.003
Residual	7	133.053	19.008		
Total	9	718.400			

	Coefficients	Standard Error	t Stat	P-value	Lower 95%	Upper 95%	Lower 95.0%	Upper 95.0%
Intercept	38.598	9.959	3.876	0.006	15.049	62.146	15.049	62.146
IQ score	0.401	0.093	4.321	0.003	0.182	0.621	0.182	0.621
X1	7.594	2.796	2.716	0.030	0.982	14.206	0.982	14.206

In our example above "P-value" is 0.006, 0.003 and 0.030, for Intercept, IQ Score, and X1 (Gender), respectively. The P-value for IQ Score, and X1 (Gender) are both statistically significant, as they are lower than 0.05. This means that IQ Score contributes significantly to the regression after the effect of X1 (Gender) is considered. Similarly, X1 (Gender) contributes significantly to the regression after the effect of IQ Score is considered.

The "Gender" categorical variable has two values (male or female), and the dummy variable (X1) was used to represent "Gender":
- $X_1 = 1$ for male candidates.
- $X_1 - 0$ for non-male candidates.

The regression coefficient for X1 (Gender) measures the difference between the dummy variable (males) and it's reference variable (females). In our table, the regression coefficient for X1 (Gender) is 7.594. This means that, after effects of IQ Score is considered, males will score 7.594 points higher for the Entry test, than it's reference group (females). And, because the P-value for X1 (Gender) is statistically significant at 0.030 (less than 0.05), it is a real effect.

	Coefficients	Standard Error	t Stat	P-value	Lower 95%	Upper 95%	Lower 95.0%	Upper 95.0%
Intercept	38.598	9.959	3.876	0.006	15.049	62.146	15.049	62.146
IQ score	0.401	0.093	4.321	0.003	0.182	0.621	0.182	0.621
X1	7.594	2.796	2.716	0.030	0.982	14.206	0.982	14.206

Coefficients

From the Summary Output, the regression line is:
Entry test score (Output)
= 38.598 + 0.401 * IQ Score + 7.594 * X1 (Gender)

Based on the above regression formula,
- For each unit increase in IQ Score, Entry test score increases with 0.401 units.
- For each unit increase in X1 (Gender), Entry test score increases with 7.594 units.

Coefficients can also be used for forecasting. For example, if IQ Score is 80 and X1 (Gender) is 1 (Male), **predicted Entry test score**
= 38.598 + 0.401 * 80 + 7.594 * 1
= 78

12.3) Multiple Regression Example 3: Predict the Output when the Inputs consist of 1 categorical variable with 3 variables

The Regression data analysis tool cannot analyze non-numeric data (Categorical data) - they need to be coded into numeric dummy variables. Each dummy variable must be coded into either 0 or 1.

Earlier we analyzed 1 categorical variable (Gender) with 2 variables (male or female). In this example, we will demonstrate how to analyze 1 categorical variable with 3 variables. We want to test if Age, Company, and Gender are predictors Salary. The independent variables, Company and Gender are categorical, while Age is quantitative.

	A	B	C	D
1	Age	Company	Gender	Salary
2	22	Company A	Male	54000
3	27	Company B	Male	47000
4	47	Company C	Male	67000
5	37	Company A	Female	59000
6	52	Company B	Female	51000
7	57	Company C	Female	51000
8	41	Company A	Male	69000
9	50	Company B	Male	66000
10	32	Company C	Male	55000
11	29	Company A	Female	51000
12	49	Company B	Female	44000
13	23	Company C	Female	31000
14	50	Company A	Male	91000
15	26	Company C	Male	51000
16	30	Company C	Female	48000
17	42	Company B	Female	37000

The data in column A to D, is converted to the data in column F to J, because Company and Gender are categorical, and needs to be coded into numeric dummy variables.

	A	B	C	D	E	F	G	H	I	J
1	Age	Company	Gender	Salary		Age	Company#1	Company#2	Gender1	Salary
2	22	Company A	Male	54000		22	1	0	1	54000
3	27	Company B	Male	47000		27	0	1	1	47000
4	47	Company C	Male	67000		47	0	0	1	67000
5	37	Company A	Female	59000		37	1	0	0	59000
6	52	Company B	Female	51000		52	0	1	0	51000
7	57	Company C	Female	51000		57	0	0	0	51000
8	41	Company A	Male	69000		41	1	0	1	69000
9	50	Company B	Male	66000		50	0	1	1	66000
10	32	Company C	Male	55000		32	0	0	1	55000
11	29	Company A	Female	51000		29	1	0	0	51000
12	49	Company B	Female	44000		49	0	1	0	44000
13	23	Company C	Female	31000		23	0	0	0	31000
14	50	Company A	Male	91000		50	1	0	1	91000
15	26	Company C	Male	51000		26	0	0	1	51000
16	30	Company C	Female	48000		30	0	0	0	48000
17	42	Company B	Female	37000		42	0	1	0	37000

There are 3 possible values for the Company variable and 2 possible values for the Gender variable. If a Categorical variable has k values, the model will require k – 1 dummy variables.

As Gender categorical variable has 2 values (Male and Female), a dummy variable, called Gender1, is enough to code Gender, as follows:

	Dummy variable
	Gender1
Male	1
Female	0

As Company categorical variable has 3 values (Company A, Company B, Company C), 2 dummy variables called Company#1 and Company#2, are needed to code Company, as follows:

	Dummy Variable	
Company	Company#1	Company#2
Company A	1	0
Company B	0	1
Company C	0	0

We can now perform regression analysis on the highlighted data with the numeric dummy variables.

	A	B	C	D	E	F	G	H	I	J
1	Age	Company	Gender	Salary		Age	Company#1	Company#2	Gender1	Salary
2	22	Company A	Male	54000		22	1	0	1	54000
3	27	Company B	Male	47000		27	0	1	1	47000
4	47	Company C	Male	67000		47	0	0	1	67000
5	37	Company A	Female	59000		37	1	0	0	59000
6	52	Company B	Female	51000		52	0	1	0	51000
7	57	Company C	Female	51000		57	0	0	0	51000
8	41	Company A	Male	69000		41	1	0	1	69000
9	50	Company B	Male	66000		50	0	1	1	66000
10	32	Company C	Male	55000		32	0	0	1	55000
11	29	Company A	Female	51000		29	1	0	0	51000
12	49	Company B	Female	44000		49	0	1	0	44000
13	23	Company C	Female	31000		23	0	0	0	31000
14	50	Company A	Male	91000		50	1	0	1	91000
15	26	Company C	Male	51000		26	0	0	1	51000
16	30	Company C	Female	48000		30	0	0	0	48000
17	42	Company B	Female	37000		42	0	1	0	37000

1) Install "Analysis ToolPak", an Excel add-in.

"Analysis ToolPak" is an add-in for Microsoft Excel that comes with Microsoft Excel. To be able to run regression using Excel, you need to first install "Analysis ToolPak", an Excel add-in program that provides data analysis tools. To load the Analysis ToolPak add-in, follow these steps:

On the File tab, click Options.

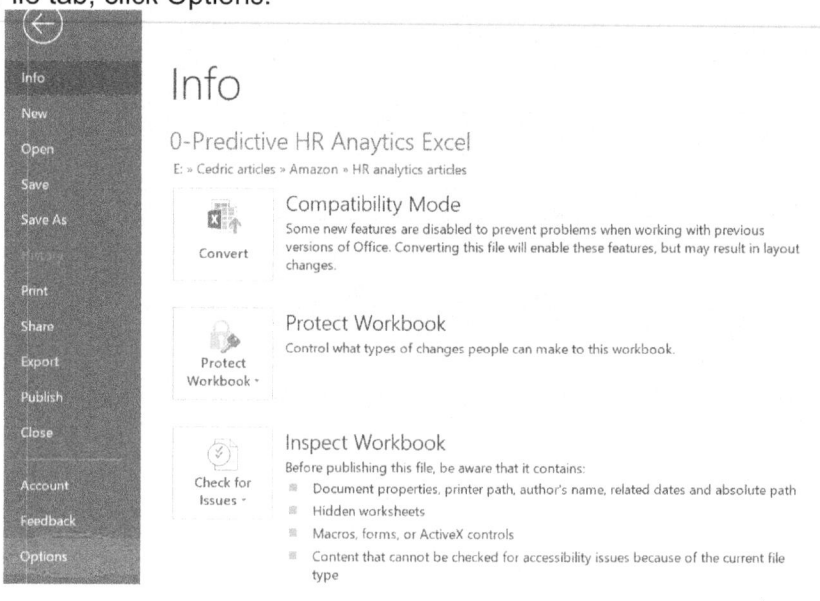

Under Add-ins, click Analysis ToolPak and click the "Go" button.

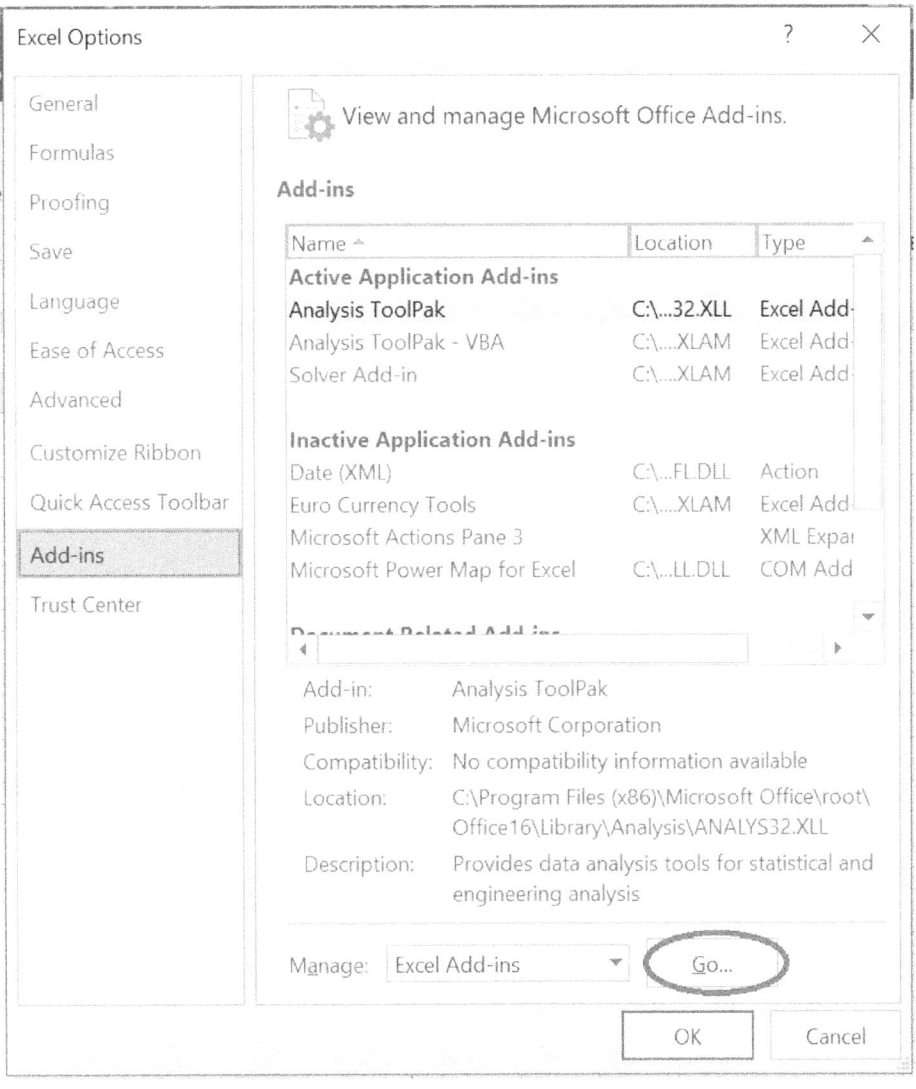

Click "Analysis ToolPak" and click on OK.

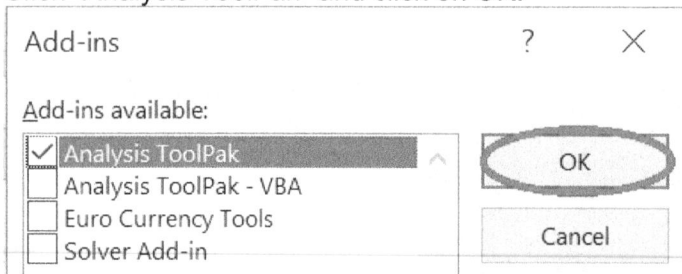

On the Data tab, in the Analysis group, you are now able to click on "Data Analysis".

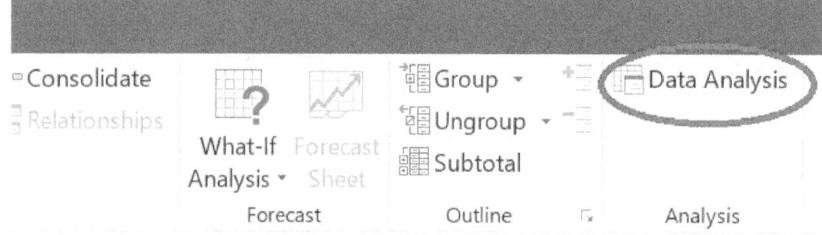

2) Select "Regression" and click "OK".

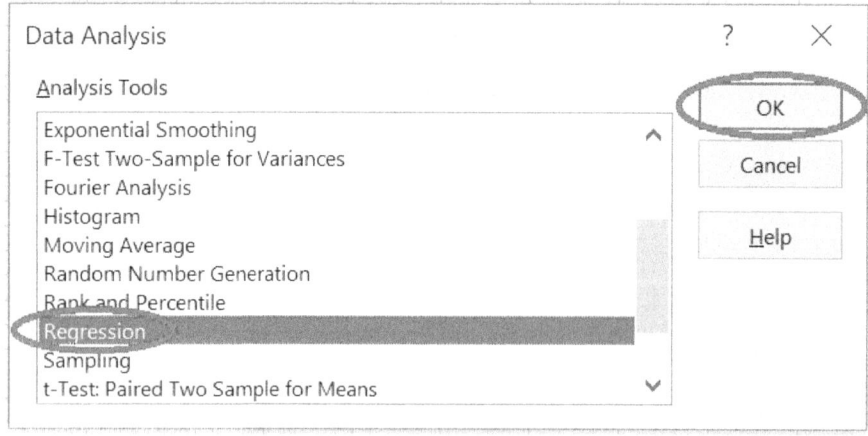

3) After you click OK in the "Data Analysis" dialog box, you will see a "Regression" dialog box.
4) For "Input Y Range", select cells (J1:J17). This is the predictor variable or dependent variable.
5) For "Input X Range", select cells (F1:I11). These are the explanatory variables or independent variables.
6) Check "Labels" box.
7) Click the "Output Range" box, and select cell A20.
8) Click "OK".

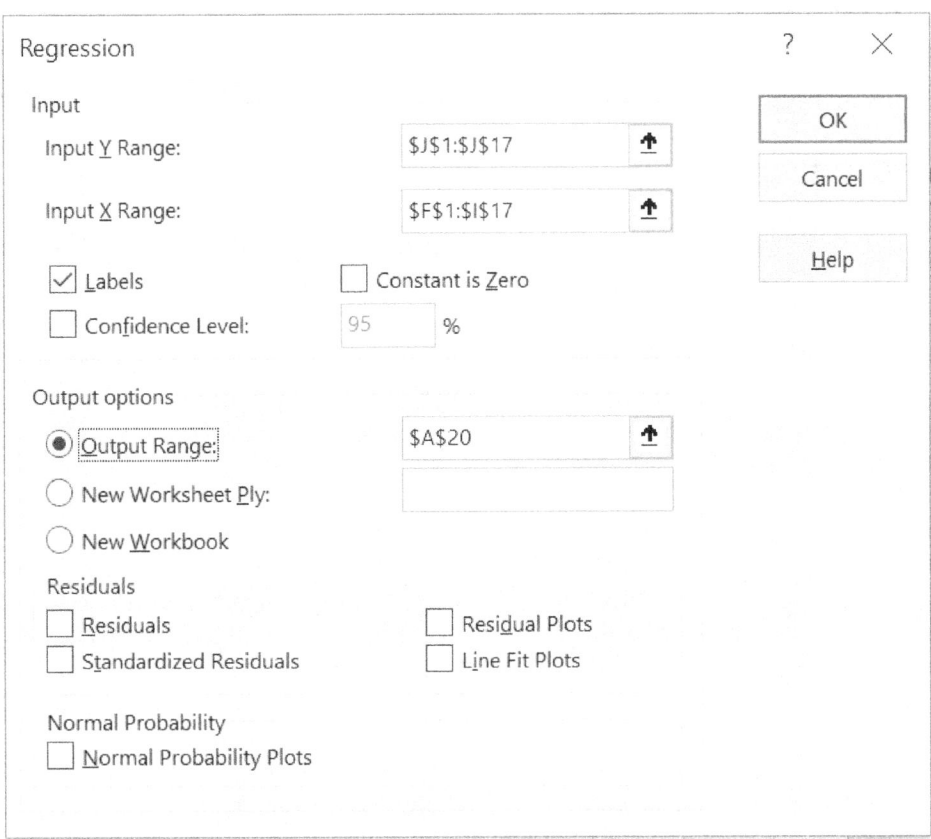

After you click "OK", Excel generates the following Summary Output. Round the numbers to 3 decimal places.

SUMMARY OUTPUT

Regression Statistics	
Multiple R	0.936
R Square	0.875
Adjusted R Square	0.830
Standard Error	5803.747
Observations	16

ANOVA

	df	SS	MS	F	Significance F
Regression	4	2597481731.310	649370432.827	19.279	0.000
Residual	11	370518268.690	33683478.972		
Total	15	2968000000.000			

	Coefficients	Standard Error	t Stat	P-value	Lower 95%	Upper 95%	Lower 95.0%	Upper 95.0%
Intercept	15289.910	5820.671	2.627	0.024	2478.699	28101.121	2478.699	28101.121
Age	760.399	138.671	5.483	0.000	455.185	1065.613	455.185	1065.613
Company#1	12732.856	3526.688	3.610	0.004	4970.668	20495.043	4970.668	20495.043
Company#2	-6117.435	3695.492	-1.655	0.126	-14251.158	2016.289	-14251.158	2016.289
Gender1	15924.911	2952.235	5.394	0.000	9427.085	22422.737	9427.085	22422.737

R Square: In the output, R Square is 0.875, which means it is a very good fit. 87% of the variation in Salary (Output) is explained by the independent variables (Inputs): Age Company#1, Company#2, Gender1. The closer R Square is to "1", the better the regression line fits the data.

Significance F and P-values: With multiple regression, there is more than one independent variable - thus we need to determine whether each independent variable contributes significantly to the regression after effects of other variables are considered. To determine if your results are statistically significant (i.e. reliable), check "Significance F" (0.001). If the value of "Significance F" is less than 0.05, it is statistically significant (i.e. reliable). If "Significance F" is bigger than 0.05, don't use this set of independent variables. Delete those variables with "P-value" that is bigger than 0.05 and run the regression again until "Significance F" drops below 0.05. Most or all of your P-values should be lower than 0.05. In our example "P-value" is 0.024, 0.000, 0.004, 0.126, and 0.0000, for Intercept, Age, Company#1, Company#2, and Gender1, respectively. The P-value for Age, Company#1, and Gender1 are statistically significant, as they are lower than 0.05. This means that Age contributes significantly to the regression after the effects of all the other variables are considered. Similarly, Company#1 and

Gender1 contributes significantly to the regression after all the other variables are considered.

Coefficients

	Coefficients	Standard Error	t Stat	P-value	Lower 95%	Upper 95%	Lower 95.0%	Upper 95.0%
Intercept	15289.910	5820.671	2.627	0.024	2478.699	28101.121	2478.699	28101.121
Age	760.399	138.671	5.483	0.000	455.185	1065.613	455.185	1065.613
Company#1	12732.856	3526.688	3.610	0.004	4970.668	20495.043	4970.668	20495.043
Company#2	-6117.435	3695.492	-1.655	0.126	-14251.158	2016.289	-14251.158	2016.289
Gender1	15924.911	2952.235	5.394	0.000	9427.085	22422.737	9427.085	22422.737

From the Summary Output, the regression line formula is:
Salary (Output)
= 15289.910 + (760.399*Age) + (12732.856* Company#1) - (6117.435* Company#2) + (15924.911*Gender1)

The regression line formula can predict the salary of a 30-year-old female with Company B, using these coding:

Age = 30,
Gender1 = 0,
Company#1= 0,
Company#2 = 1

From the below formula, you can see that the model forecasts that **this person would have a salary of 31,984:**

Salary (Output)
= 15289.910 + (760.399*Age) + (12732.856* Company#1) - (6117.435* Company#2) + 15924.911*Gender1)
= 15289.910 + (760.399*30) + (12732.856* 0) - (6117.435* 1) + (15924.911*0)
= 31,984

Similarly, the regression line formula can predict the salary of a 45-year-old male with Company C, using these coding:

Age = 45,
Gender1 = 0,
Company#1 = 0,
Company#2 = 0

From the below formula, you can see that the model forecasts that **this person would have a salary of 49,507:**

Salary (Output)
= 15289.910 + (760.399*Age) + (12732.856* Company#1) - (6117.435* Company#2) + (15924.911*Gender1)
= 15289.910 + (760.399*45) + (12732.856*0) - (6117.435* 0) + (15924.911*0)
= 49,507

12.4) Logistic Regression: Predict the value of a categorical output

Excel multiple linear regression cannot analyze dependent non-numeric data (categorical data) - they must be coded into numeric dummy variables (0 or 1). Unlike a linear regression that predicts values like sales amount, the logistic regression equation predicts probabilities. Compared with multiple linear regression, Logistic regression is more complicated for people without statistics background to grasp. In the data below, we want to predict the non-numeric dependent variable, customer purchase (i.e. whether the customer "buy" or "didn't buy") using Excel Logistic regression. Our data represents events that have already happened, thus, in the "Buy" column,
- "didn't buy" = 0
- "buy" = 1

Buy	Income	Age
0	85	49
1	85	57
0	86	51
0	62	52
0	86	52
0	91	55
1	71	49
0	98	54
1	98	55
0	99	53
0	86	44
0	80	54

Logistic regression is used to predict the binomial (Yes/No, Buy/Didn't buy, Resign/Stay, 1/0, etc) outcome of a response (dependent) variable, using one or several predictor (independent) variables. The predictors can be binomial, categorical, or numerical. Similar to multiple linear regression, we need a function that connects the independent variables to the dependent variable. But the difference here is that the dependent variable can only take two values: 1/0 (Yes/No).

1) This is where we create our logistic regression equation. We need to map a continuous function of independent variables to a binary outcome. Our objective is to create an equation with coefficients b_0 to b_2 and then enter values for Income and Age to predict customer purchase (buy or didn't buy). We have 3 coefficients (1 coefficient for constant, 1 coefficient for income, 1 coefficient for Age). The logistic equation looks similar to linear regression equation:

$$\text{Logit(Purchase)} = b_0 + b_1 * \text{Income} + b_2 * \text{Age}$$

2) Logit is a function that takes a probability of an event as input and returns the logarithm of the odds of that event as output. We need to first assign an arbitrary value (e.g. 0.000) for these coefficients (b0, b1, b2, b3) - Later you will be shown how to use the Excel Solver to replace these starting arbitrary coefficients (e.g. 0.000) with optimized coefficients to create an equation to predict probability.

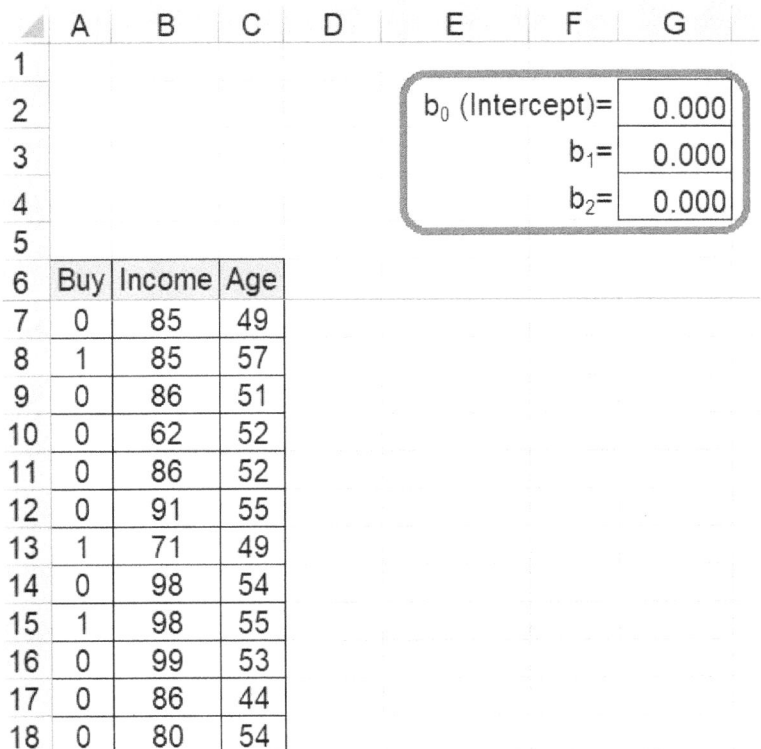

3) Here, you need to calculate a Logit for each record. Enter the Logit formula below for all the data records, in your Excel spreadsheet:

$$\text{Logit(Purchase)} = b_0 + b_1 * \text{Income} + b_2 * \text{Age}$$

| N | | | | fx | =$G$2+$G$3*B7+$G$4*C7 |

	A	B	C	D	E	F	G	H	I
1									
2						b_0 (Intercept)=	0.000		
3						b_1=	0.000		
4						b_2=	0.000		
5									
6		Buy	Income	Age	Logit (L)				
7		0	85	49	=$G$2				
8		1	85	57	0.000				
9		0	86	51	0.000				
10		0	62	52	0.000				
11		0	86	52	0.000				
12		0	91	55	0.000				
13		1	71	49	0.000				
14		0	98	54	0.000				
15		1	98	55	0.000				
16		0	99	53	0.000				
17		0	86	44	0.000				
18		0	80	54	0.000				

4) Here, you need to calculate e^L for each record. The number e is the base of the natural logarithm. It is approximately equal to 2.71828163. e^L must be calculated for each record. Enter the Exponential formula below, in your Excel spreadsheet:

	A	B	C	D	E	F	G
1							
2					b_0 (Intercept)=		0.000
3					b_1=		0.000
4					b_2=		0.000
5							
6	Buy	Income	Age	Logit (L)	e^L		
7	0	85	49	0.000	=EXP(D7)		
8	1	85	57	0.000	1.000		
9	0	86	51	0.000	1.000		
10	0	62	52	0.000	1.000		
11	0	86	52	0.000	1.000		
12	0	91	55	0.000	1.000		
13	1	71	49	0.000	1.000		
14	0	98	54	0.000	1.000		
15	1	98	55	0.000	1.000		
16	0	99	53	0.000	1.000		
17	0	86	44	0.000	1.000		
18	0	80	54	0.000	1.000		

Cell N: f_x =EXP(D7)

5) Here, you need to calculate P(X) for each record. P(X) is the probability of event X occurring. Enter the formula for probability of the event (i.e. Buy) in your Excel spreadsheet, using the formula below:

$$P(X) = e^L / (1 + e^L)$$

	A	B	C	D	E	F	G
1							
2						b_0 (Intercept)=	0.000
3						b_1=	0.000
4						b_2=	0.000
5							
6	Buy	Income	Age	Logit (L)	e^L	P (X)	
7	0	85	49	0.000	1.000	=E7/(1+E7)	
8	1	85	57	0.000	1.000	0.500	
9	0	86	51	0.000	1.000	0.500	
10	0	62	52	0.000	1.000	0.500	
11	0	86	52	0.000	1.000	0.500	
12	0	91	55	0.000	1.000	0.500	
13	1	71	49	0.000	1.000	0.500	
14	0	98	54	0.000	1.000	0.500	
15	1	98	55	0.000	1.000	0.500	
16	0	99	53	0.000	1.000	0.500	
17	0	86	44	0.000	1.000	0.500	
18	0	80	54	0.000	1.000	0.500	

Formula bar: =E7/(1+E7)

6) Here, you need to calculate LL, the Log-Likelihood Function. The log-likelihood function computes a probability based on the input variables values. Enter the log-likelihood formula below, in your Excel spreadsheet:

N   fx =A7*LN(F7)+(1-A7)*(LN(1-F7))

	A	B	C	D	E	F	G	H
1								
2						b_0 (Intercept)=	0.000	
3						b_1=	0.000	
4						b_2=	0.000	
5								
6		Buy	Income	Age	Logit (L)	e^L	P (X)	Log-Likelihood (LL)
7	0	85	49	0.000	1.000	0.500	=A7*LN(F7	
8	1	85	57	0.000	1.000	0.500	-0.693	
9	0	86	51	0.000	1.000	0.500	-0.693	
10	0	62	52	0.000	1.000	0.500	-0.693	
11	0	86	52	0.000	1.000	0.500	-0.693	
12	0	91	55	0.000	1.000	0.500	-0.693	
13	1	71	49	0.000	1.000	0.500	-0.693	
14	0	98	54	0.000	1.000	0.500	-0.693	
15	1	98	55	0.000	1.000	0.500	-0.693	
16	0	99	53	0.000	1.000	0.500	-0.693	
17	0	86	44	0.000	1.000	0.500	-0.693	
18	0	80	54	0.000	1.000	0.500	-0.693	
19						sum of LL:	-8.318	

7) The sum that we wish to maximize is the total of log-likelihood (LL):

G19   fx =SUM(G7:G18)

	A	B	C	D	E	F	G
1							
2						b_0 (Intercept)=	0.000
3						b_1=	0.000
4						b_2=	0.000
5							
6		Buy	Income	Age	Logit (L)	e^L	P (X)	Log-Likelihood (LL)
7	0	85	49	0.000	1.000	0.500	-0.693
8	1	85	57	0.000	1.000	0.500	-0.693
9	0	86	51	0.000	1.000	0.500	-0.693
10	0	62	52	0.000	1.000	0.500	-0.693
11	0	86	52	0.000	1.000	0.500	-0.693
12	0	91	55	0.000	1.000	0.500	-0.693
13	1	71	49	0.000	1.000	0.500	-0.693
14	0	98	54	0.000	1.000	0.500	-0.693
15	1	98	55	0.000	1.000	0.500	-0.693
16	0	99	53	0.000	1.000	0.500	-0.693
17	0	86	44	0.000	1.000	0.500	-0.693
18	0	80	54	0.000	1.000	0.500	-0.693
19						sum of LL:	-8.318

The objective of Logistic Regression is find the coefficients of the Logit (b_0, b_1, b_2 + ...+ b_k) that maximize LL, the Log-Likelihood Function in cell G19, to produce the Maximum Log-Likelihood (MLL) Function. The only values we can change are the guesses for the coefficient b_0 through b_3, which we have assigned an arbitrary value of 0.000. We don't have to optimize them ourselves, as we can use Solver, an Excel add-in, that adjusts the coefficient to maximize or minimize the value in the cell.

8) The Excel Solver is an add-in that is included in Excel. But it must be manually activated by you before it can be utilized for the first time. To use Solver, on the File tab, click Options.

9) Under Add-ins, click Excel Add-ins and click the "Go" button.

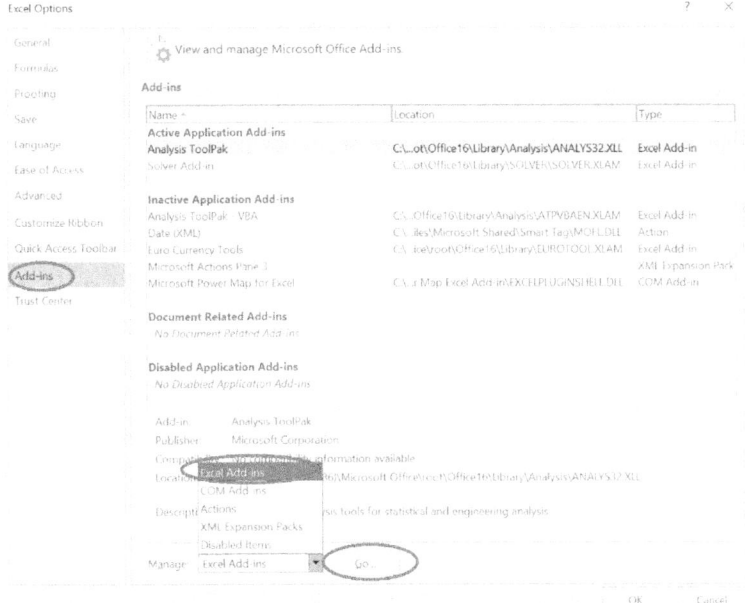

10) Check "Solver Add-In" and click on OK.

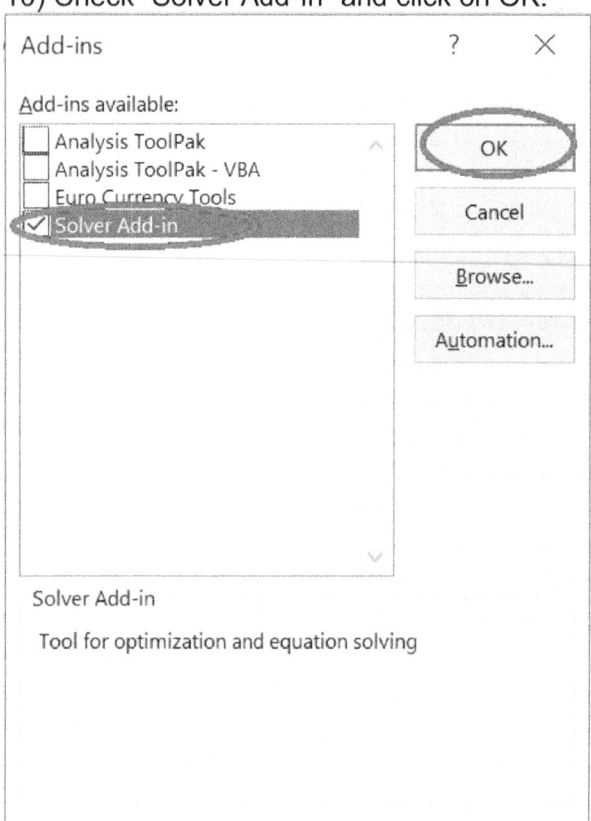

11) On the Data tab, in the Analysis group, you are now able to click on "Solver". Click the Solver button.

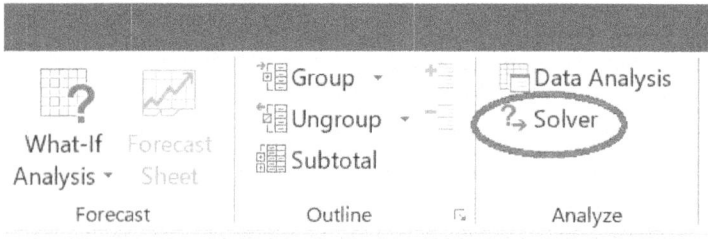

12) The objective is to maximize the sum of the log-likelihood column (LL), by changing the values in G2:G4, representing coefficients b_0-b_2.
- For "Set Objective", select cell (G19), the sum of the log-likelihood column (LL).
- Uncheck the box labeled "Make Unconstrained Variables Non-Negative".
- For "Select a Solving Method", select "GRG Nonlinear" because we are not performing a linear optimization.
- Click "Solve"

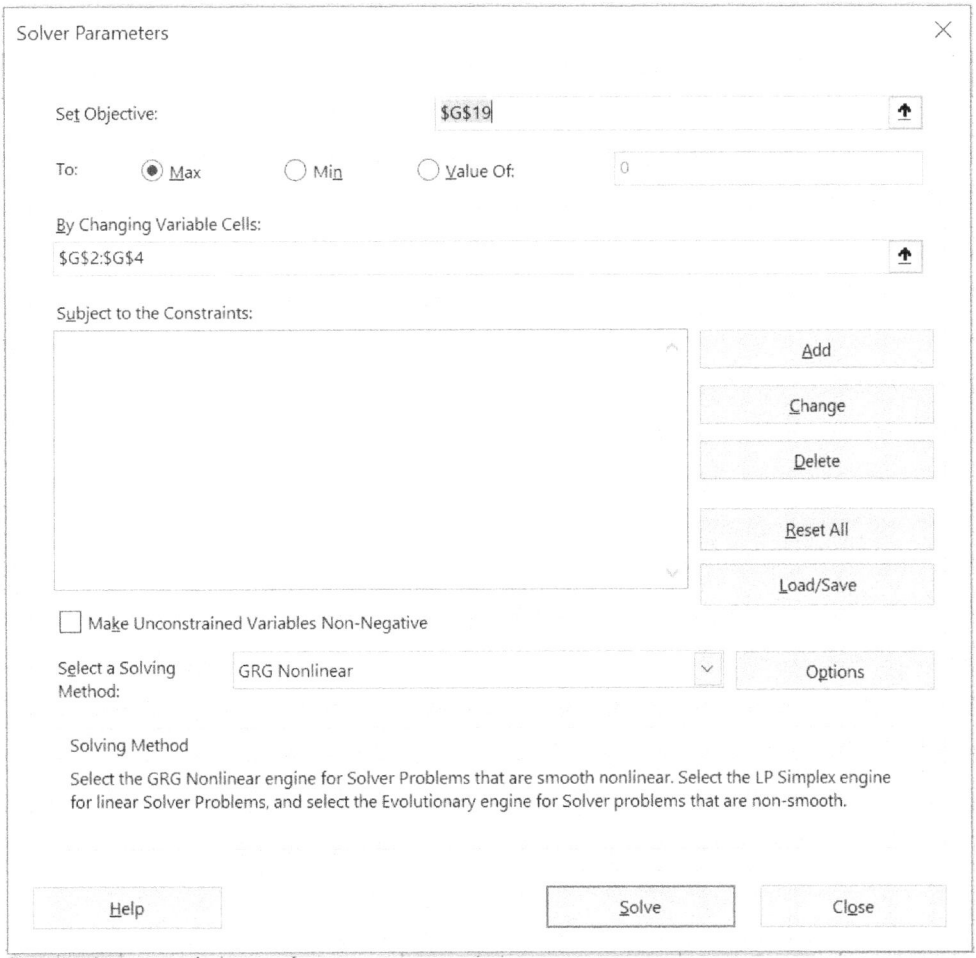

13) After clicking "Solve", you will see the screen "Solver Results". Check "Keep Solver Solution".

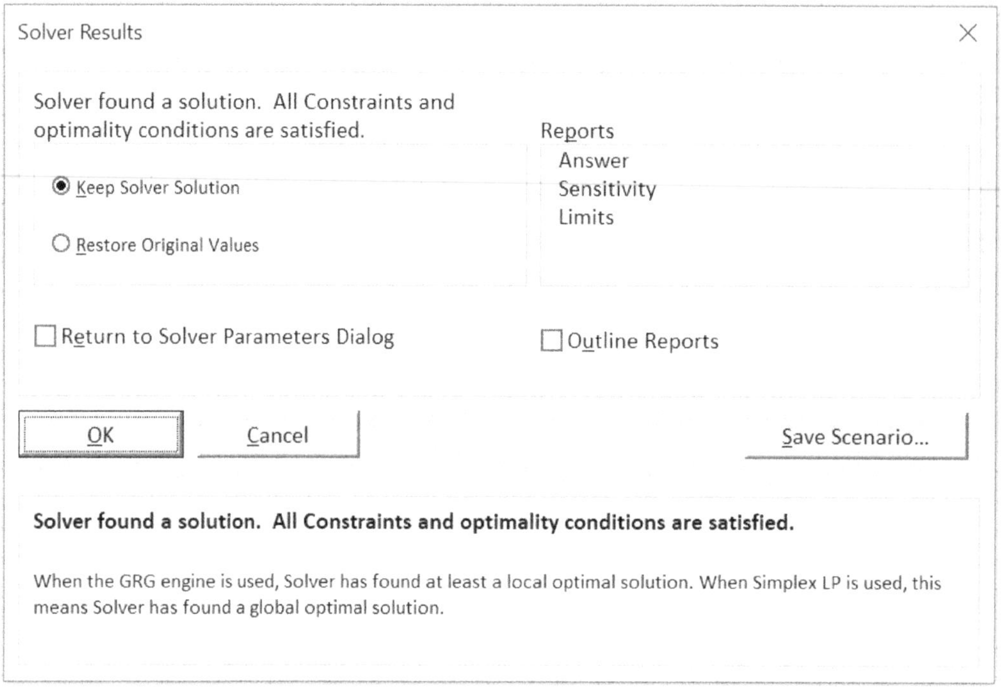

Solver Results

Solver found a solution. All Constraints and optimality conditions are satisfied.

● Keep Solver Solution

○ Restore Original Values

Reports
Answer
Sensitivity
Limits

☐ Return to Solver Parameters Dialog   ☐ Outline Reports

OK   Cancel   Save Scenario...

Solver found a solution. All Constraints and optimality conditions are satisfied.

When the GRG engine is used, Solver has found at least a local optimal solution. When Simplex LP is used, this means Solver has found a global optimal solution.

14) After clicking "Solve" you get new values in G2:G4, representing coefficients b_0-b_2.

	A	B	C	D	E	F	G
1							
2						b_0 (Intercept)=	-14.077
3						b_1=	-0.037
4						b_2=	0.306
5							
6		Buy	Income	Age	Logit (L)	e^L	P (X)	Log-Likelihood (LL)
7	0	85	49	-2.234	0.107	0.097	-0.102
8	1	85	57	0.210	1.234	0.552	-0.593
9	0	86	51	-1.660	0.190	0.160	-0.174
10	0	62	52	-0.470	0.625	0.385	-0.485
11	0	86	52	-1.354	0.258	0.205	-0.230
12	0	91	55	-0.622	0.537	0.349	-0.430
13	1	71	49	-1.719	0.179	0.152	-1.884
14	0	98	54	-1.185	0.306	0.234	-0.267
15	1	98	55	-0.880	0.415	0.293	-1.227
16	0	99	53	-1.528	0.217	0.178	-0.196
17	0	86	44	-3.799	0.022	0.022	-0.022
18	0	80	54	-0.522	0.593	0.372	-0.466
19						sum of LL:	-6.075

15) To predict whether a customer whose income is 90 and age is 60 will buy, copy those figures that are circled in your spreadsheet.

D22					f_x	=$G$2+$G$3*B22+$G$4*C22		
	A	B	C	D	E	F	G	H	I

	A	B	C	D	E	F	G
1							
2						b_0(Intercept)=	-14.078
3						b_1=	-0.037
4						b_2=	0.306
5							
6	Buy	Income	Age	Logit (L)	e^L	P (X)	Log-Likelihood (LL)
7	0	85	49	-2.234	0.107	0.097	-0.102
8	1	85	57	0.210	1.234	0.552	-0.593
9	0	86	51	-1.660	0.190	0.160	-0.174
10	0	62	52	-0.470	0.625	0.385	-0.485
11	0	86	52	-1.354	0.258	0.205	-0.230
12	0	91	55	-0.622	0.537	0.349	-0.430
13	1	71	49	-1.719	0.179	0.152	-1.884
14	0	98	54	-1.185	0.306	0.234	-0.267
15	1	98	55	-0.880	0.415	0.293	-1.227
16	0	99	53	-1.528	0.217	0.178	-0.196
17	0	86	44	-3.799	0.022	0.022	-0.022
18	0	80	54	-0.522	0.593	0.372	-0.466
19						sum of LL:	-6.075
20							
21		Income	Age	Logit (L)	e^L	P (X)	
22		90	60	0.943			
23							

You will notice that the formula is cell D22 is:

Logit(L)
=$G$2+$G$3*B22+$G$4*C22
= b_0 + b_1*Income + b_2*Age
= -14.077 – 0.037*Income + 0.306*Age

Just like a multiple linear regression, you need to enter the new b_0, b_1, and b_2 coefficient values into your logistic regression equation to predict a value. But unlike a linear regression that predicts values like sales amount, the logistic regression equation predicts probabilities.

16) Next, you need to calculate e^L. The number e is the base of the natural logarithm. It is approximately equal to 2.71828163. Enter the Exponential formula below, in your Excel spreadsheet:

E22   fx =EXP(D22)

	A	B	C	D	E	F	G
1							
2						b_0(Intercept)=	-14.078
3						b_1=	-0.037
4						b_2=	0.306
5							
6		Buy	Income	Age	Logit (L)	e^L	P (X)	Log-Likelihood (LL)
7	0	85	49	-2.234	0.107	0.097	-0.102
8	1	85	57	0.210	1.234	0.552	-0.593
9	0	86	51	-1.660	0.190	0.160	-0.174
10	0	62	52	-0.470	0.625	0.385	-0.485
11	0	86	52	-1.354	0.258	0.205	-0.230
12	0	91	55	-0.622	0.537	0.349	-0.430
13	1	71	49	-1.719	0.179	0.152	-1.884
14	0	98	54	-1.185	0.306	0.234	-0.267
15	1	98	55	-0.880	0.415	0.293	-1.227
16	0	99	53	-1.528	0.217	0.178	-0.196
17	0	86	44	-3.799	0.022	0.022	-0.022
18	0	80	54	-0.522	0.593	0.372	-0.466
19						sum of LL:	-6.075
20							
21		Income	Age	Logit (L)	e^L	P (X)	
22		90	60	0.943	2.568		

17) Next, you need to calculate P(X). P(X) is the probability of event X occurring. Enter the formula for probability of the event (i.e. Buy) in your Excel spreadsheet, using the formula below:

$$P(X) = e^L / (1 + e^L)$$

F22   fx   =E22/(1+E22)

	A	B	C	D	E	F	G
1							
2						b_0(Intercept)=	-14.078
3						b_1=	-0.037
4						b_2=	0.306
5							
6		Buy	Income	Age	Logit (L)	e^L	P (X)	Log-Likelihood (LL)
7	0	85	49	-2.234	0.107	0.097	-0.102
8	1	85	57	0.210	1.234	0.552	-0.593
9	0	86	51	-1.660	0.190	0.160	-0.174
10	0	62	52	-0.470	0.625	0.385	-0.485
11	0	86	52	-1.354	0.258	0.205	-0.230
12	0	91	55	-0.622	0.537	0.349	-0.430
13	1	71	49	-1.719	0.179	0.152	-1.884
14	0	98	54	-1.185	0.306	0.234	-0.267
15	1	98	55	-0.880	0.415	0.293	-1.227
16	0	99	53	-1.528	0.217	0.178	-0.196
17	0	86	44	-3.799	0.022	0.022	-0.022
18	0	80	54	-0.522	0.593	0.372	-0.466
19						sum of LL:	-6.075
20							
21		Income	Age	Logit (L)	e^L	P (X)	
22		90	60	0.943	2.568	0.720	
23							

From the spreadsheet, if a customer income is 90 and age is 60, the probably that he will buy is 0.72, which is closer to 1 than to 0. Closer to 1 means it is probably "buy", while closer to 0 means it is probably "didn't buy".

12.5) Logistic Regression: Fixing Solver Error Messages

For simple problems, the Excel's Solver is able to quickly find the optimal Solver variable values for the objective function. However, sometimes, Solver has difficulty finding the Solver variable values that optimize the objective function, and displays an error message. Here are some Solver messages that you will encounter: [1]

- **Solver has found a solution**

This message tells you that Solver found a set of variable values that meets your constraints.

- **Solver has converged to the current solution**

This message tells you that Excel found a solution but is not confident. To get a better solution, adjust the Convergence setting in the Solver Options dialog box so that Excel works at a higher level of precision. In the Convergence box, enter the amount of relative change that you want to allow before Solver stops with the message "Solver converged to the current solution." Smaller values mean that Solver will take more time, but will stop at a point closer to the optimal solution. [2]

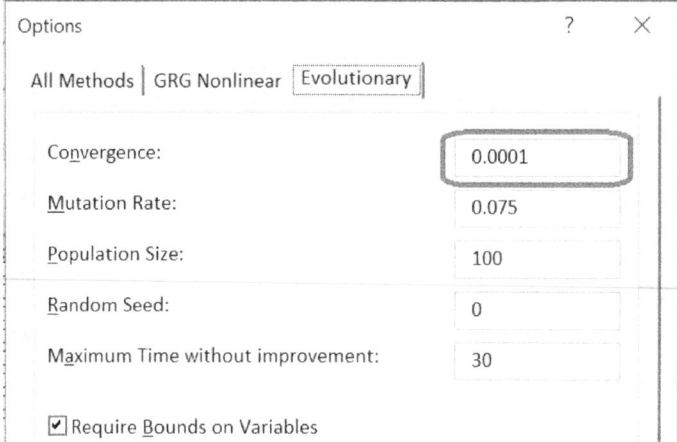

- **Solver cannot improve the current solution**

This message tells you that, Excel has calculated a rough solution. To get a better solution, increase the precision setting using the Solver Options dialog box.

- **Stop chosen when maximum time limit was reached**

This message tells you that Excel ran out of time. To fix this, use a larger Max Time setting. [1]

- **Stop chosen when maximum iteration limit was reached**

This message tells you that Excel ran out of iterations before it can find the optimal solution. To fix this, set larger iterations value in the Solver Options dialog box.

- **Objective Cell values do not converge**

This message tells you that the objective function doesn't have an optimal value. It keeps getting a better objective function value with every iteration, but it doesn't get closer to a final objective function value. You may need to check your objective formula, and constraint formulas.

- **Solver could not find a feasible solution**

This message tells you that there isn't any answer to your optimization modeling problem.

- **Linearity conditions required by this LP Solver are not satisfied**

The message tells you that you chose the "Simplex LP" solving method when your model isn't linear. To fix this, choose the "GRG Nonlinear" solving method.

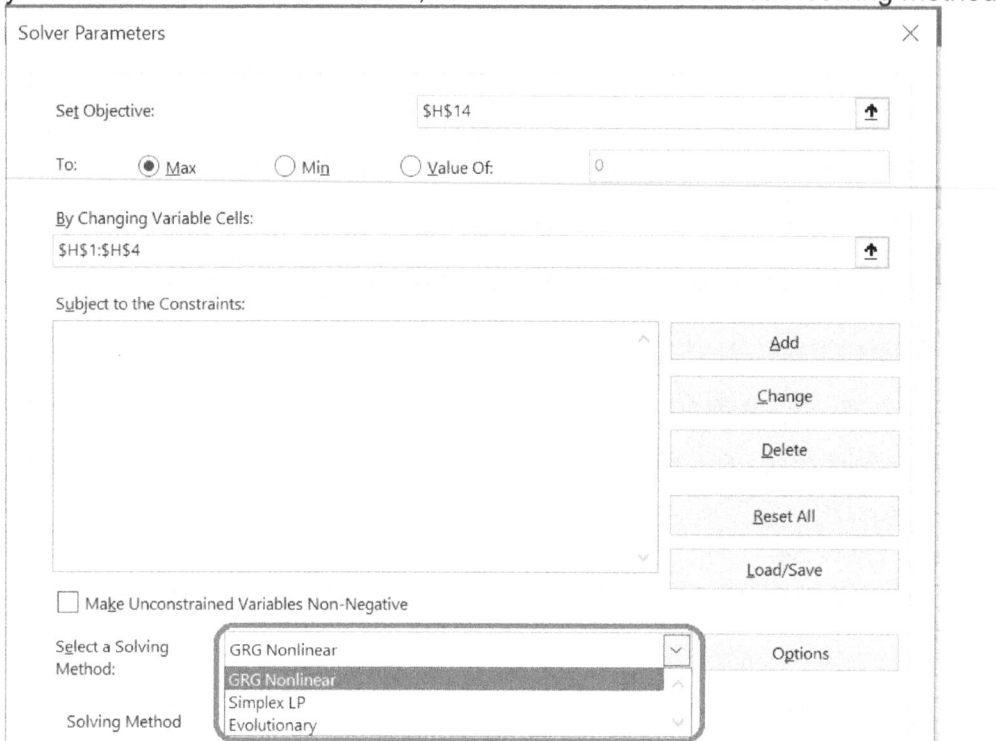

- **The problem is too large for Solver to handle**

This message tells you that your problem is too big for solver because you've got more than 100 decision variable or constraints. To fix this, reduce the number of variables or constraints.

- **Solver encountered an error value in objective or constraint cell**

This message means that you need to fix your formula or constraint.

- **Error in model. Please verify that all cells and constraints are valid**

This message means that you need to check your formulas and your input values as there is mistake. Check that you did not use the word "solver" in your variables, as it can confuse Solver.

References:
(1) Stephen L. Nelson, E. C. Nelson (2019) What do the Solver Error Messages in Excel Mean?, https://www.dummies.com/software/microsoft-office/excel/what-do-the-solver-error-messages-in-excel-mean/ (6 March 2019)
(2) FrontlineSolvers (2019) Excel Solver – Change Options for CRG Nonlinear Solving Method https://www.solver.com/excel-solver-change-options-grg-nonlinear-solving-method (6 March 2019)

13) Text Mining (Text Analytics)

According Bernard Marr, 90% of the world's data was generated in the last two years alone. A lot of these data are often untapped unstructured data in the form of text, audio, images, and videos – not numbers. For organizations the most common type of unstructured data is usually text. Text analytics provides enormous insights for organizations, because text data is often not been analyzed, and text is the largest and richest source of data in most organizations. [1]

Text mining (also known as text analytics) is the process of analyzing unstructured text to identify patterns. Mining and analyzing text help companies uncover valuable insights through documents, emails, call center logs, employee survey comments, social network posts, Glassdoor, and screening job candidates based on their resume words.

There are three common approaches of use text analytics: [2]

(i) Most Frequent Words and Phrases
Frequently mentioned words usually represent topics that require further analysis, as "more frequent = more important" because a large number of employees mentioned it. Determining which words and phrases are most frequently mentioned in comments requires a count of meaningful words within comments. Words like "the", "is", and "that" are not meaningful. But words like "happy", and "sad" are meaningful. But, it's not sufficient to consider individuals words because phrases such as "Work from home" provide better context about employees' sentiment.

(ii) Themes
After ascertaining which words and phrases are frequently used, place them into groups of related topics or themes. Groupings allow for a much more pinpointed analysis than if the highly frequent words considered in isolation. The words and phrases "stress", "sickness", and "work life balance" relates to the "Wellbeing" theme. The words and phrases "career aspiration", "job rotation", and "upgrade" all relate to the "Learning & Development" theme.

(iii) Sentiment

After determining the frequently mentioned words and grouping them into themes, the next step is sentiment analysis. Sentiment is an opinion about something, categorized into negative, positive, and neutral sentiment. Sentiment analysis (also known as opinion mining) mines text from emails, online reviews, social networks, Glassdoor, and call center logs to identify positive or negative feelings by scoring text to get a probability score.

References:
(1) Bernard Marr (2018), How Much Data Do We Create Every Day? The Mind-Blowing Stats Everyone Should Read. https://www.forbes.com/sites/bernardmarr/2018/05/21/how-much-data-do-we-create-every-day-the-mind-blowing-stats-everyone-should-read/#751cb49460ba (3 June 2019)
(2) Dan Harris, Ph.D. (2018), 3 Ways Managers Can Use Text Analytics to Drive Engagement. https://www.quantumworkplace.com/future-of-work/3-ways-managers-can-use-text-analytics-to-drive-engagement (6 June 2019)

13.1) Count frequently mentioned words with VBA in Microsoft Word

This example, shows you how to run a Marco to count the occurrences of each word.

1) Copy and paste the text below, into your word document.

> LG Electronics, Samsung, Maruti Suzuki and Hero Honda, riding the current economic recovery, are set to celebrate this festival season with their employees by paying higher bonuses than last year when sales were sluggish and future uncertain. But, unlike past years, most of this money may not come back into the consumer market this time, as some employees ET spoke to are not in a mood to celebrate: they would rather invest their bonuses.

Source: Mahima Puri (2009), Higher bonuses set to spread cheer this festive season. https://economictimes.indiatimes.com/jobs/higher-bonuses-set-to-spread-cheer-this-festive-season/articleshow/5044436.cms (31 May 2019)

2) Right click on a blank area at the top of your Word document, and choose "Customize the Ribbon".

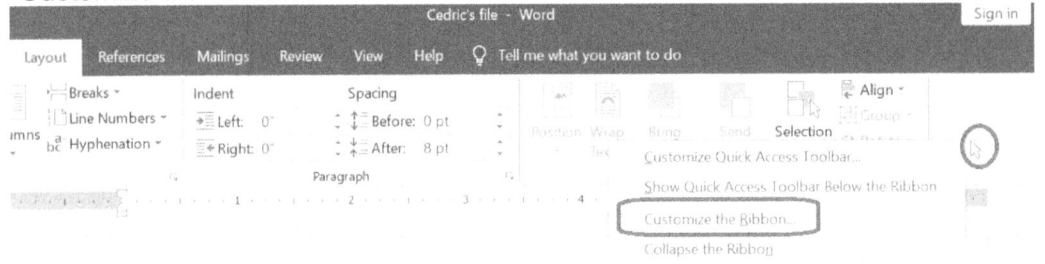

3) Click "developer" and click "OK", and you will see a Developer tab.

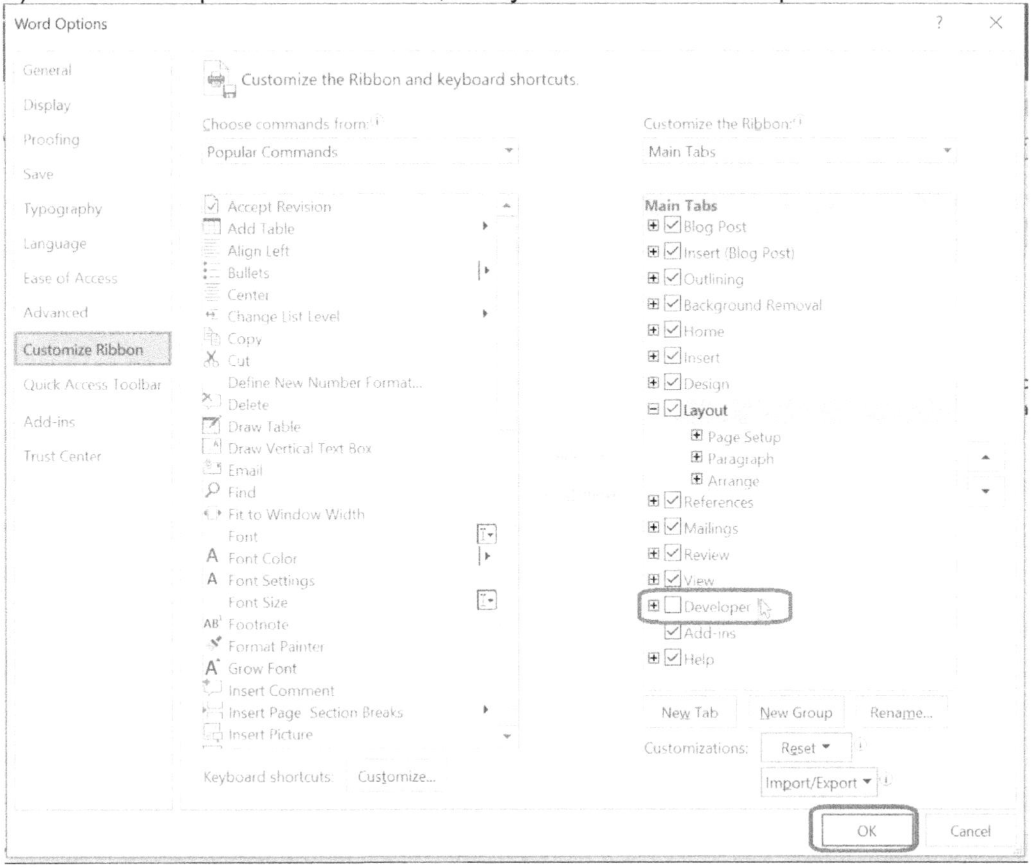

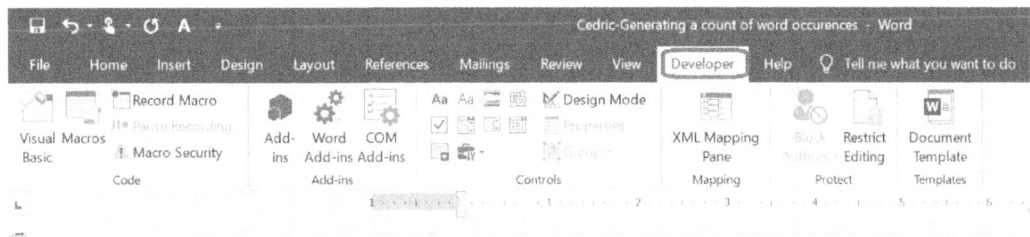

4) Allen Wyatt developed a VBA macro to the occurrences of each word in your document. To use his VBA macro, go to this website (https://word.tips.net/T001833_Generating_a_Count_of_Word_Occurrences.html), to copy his VBA macro codes. Here's a partial screenshot of the VBA macro codes that you need to copy.

5) At the Developer tab. Click "Visual Basic" to open the VBA editor.

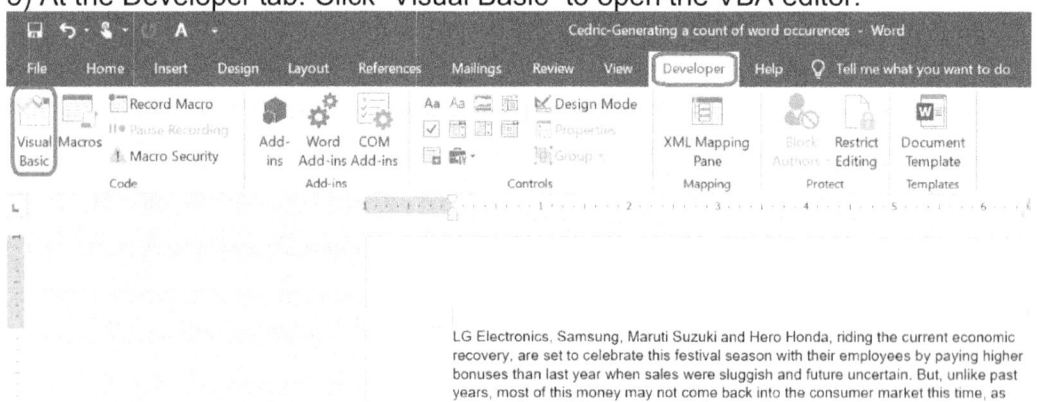

6) After you click "Visual Basic", you will see this screen:

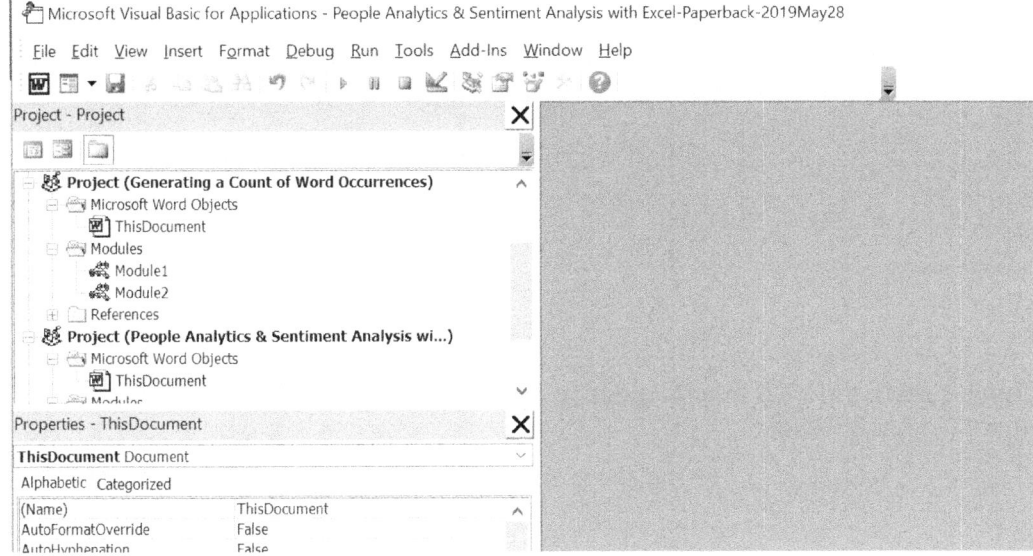

7) Click "Insert", and click "Module".

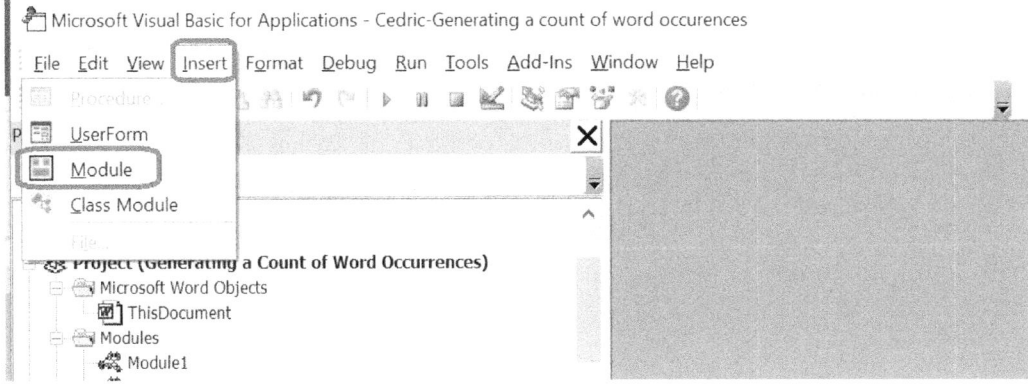

8) Use "Control V" to paste the VBA macro codes that you copied earlier here:

9) After you paste the VBA macro codes, go to "File", and click "Close and Return to Microsoft Word".

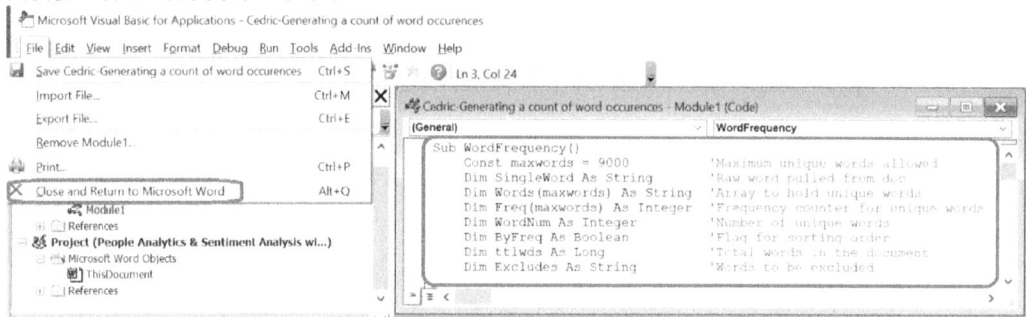

10) Go to "Developer", click "Macros", click "WordFrequency", and click "Run".

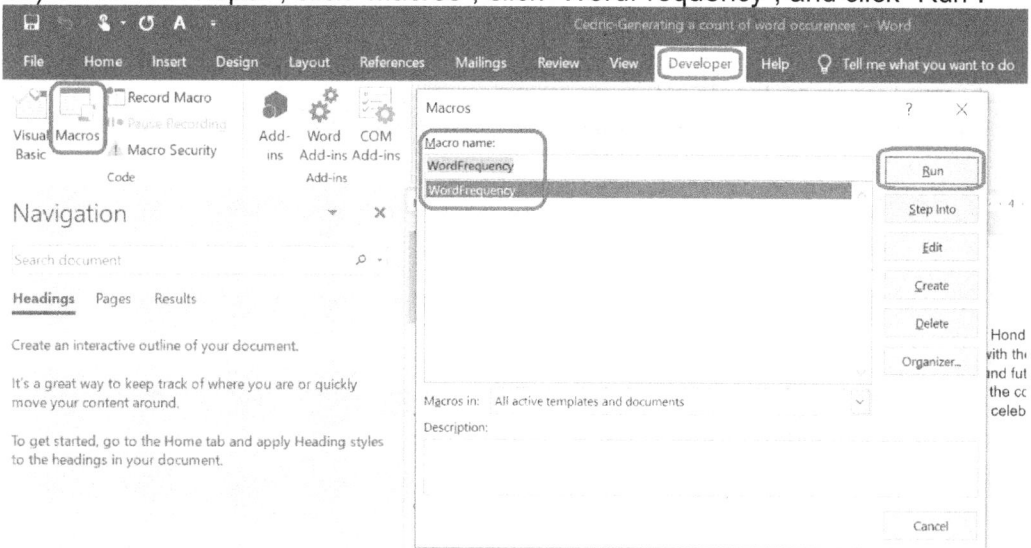

11) At the "Word order" window, click "OK".

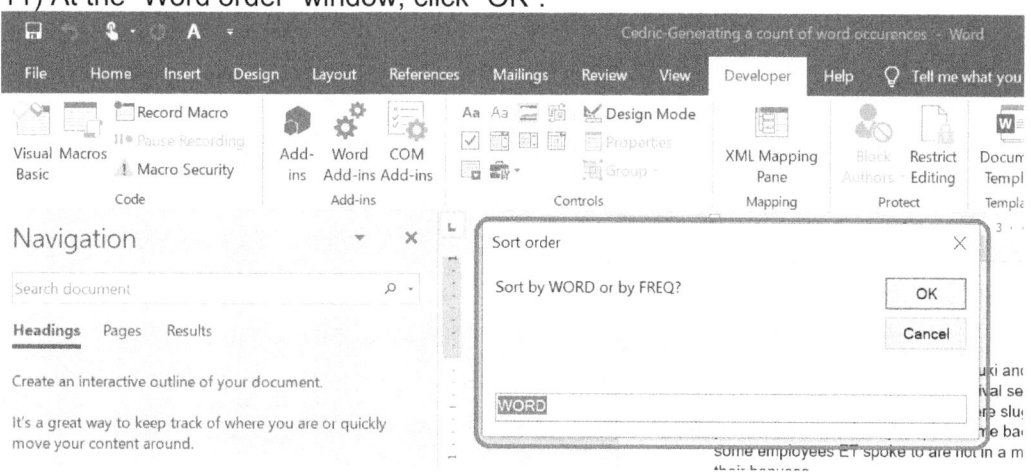

12) After you click "OK", a new word document appears, with a window saying "There are 55 different words". Click "OK".

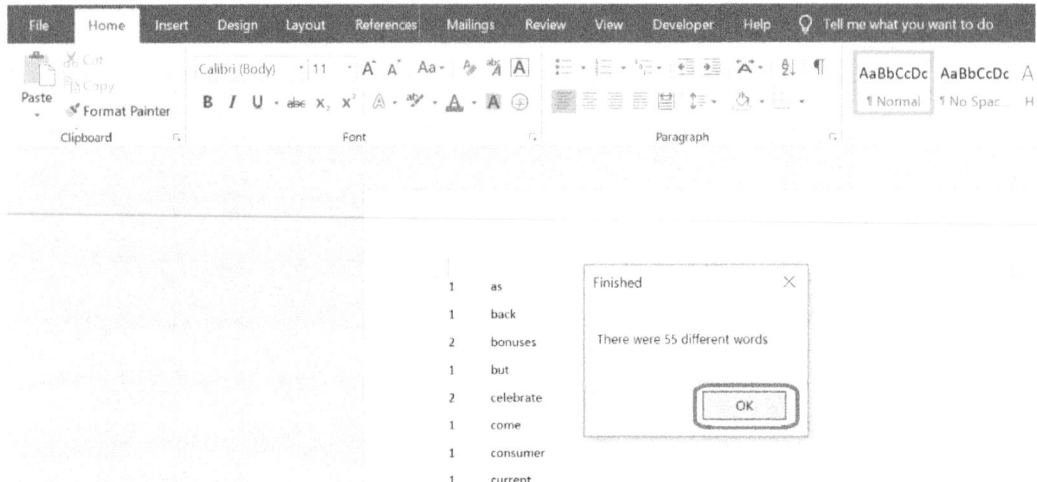

13) Select all the text, go to "Insert" tab, click "convert text to table"

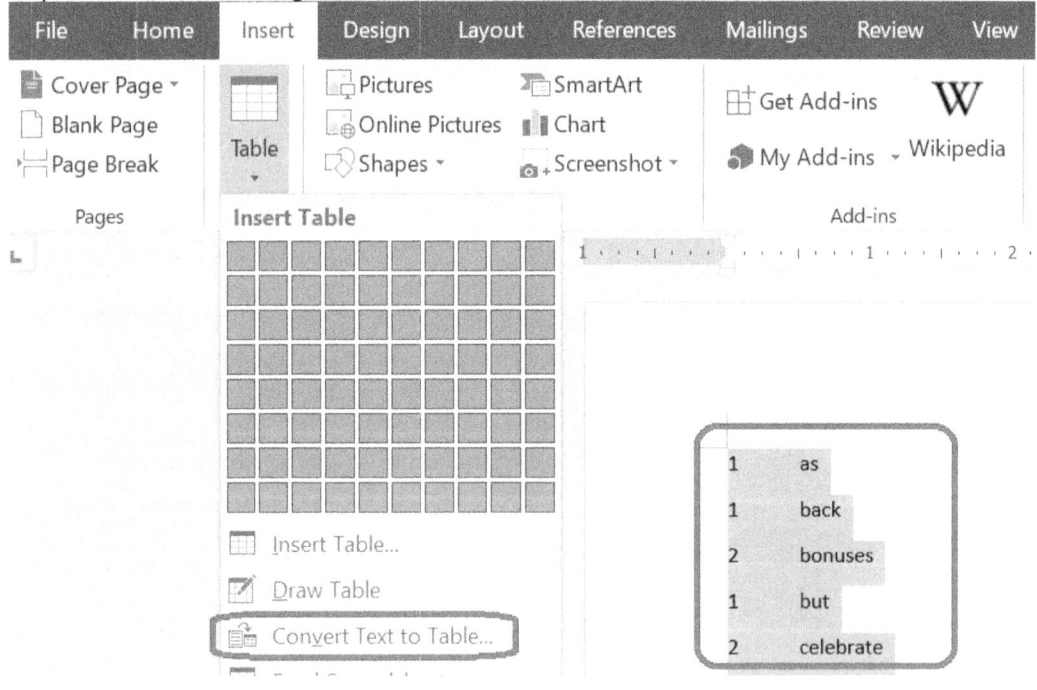

14) Click "OK"

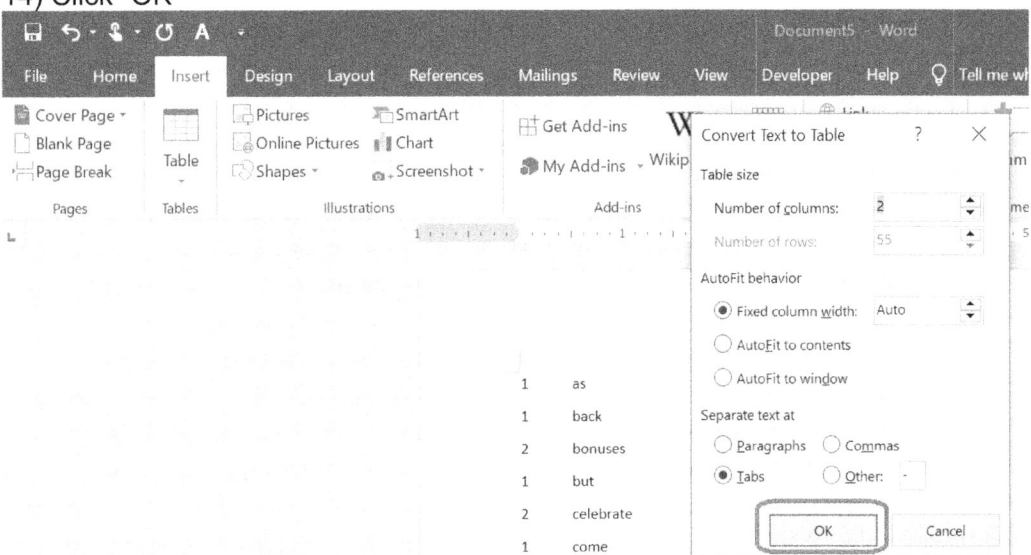

15) After you click "OK", you will see that your text are now contained in tables.

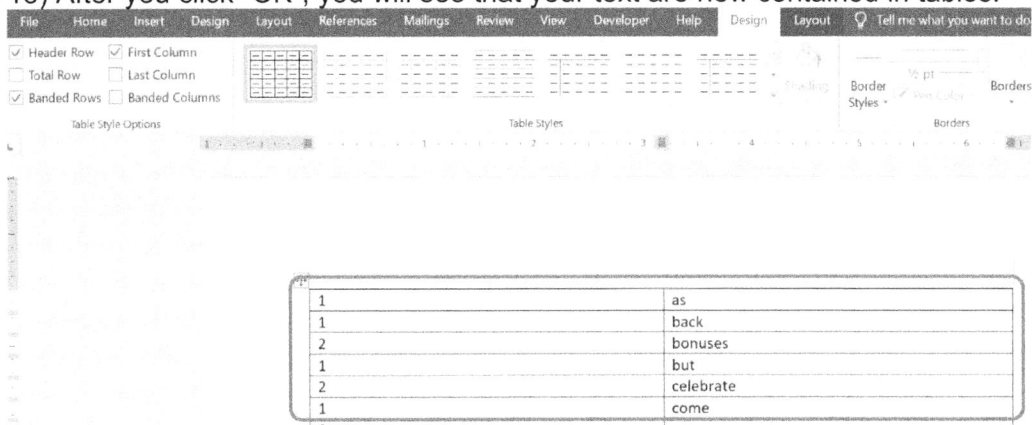

16) "Right click" the first row, click "Insert", then click "Insert Rows Above".

1	as	
1	back	
2	bonuses	
1	but	
2	celebrate	
1	come	
1	consume	
1	current	
1	economy	
1	electron	
2	employ	
1	et	
1	festival	
1	future	
1	hero	

17) At the top row, type "Occurrence" and "Word", and then "Bold" it as follows:

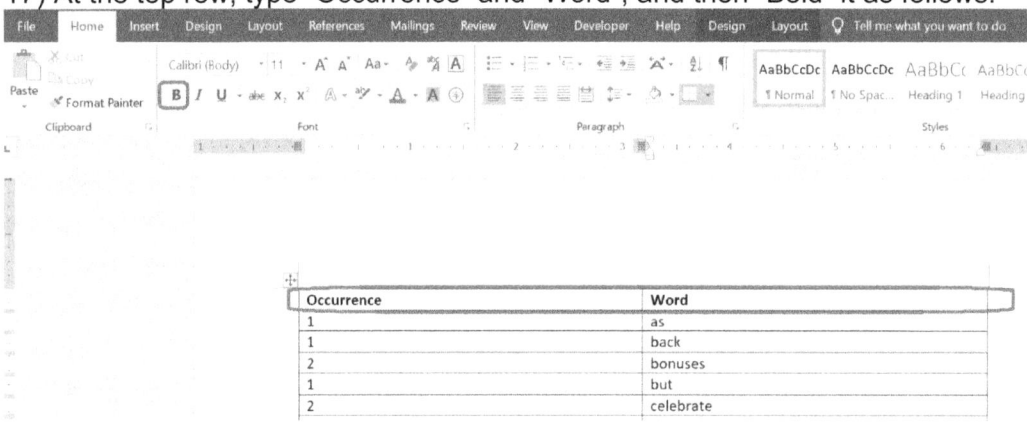

Occurrence	Word
1	as
1	back
2	bonuses
1	but
2	celebrate

18) Click the **left column**, at "Layout" tab, click "Sort", click "Descending", and click "OK".

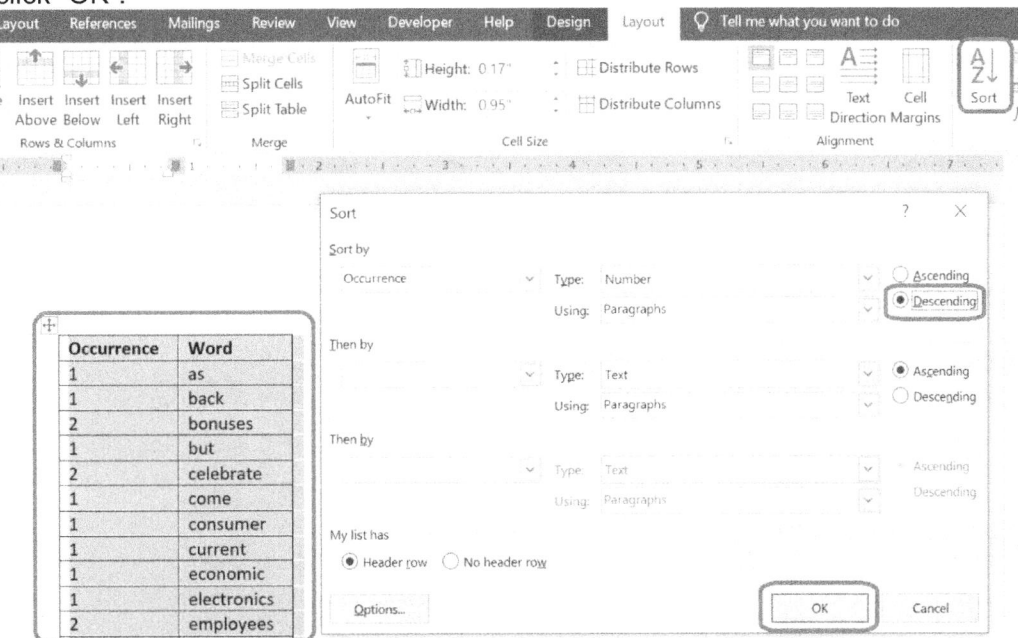

19) Your text is now sorted in descending order by occurrences.

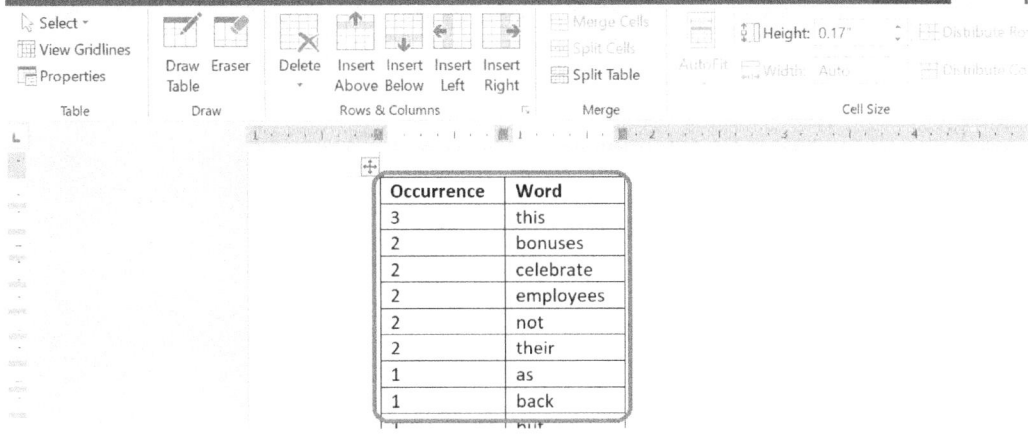

20) To visualize your word occurrences, copy these texts and paste in Excel, then in graphs, select "Clustered Column".

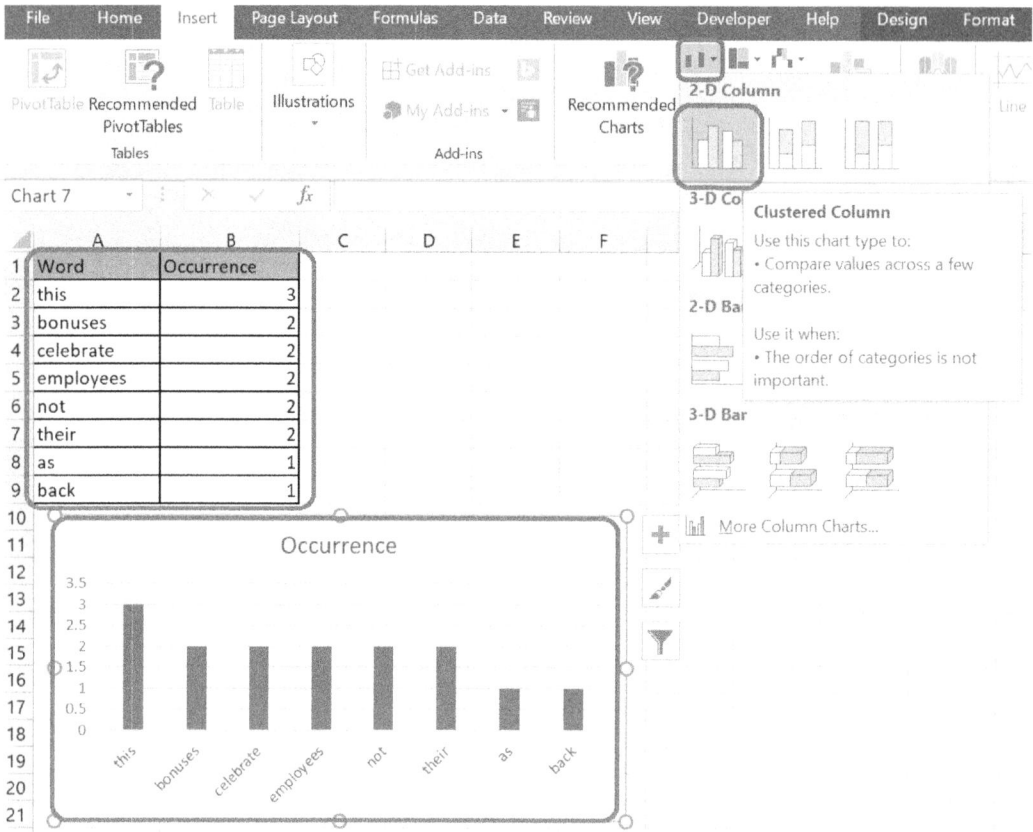

13.2) Count frequently mentioned words online

You can also use online word counters (https://www.textfixer.com/tools/online-word-counter.php) to count each word occurrences. These online word counters can exclude common words from the word frequency count.

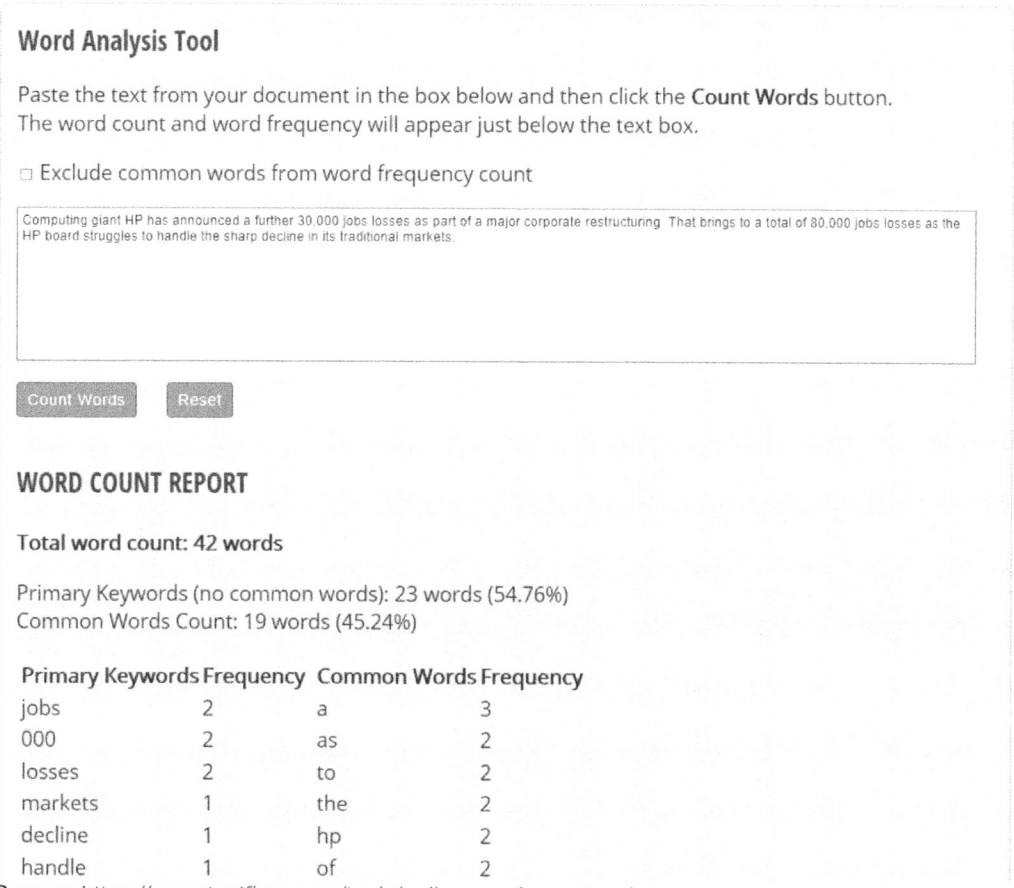

Source: https://www.textfixer.com/tools/online-word-counter.php

13.3) Create a "Word Cloud" to visualize frequently mentioned words using "Pro Word Cloud" Microsoft Word Add-In.

In this section, we will explain the Text mining methods to highlight frequently used keywords in a paragraph of texts, and explain how to generate "word clouds" using the Microsoft Word. Word cloud (also called text cloud or tag cloud) is a visual representation of text data. Word clouds are more visually engaging than a table data - it is easy to understand and present, and is impactful as the most used keywords stands out graphically. Social media sites use word clouds to collect, analyze and share user sentiments. Marketers use word clouds to highlight the needs and pain points of customers. Data scientists use word clouds to report qualitative data.

Word Clouds is a popular way to visualize a message. In Word Clouds, the size of each word indicates its frequency and importance. Pro Word Cloud is a free Microsoft add-in to create Word Clouds.

1) Copy and paste the text below, into your word document.

> Computing giant HP has announced a further 30,000 jobs losses as part of a major corporate restructuring. That brings to a total of 80,000 jobs losses as the HP board struggles to handle the sharp decline in its traditional markets.

Source: Jim Riley (2015), More Retrenchment at HP. https://www.tutor2u.net/business/blog/more-retrenchment-at-hp (8 March 2019)

2) Right click on a blank area at the top of your Word document, and choose "Customize the Ribbon".

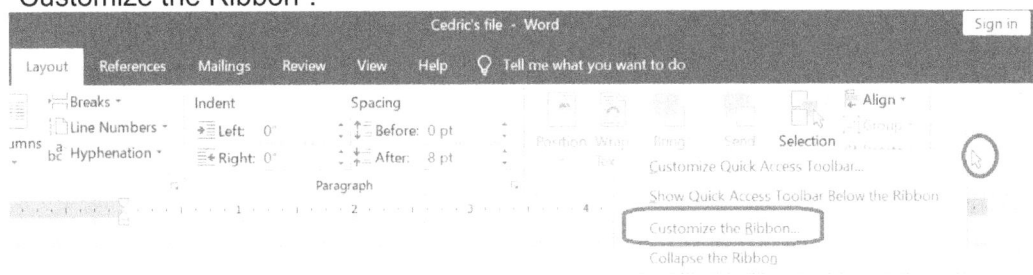

Computing giant HP has announced a further 30,000 jobs losses as part of a major corporate restructuring. That brings to a total of 80,000 jobs losses as the HP board struggles to handle the sharp decline in its traditional markets.

3) Tick "developer" and click ok, and you will see a Developer tab.

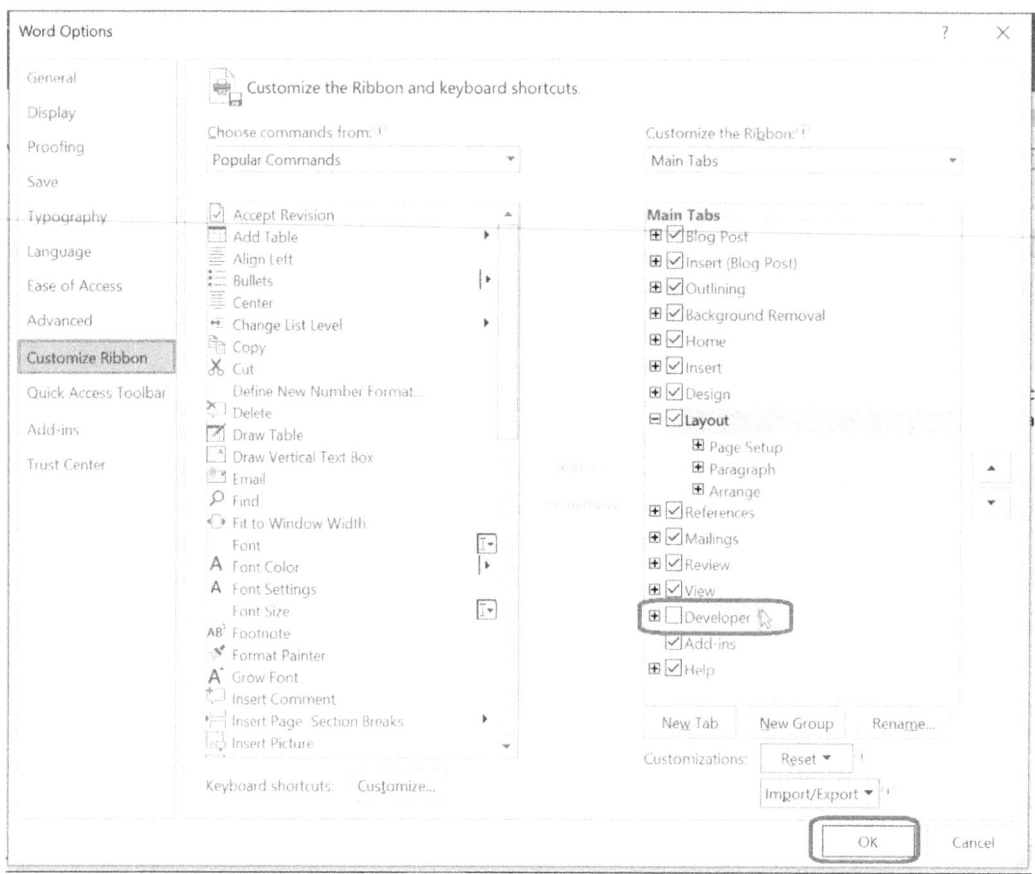

4) At the Developer tab, click the "Add-ins" button:

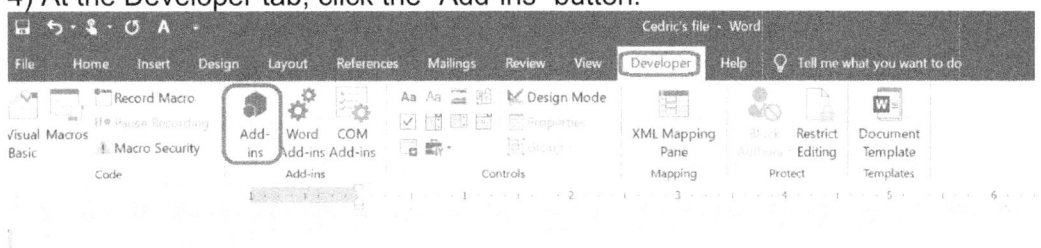

Computing giant HP has announced a further 30,000 jobs losses as part of a major corporate restructuring. That brings to a total of 80,000 jobs losses as the HP board struggles to handle the sharp decline in its traditional markets.

5) Click "Store":

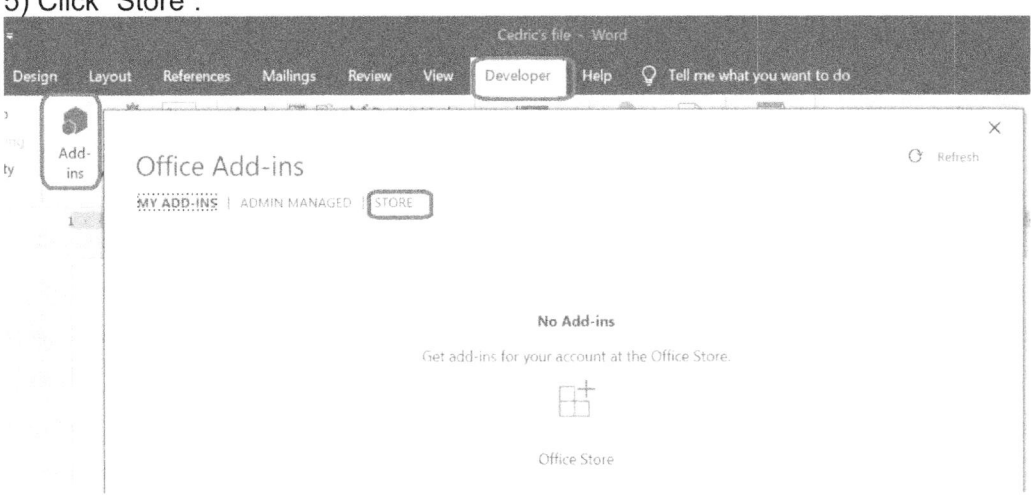

6) Find "pro word cloud", and click "Add".

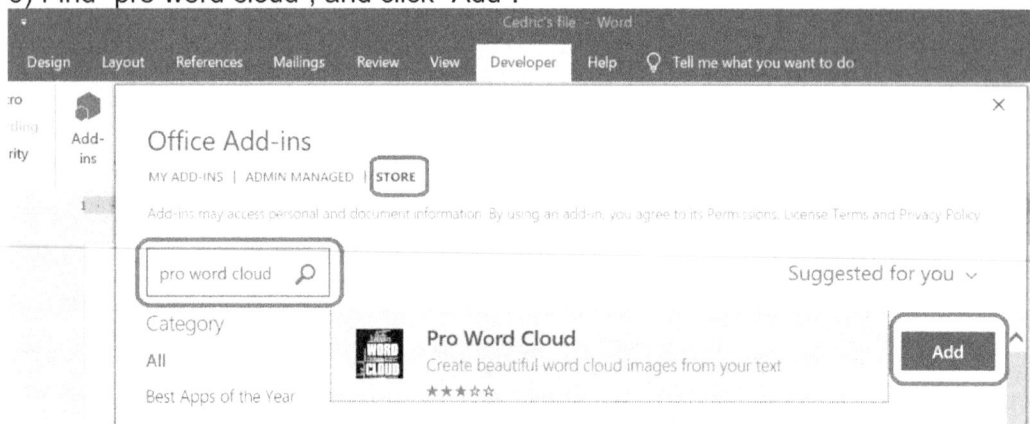

7) After you click "Add", a "Pro Word Cloud" panel appears on the right side of your Word document.

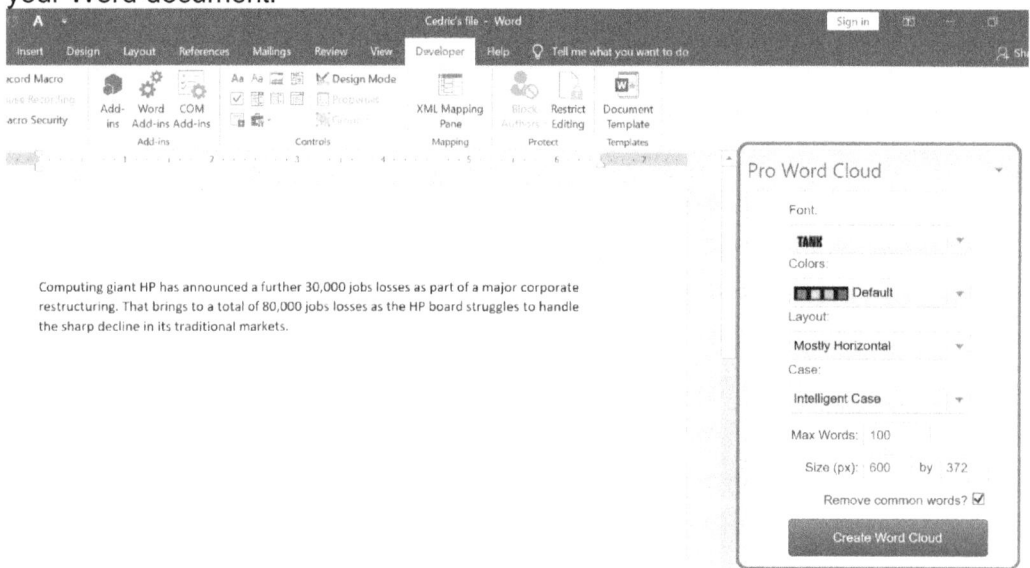

8) To create a Word Cloud, highlight the text that you want to turn into a Word Cloud, and click "Create Word Cloud".

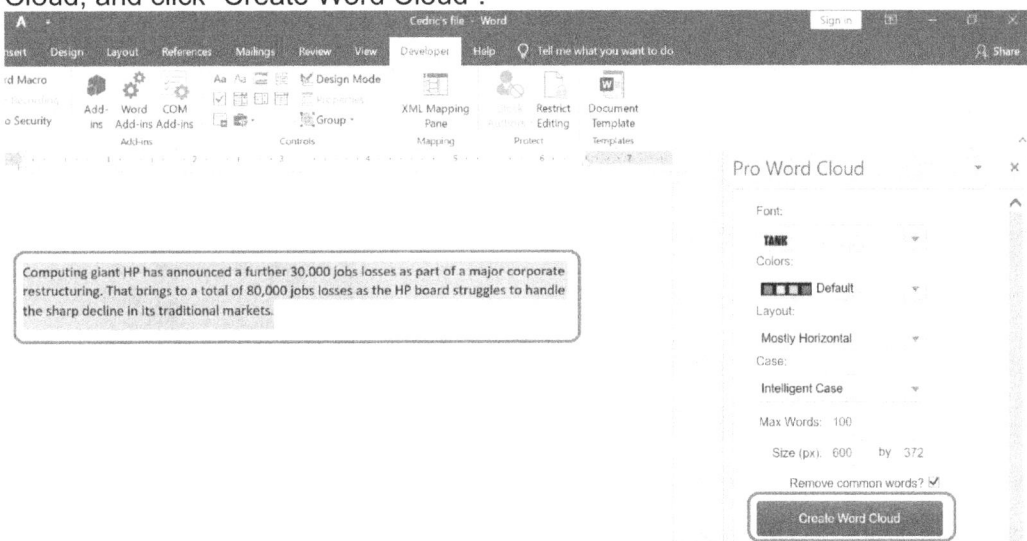

9) After you click "Create Word Cloud", a Word Cloud appears at the top right-hand corner of your Word document.

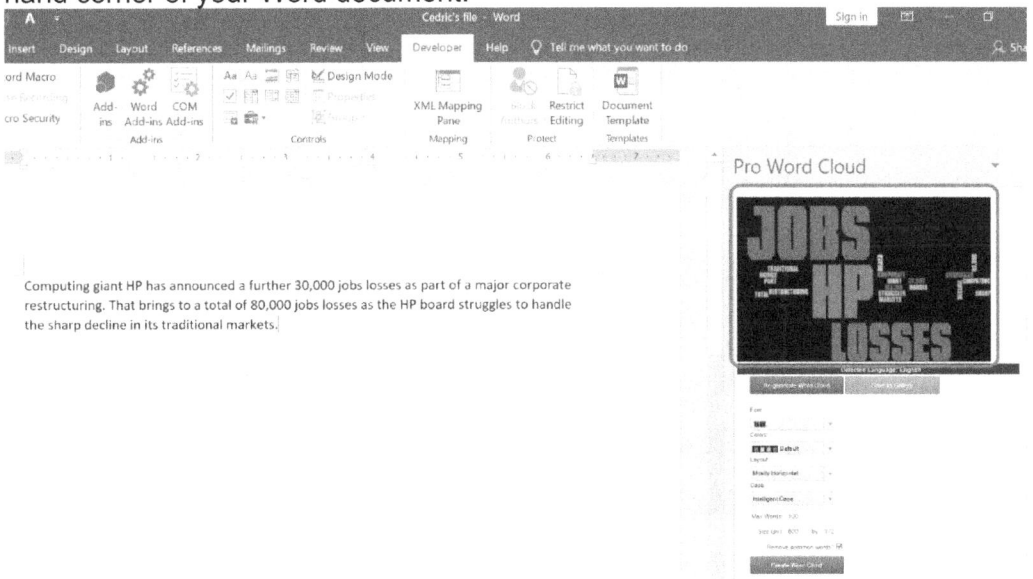

10) To copy the Word Cloud, you can either:
(i) Right click on the Word Cloud, and click "Copy"; or
(ii) Highlight the Word Cloud, and use "Control+C" to copy it

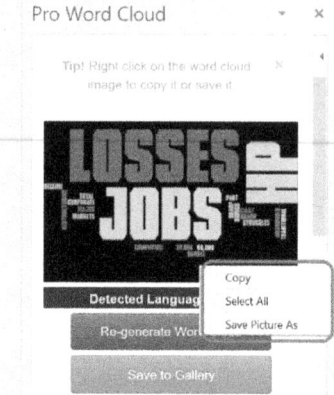

11) Paste the Word Cloud on your Word Document.

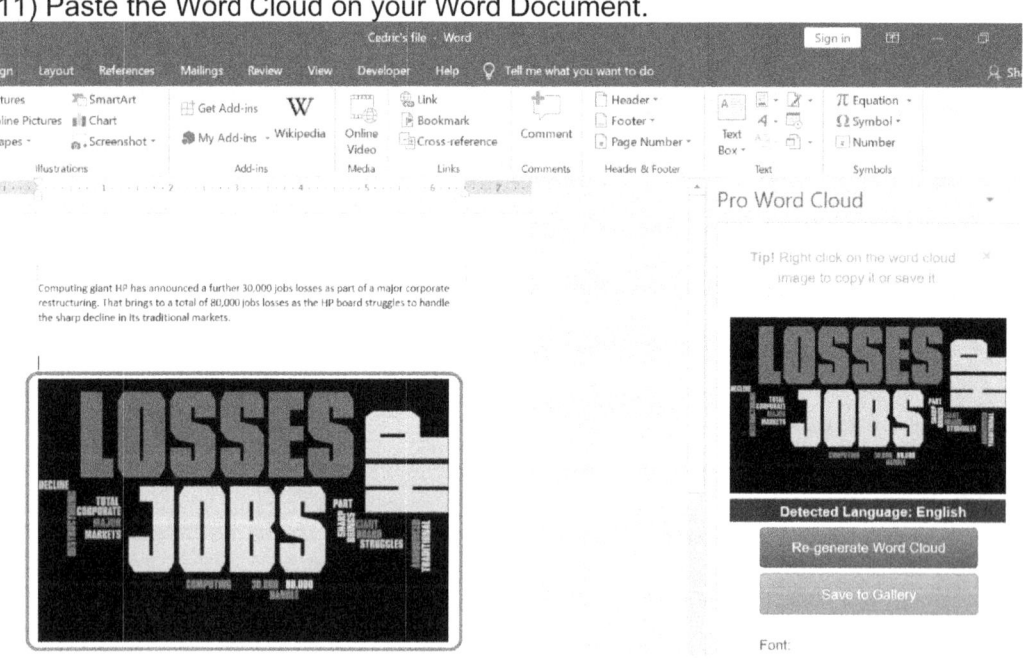

12) To format your Word Cloud, left click on your Word Cloud, click the "Format" tab, and you will see a list of formatting options.

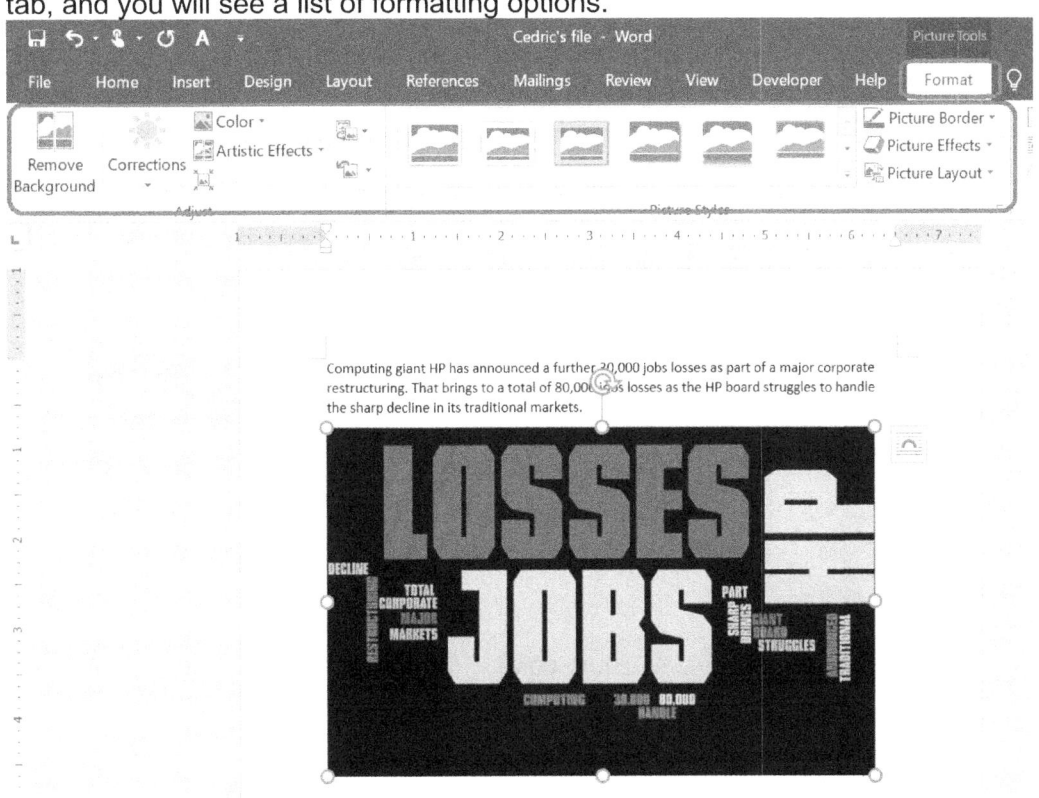

13) Click "Blue, Accent color 1 Light"

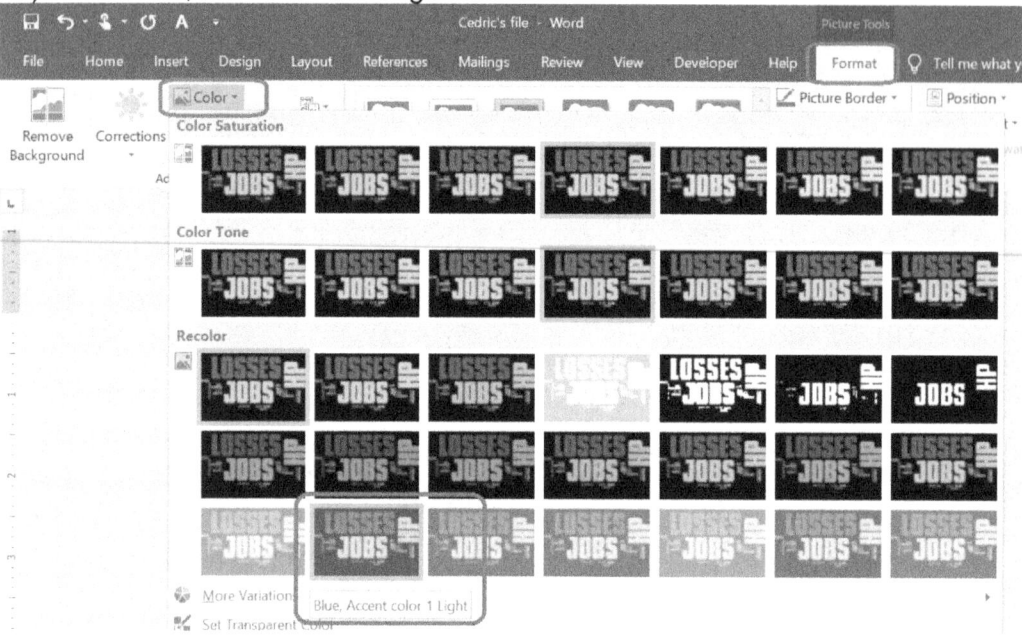

14) Your Word Cloud changes to Blue color. The above word cloud clearly shows that "HP", "Jobs", and "Losses" are the three most important words in the article.

14) Sentiment Analysis

Sentiment Analysis (also known as opinion mining) is the process of mathematically categorizing opinions expressed in text, to determine whether the attitude towards a company, product, or topic, is positive, negative, or neutral. Text information is constantly growing in review sites, forums, and social media. By using sentiment analysis, you gauge how people feel about your company without having to read through thousands of comments.

Source: Symeon Symeonidis (2018) 5 Things You Need to Know about Sentiment Analysis and Classification. https://www.kdnuggets.com/2018/03/5-things-sentiment-analysis-classification.html (13 June 2019)

People's engagement with business and brand perception depends heavily on public opinion. According to a survey by Podium, 93 percent of consumers say that online reviews influence their buying decisions. Sentiment analysis allows companies to monitor and measure people's attitude towards an organization so that they can address it timely. [1]

Customer Feedback text	Sentiment
This restaurant is excellent! The waiters are very friendly and the steak is delicious!	Positive
I will not recommend this restaurant to anyone. The steak is horrible and expensive, and the service is poor.	Negative

How Sentiment Analysis works

Rules-based sentiment analysis technique and uses a dictionary of words labelled by sentiment to determine the sentiment (-1 = Negative, +1 = Positive) of a sentence: [2]

- **Define two lists of polarized words.**
  E.g. Negative words such as bad, worst, ugly, etc. Positive words such as excellent, best, beautiful, etc.
- **Count the number of positive and negative words.**
  Given a text, count the number of positive words, and negative words. If the number of positive word appearances is greater than the number of negative word appearances return a positive sentiment, conversely, return a negative sentiment. Otherwise, return neutral.

Sentence	Sentiment	Score
great colleagues	positive	0.82
nasty bosses, long hours	negative	0.42

Sentiment Analysis limitations

Sentiment Analysis has a few limitations:

- **Context and Polarity**
  Analyzing sentiment without context is difficult because of changes in polarity. If the responses to a survey question, "What did you like about the event?", is "Everything of it" and "Absolutely nothing!", the first response would be positive and the second one would be negative. But, if the responses come from answers to the question, "What did you Dislike about the event?", the negative in the question will make sentiment analysis change altogether! [2]

- **Irony and Sarcasm**
  If the response to a survey question, "Have you had a nice customer experience with us?", is "Yeah, sure.", this likely to be classified as negative because "yeah and sure "belong to positive or neutral texts. But, in reality it might be a negative sentiment from a customer using irony and sarcasm. [2]

- **Comparisons**
  How to treat comparisons such as "This is better than old tools" in sentiment analysis is another challenge, where context makes a difference. Would you classify them as neutral or positive? At first glance, it seems to be a positive sentiment. But if the old tools are useless, then it is a neutral sentiment. [2]

Sentiment Word Cloud

To visualize the results of Sentiment Analysis, you can use graphs, histograms, and Word Cloud.

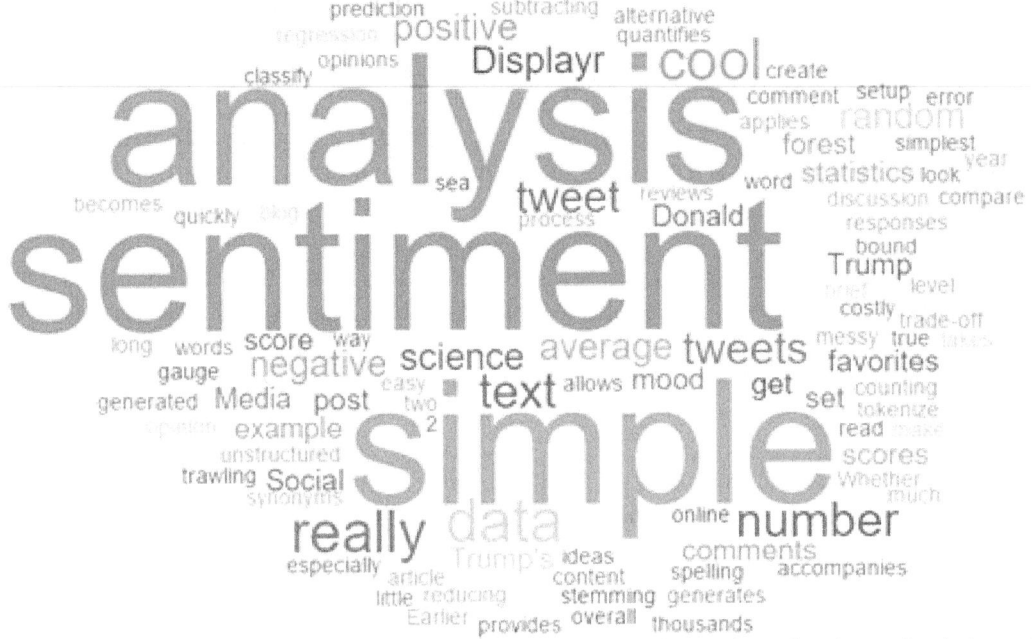

Source: Symeon Symeonidis (2018) 5 Things You Need to Know about Sentiment Analysis and Classification. https://www.kdnuggets.com/2018/03/5-things-sentiment-analysis-classification.html (13 June 2019)

References:
(1) 2017 State of Online Reviews (2017) Consumers Get "Buy" with a little help from their friends. http://learn.podium.com/rs/841-BRM-380/images/2017-SOOR-Infographic.jpg (13 June 2019)
(2) Monkeylearn (2019) Sentiment Analysis - Nearly Everything You Need to Know. https://monkeylearn.com/sentiment-analysis/ (13 June 2019)

14.1) Real-World Impact of Sentiment Analysis

Research have shown that Employer Branding and Sentiment Analysis has an impact on employee salaries, number of job applicants, cost per hire, retention, and employee turnover:

- Glassdoor found that workplace culture matters for employee retention. When employees switch employers, Glassdoor found they usually move to companies with higher Glassdoor ratings. In particular, Glassdoor found that raising a company's overall rating on Glassdoor by one star (on a one-to-five scale) was associated with a four-percentage-point higher chance that employees would stay for their next role. [1]

- Companies with weak brand overpay salaries by 10 percent. [2]

- 69 percent of candidates wouldn't take a job at a bad company even if they were unemployed. [2]

- A strong employer brand can lead to a 50 percent decrease in cost/hire. [2]

- A strong employer brand can lead to a 28 percent increase in retention. [2]

- A strong employer brand can lead to 50 percent more qualified applicants. [2]

- 92 percent of candidates would consider leaving their jobs if a company with an excellent corporate reputation offered them. [2]

- Companies with a strong talent brand get up to 2.5 times more applicants per job post on LinkedIn. [2]

References:
(1) Dr. Andrew Chamberlain (2017) Why Do Employees Stay? A Clear Career Path and Good Pay, for Starters. https://www.glassdoor.com/research/why-do-employees-stay-a-clear-career-path-and-good-pay-for-starters/ (26 November 2018)
(2) getfive (2018) A Cautionary Tale About Bad Glassdoor Ratings. https://getfive.com/blog/a-cautionary-tale-about-bad-glassdoor-ratings/ (26 November 2018)

14.2) Glassdoor Company Ratings and Reviews

Employee's sentiment about a company can be assessed from the company's rating at Glassdoor, where employees can rate what it's really like to work inside, using a 5-point scale ranging from "Very Dissatisfied" to "Very Satisfied".

Glassdoor (https://www.glassdoor.sg/) is a website whereby you can search the ratings and reviews of over 600,000 companies worldwide. At Glassdoor's website, you can find out what it's really like to work inside any company, from people who've actually worked there. Company ratings on Glassdoor are determined by recent employee feedback. [1]

Company ratings are based on a 5-point scale:

- 0.00 - 1.50 Employees are "Very Dissatisfied"
- 1.51 - 2.50 Employees are "Dissatisfied"
- 2.51 - 3.50 Employees say it's "OK"
- 3.51 - 4.00 Employees are "Satisfied"
- 4.01 - 5.00 Employees are "Very Satisfied"

Example of a Glassdoor Company Ratings and Reviews,

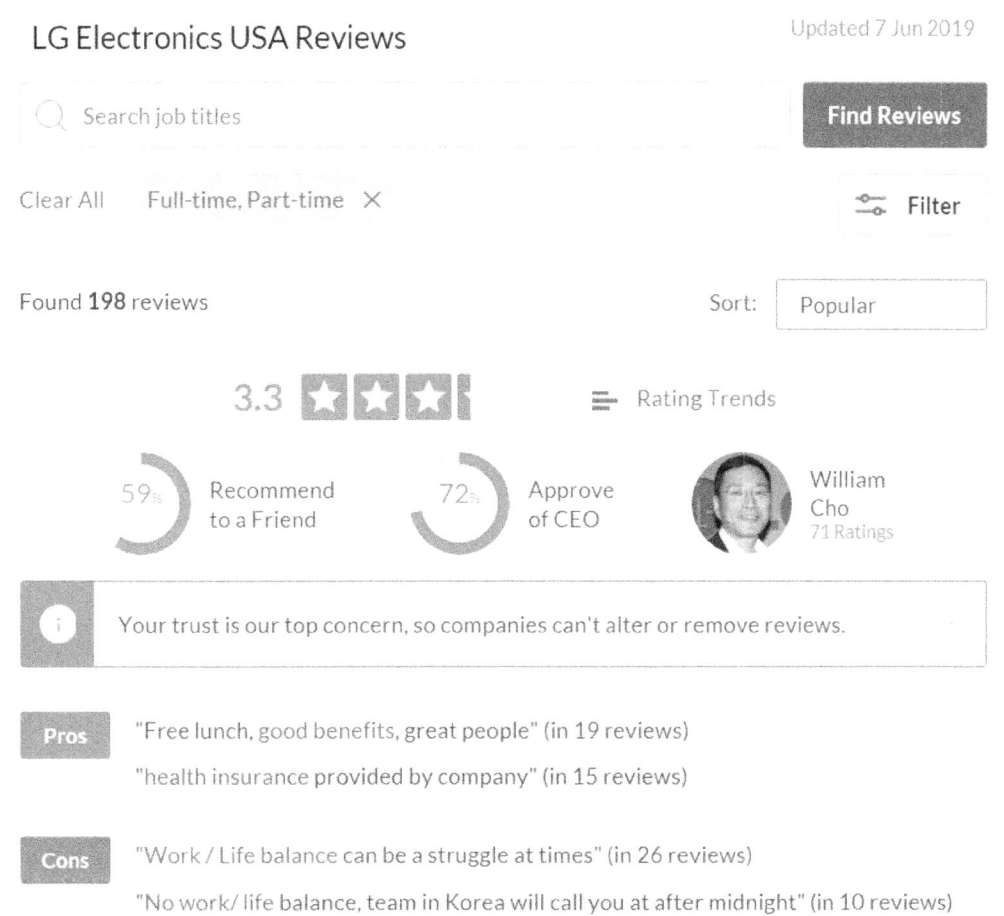

Source: Glassdoor (2019) LG Electronics USA Reviews https://www.glassdoor.sg/Reviews/LG-Electronics-USA-Reviews-E276566.htm (15 June 2019)

References:
(1) Glassdoor (2019) Ratings on Glassdoor. https://help.glassdoor.com/article/Ratings-on-Glassdoor/en_US/ (15 June 2019)

14.3) Run Sentiment Analysis in Excel with Azure Machine Learning

It would be tedious if you have to go through hundreds of survey comments to assess employee sentiment. There is a free add-in from Microsoft that helps you to do sentiment analysis in Excel, and it can compute a probability showing how positive or negative each comment is. It has a dictionary of 5,097 negative and 2,533 positive words, and each word is assigned a strong or weak polarity. [1]

1) Copy the text below into cells A1 to A9 of your Excel spreadsheet:

	A
1	**Glassdoor review**
2	pay is ok, friendly colleagues
3	nice colleagues
4	good learning ground for fresh graduates
5	good worklife
6	not much to learn
7	slow progression, low pay and a lot of work
8	entry level salary is not competitive
9	bad management, politics and work culture

2) Go to "Insert" tab, click "Get Add-ins":

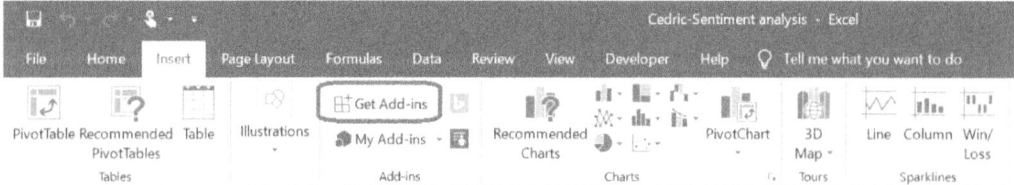

3) Search for "Azure Machine", click add:

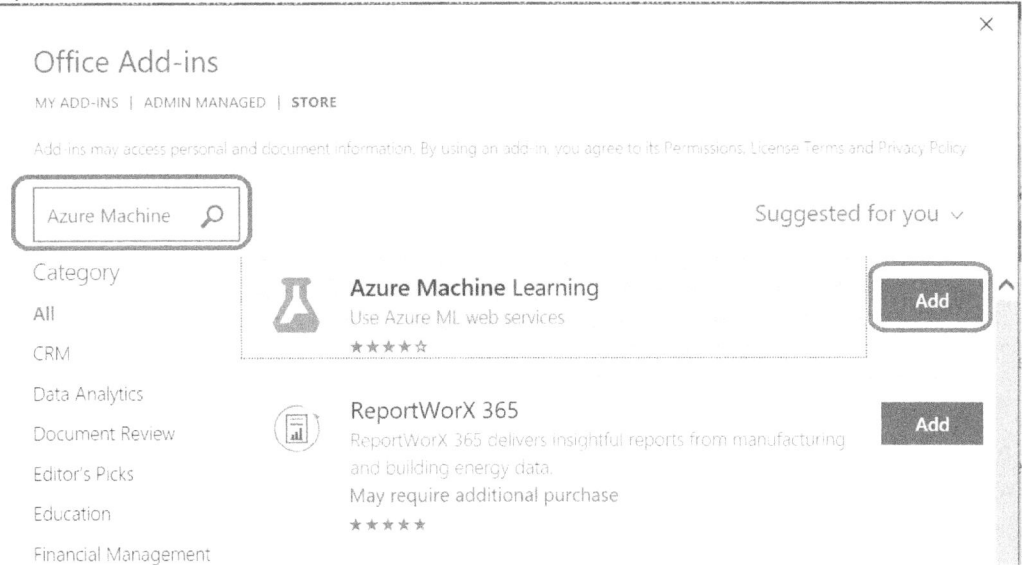

4) Click "Text Sentiment Analysis":

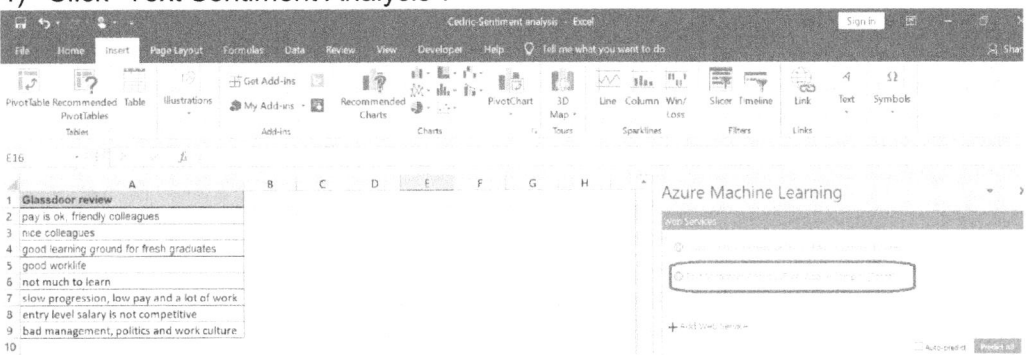

5) Click "View Schema":

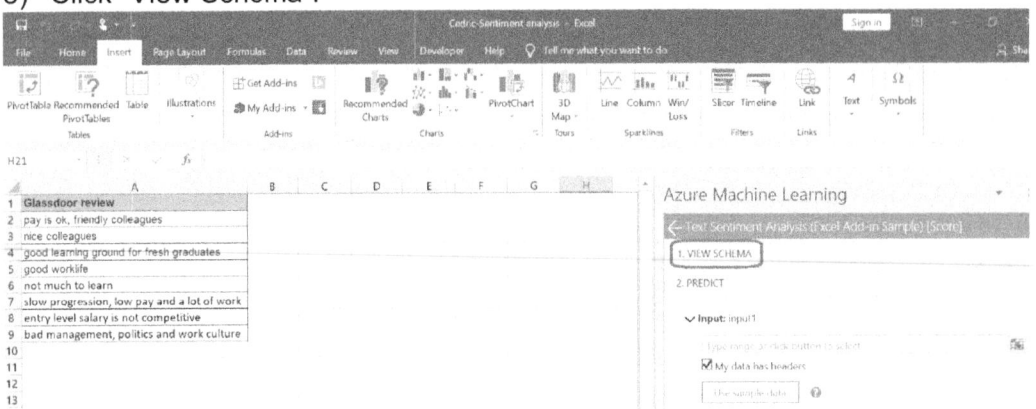

6) Replace your "Glassdoor review" heading word with the Schema word "tweet_text":

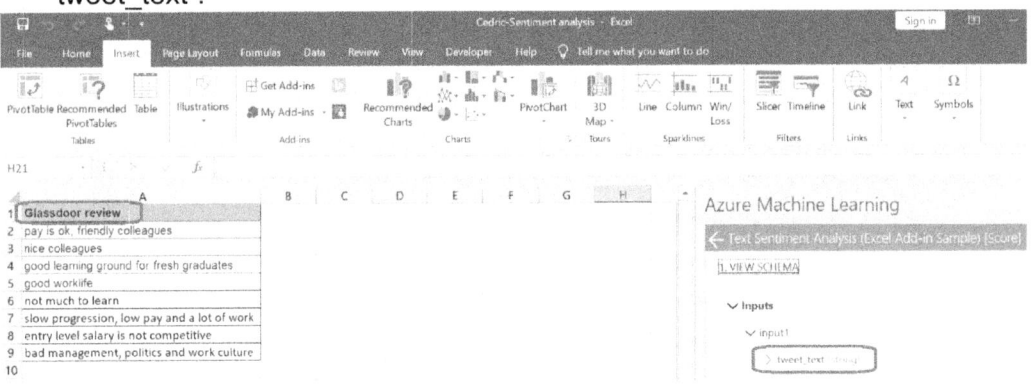

7) Replace the header "Glassdoor review" with the word "tweet_text". Take note that the word "tweet_text" has to be case sensitive.

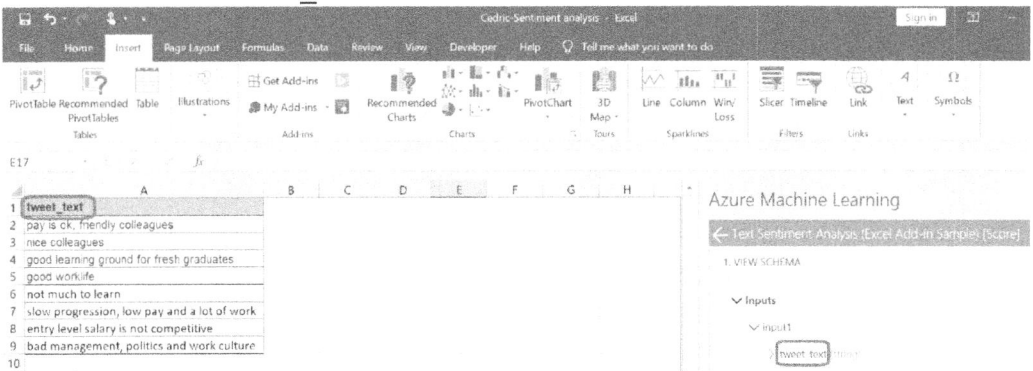

8) Click "View Scema" to close it.

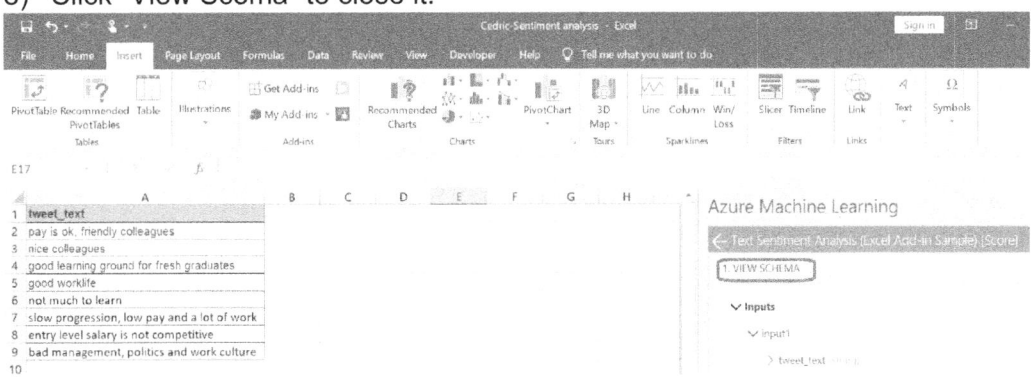

9) Select cells "A1:A9" and tick "My data has headers" for "Input". Input "B1" and tick "Include headers" for "Output". Make sure you have 2 blank columns beside column A. Then click "Predict".

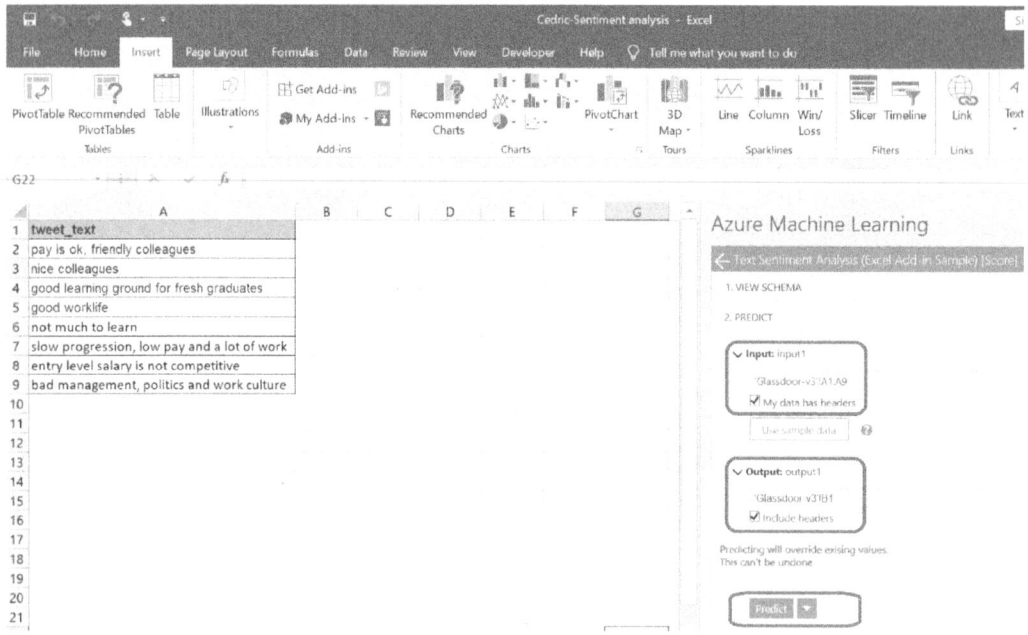

10) After you click "Predict", the "Sentiment" and "Score" column appears. Select Column C, the click "%" at the home bar to convert "Score" to percentages.

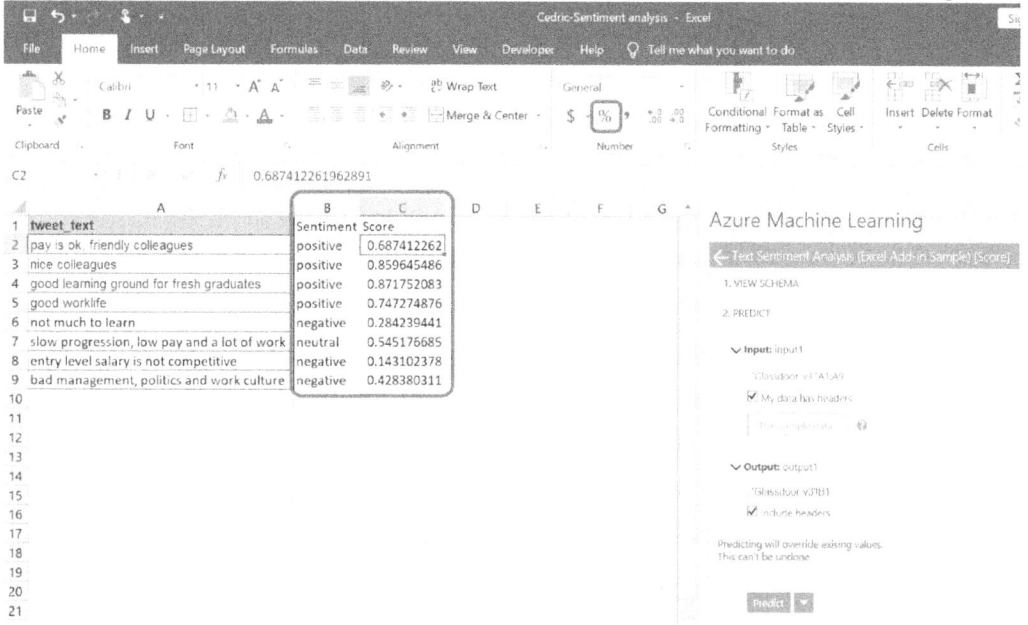

11) Sentiment ranges from 0% to 100%. 100% means very positive sentiment. 0% means very negative sentiment.

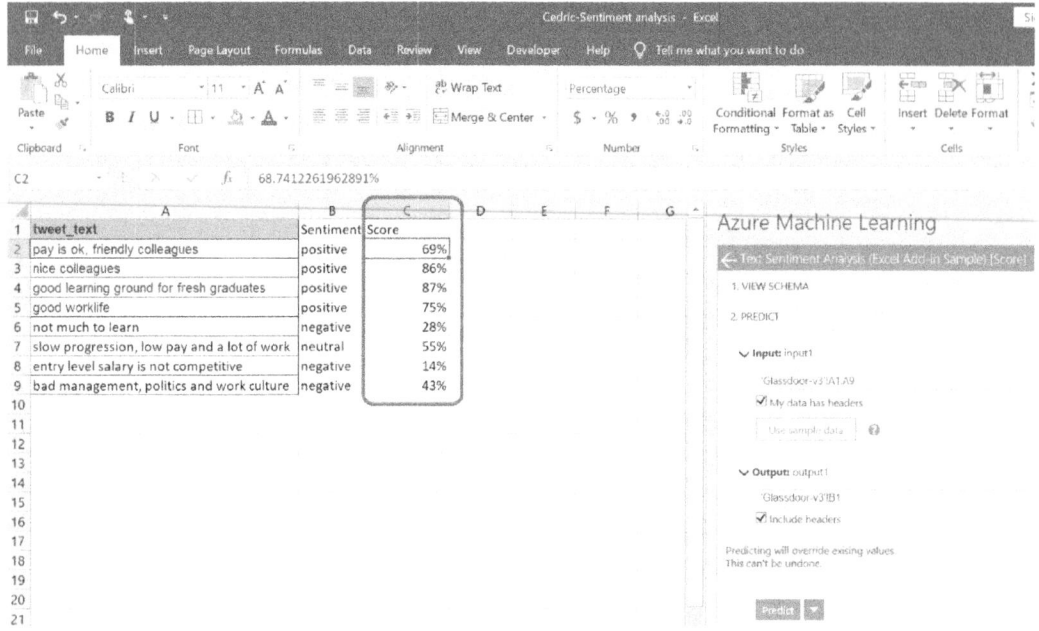

12) Click cell C2, then at "data tab" select sort "Z to A" to sort the "Score from highest to lowest.

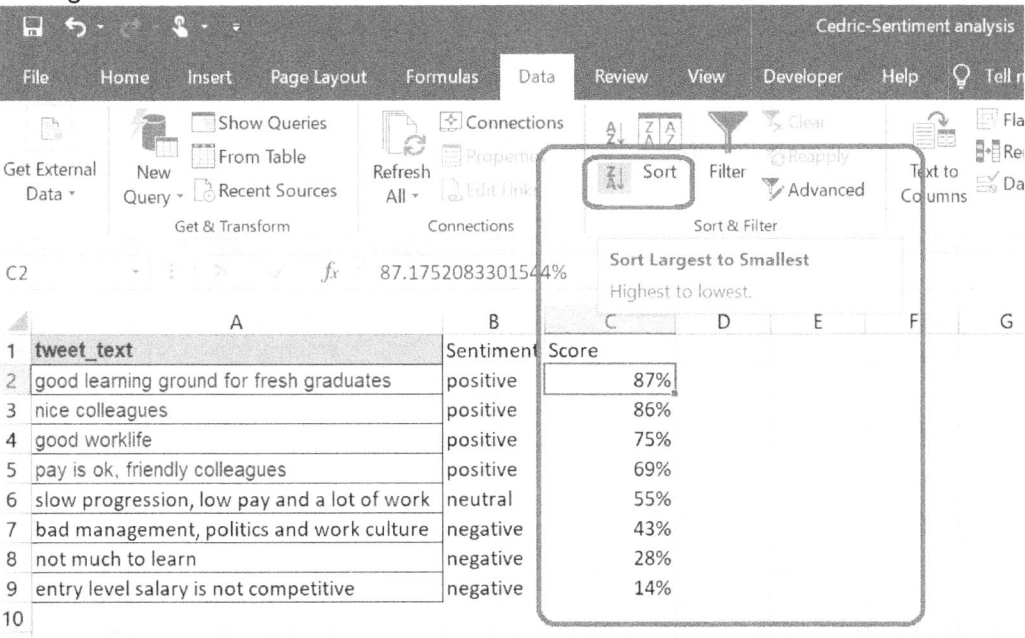

13) To Insert pivot table, to "Insert" tab, click "PivotTable". Under "Select a table or range", select cells A1 to C9. Under "Existing worksheet" enter cell E2 for "location", and click "OK".

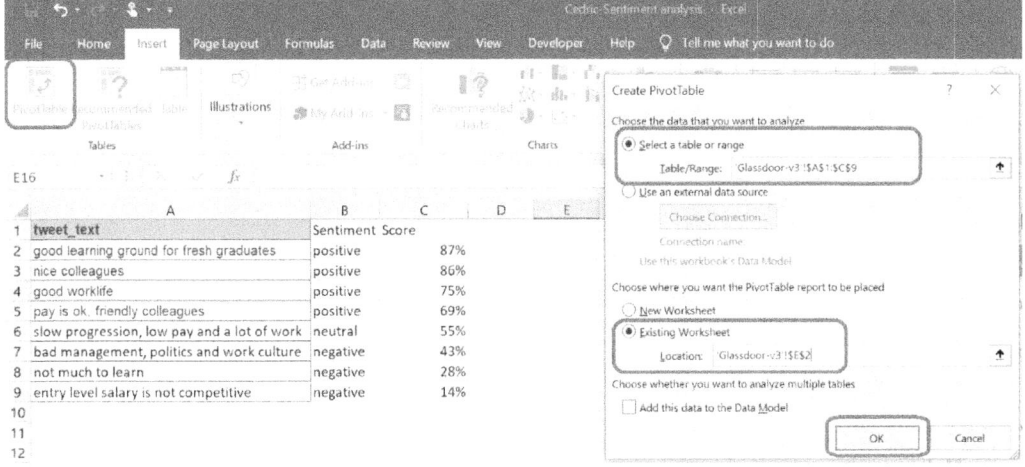

14) At the "PivotTable Fields", click "Sentiment" and "Score".

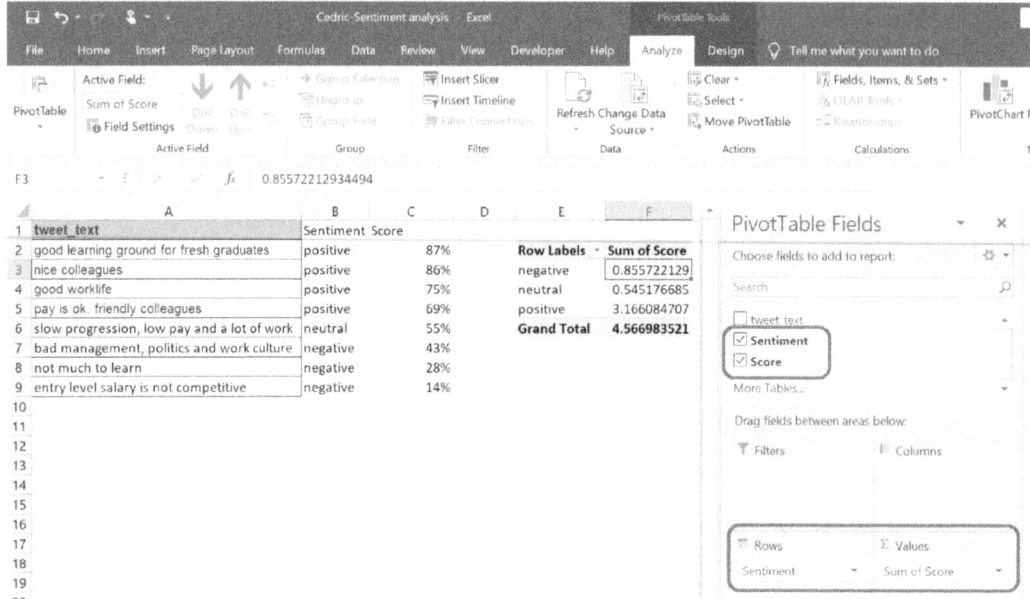

15) Click "Sum of Score", then click "Value Field Settings".

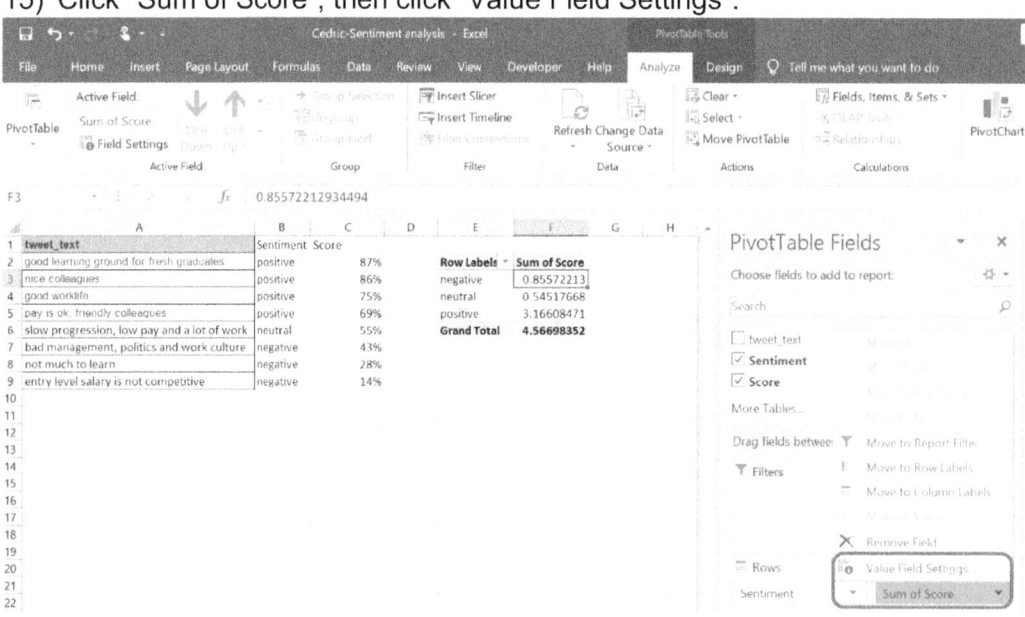

16) At the "Value Field Settings" tab, click "Average", then click "OK".

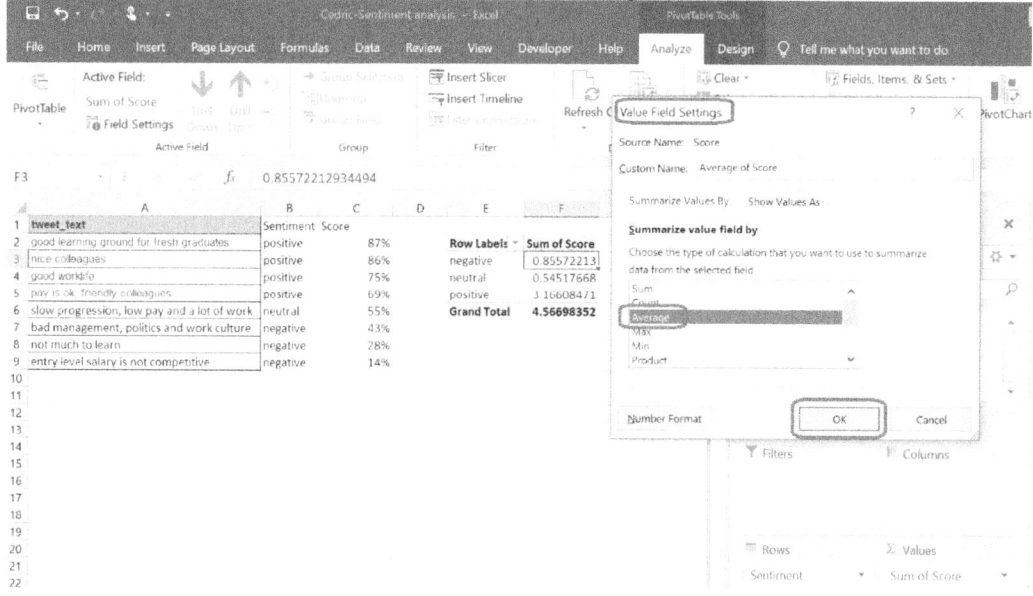

17) Check "tweet_text" and move it to "Values". You'll see "Count of tweet_text".

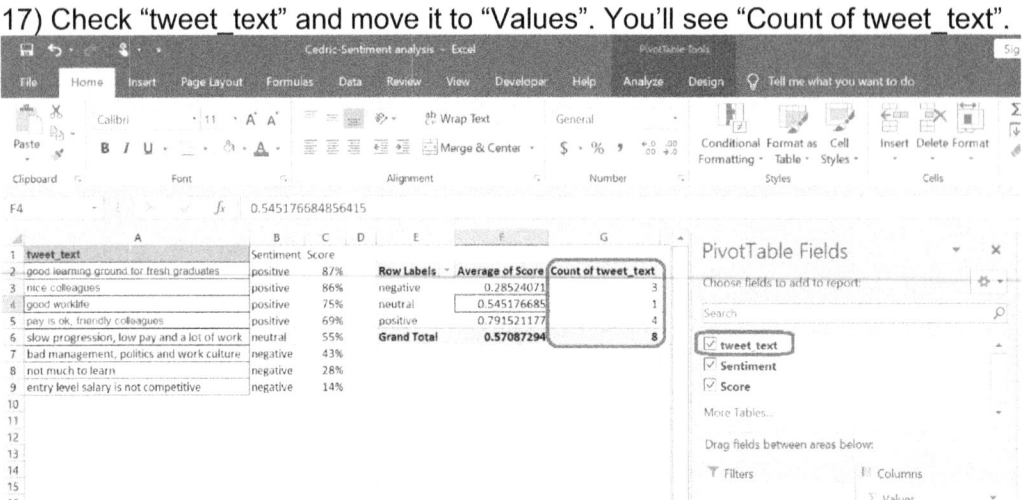

References:
(1) Bill Jelen (2017), Sentiment Analysis. https://www.mrexcel.com/excel-tips/sentiment-analysis/ (3 June 2019)

14.4) Correlation Example: Is there a relationship between "Glassdoor Company Ratings" and "Company Attrition Rate"?

In this example, we want to use correlation to find out which of the following external factors affects a Company's Attrition rate.:
- Unemployment %
- GDP growth %
- Inflation %
- Glassdoor Company ratings

Year	Unemployment %	GDP growth %	Inflation %	Company ABC's Glassdoor rating	Company ABC's attrition rate
2009	2.3	4.1	2.3	1	20
2010	2.4	3.9	2.3	1	21
2011	2.3	4.3	2.3	2	18
2012	2.5	3.5	2.3	2	15
2013	2.6	2.8	2.2	2	17
2014	2.8	2.4	2.3	3	13
2015	3.1	2.6	2.2	3	13
2016	2.9	2.5	2.3	3	12
2017	3.2	2.1	2.2	4	12
2018	3.3	2.3	2.3	4	11
2019	3.4	2.2	2.3	4	11

1) Install "Analysis ToolPak", an Excel add-in

"Analysis ToolPak" is an add-in for Microsoft Excel that comes with Microsoft Excel. To be able to run regression using Excel, you need to first install "Analysis ToolPak", an Excel add-in program that provides data analysis tools. To load the Analysis ToolPak add-in, follow these steps:

- On the File tab, click Options.

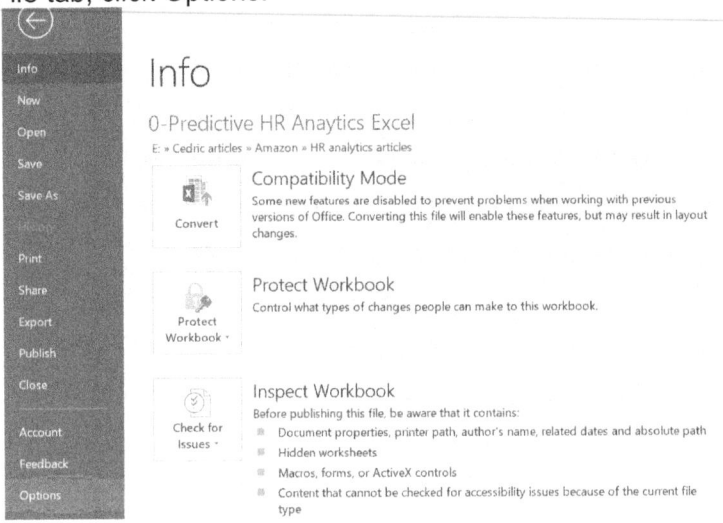

- Under Add-ins, click Analysis ToolPak and click the "Go" button.

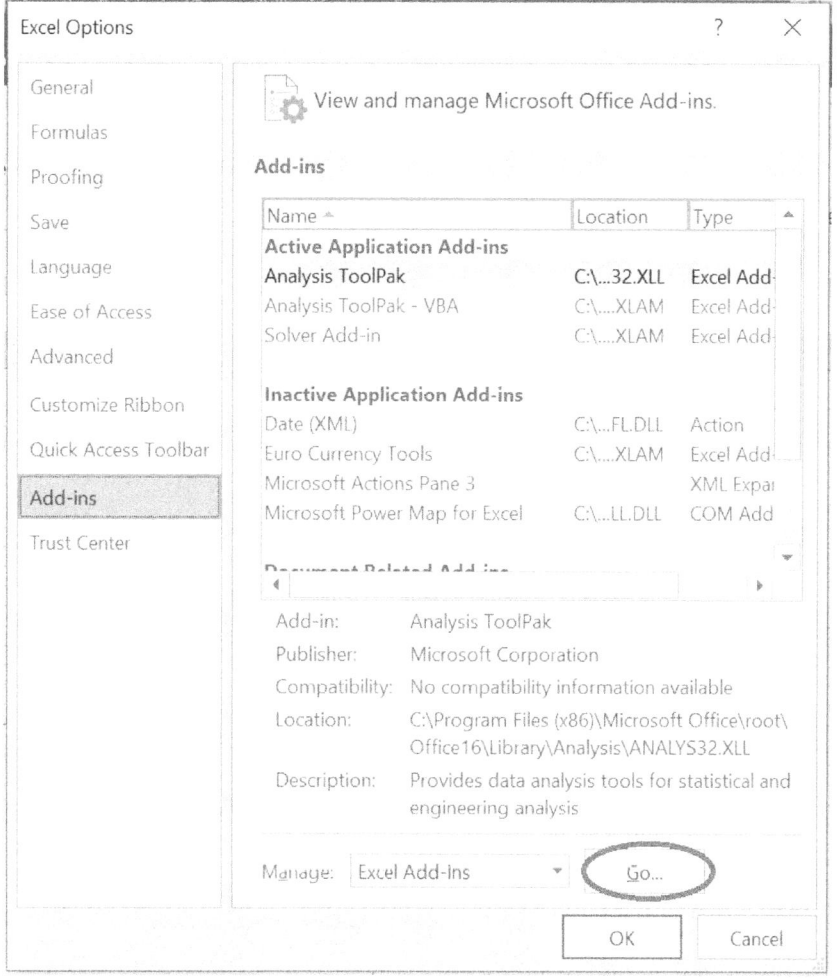

- Click "Analysis ToolPak" and click on OK.

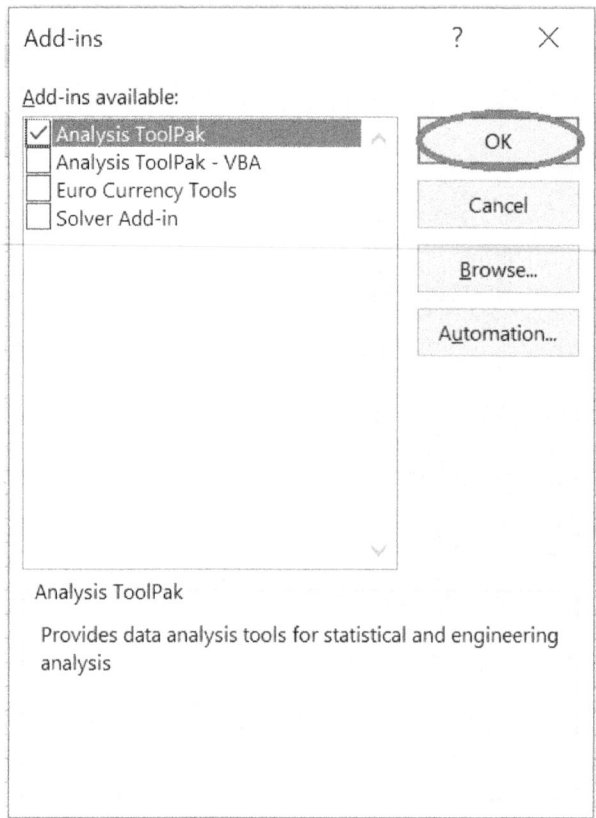

- On the Data tab, in the Analysis group, you are now able to click on "Data Analysis".

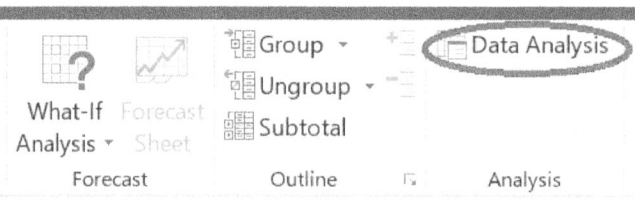

2) Copy the example data in the following table, and paste it in cell A1 of a new Excel worksheet.

	A	B	C	D	E	F
1	Year	Unemployment %	GDP growth %	Inflation %	Company ABC's Glassdoor rating	Company ABC's attrition rate
2	2009	2.3	4.1	2.3	1	20
3	2010	2.4	3.9	2.3	1	21
4	2011	2.3	4.3	2.3	2	18
5	2012	2.5	3.5	2.3	2	15
6	2013	2.6	2.8	2.2	2	17
7	2014	2.8	2.4	2.3	3	13
8	2015	3.1	2.6	2.2	3	13
9	2016	2.9	2.5	2.3	3	12
10	2017	3.2	2.1	2.2	4	12
11	2018	3.3	2.3	2.3	4	11
12	2019	3.4	2.2	2.3	4	11

3) Select "Correlation" and click "OK".

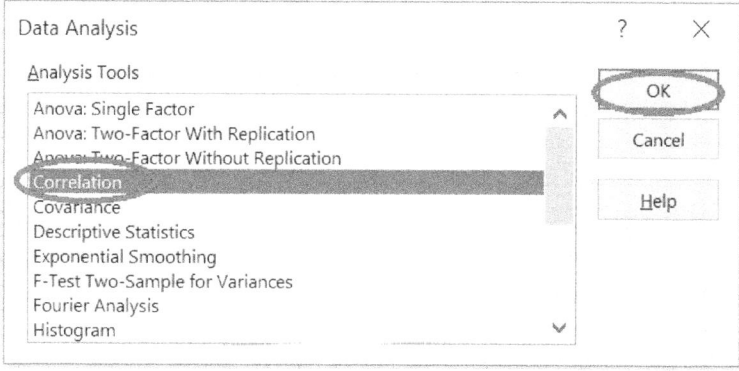

4) After you click OK in the "Data Analysis" dialog box, you will see a "Correlation" dialog box.
5) For "Input Range", select cells (B1:F12).
6) Check "Labels in first row".
7) For "Output Range", select cells (A14).
8) Click "OK".

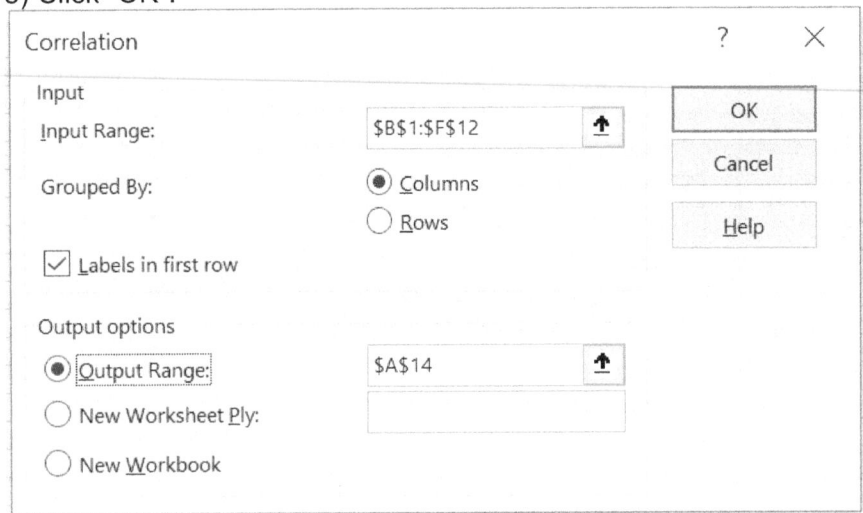

After you click "OK", Excel generates the following Correlation analysis.

	Unemployment %	GDP growth %	Inflation %	Company ABC's Glassdoor rating	Company ABC's attrition rate
Unemployment %	1				
GDP growth %	-0.91	1			
Inflation %	-0.26	0.37	1		
Company ABC's Glassdoor rating	0.94	-0.88	-0.21	1	
Company ABC's attrition rate	-0.90	0.89	0.14	-0.95	1

A negative correlation coefficient means that an increase in X is associated with a decrease in Y. Similar to a positive correlation, a negative correlation shows a connection between two variables, and the relative strengths are the same. In other words, a correlation coefficient of 0.85 has the same strength as a correlation coefficient of -0.85. Correlation coefficients are always values between -1 and 1, where "-1" means that there is a perfect linear negative correlation, while "1" shows a perfect linear positive correlation. A correlation coefficient of zero, or near to zero, means that there is no meaningful relationship between variables. Correlation coefficient of 0.91 or -0.92 shows a very strong positive and negative correlation respectively. However, correlation does not mean causation. An example of negative correlation is the amount of snowfall and the temperature. As the temperature increases, the amount of snowfall decreases. An example of positive correlation is the relationship between temperature and ice cream sales. As temperature increases, so do ice cream sales.

9) Observations from the above Excel Correlation analysis:

	Unemployment %	GDP growth %	Inflation %	Company ABC's Glassdoor rating	Company ABC's attrition rate
Unemployment %	1				
GDP growth %	-0.91	1			
Inflation %	-0.26	0.37	1		
Company ABC's Glassdoor rating	0.94	-0.88	-0.21	1	
Company ABC's attrition rate	-0.90	0.89	0.14	-0.95	1

From the Excel Correlation analysis, these variables are good predictors of Company ABC's attrition rate as they have strong correlation of below -0.75 and above 0.75:
- Unemployment %: -0.90 Correlation coefficient with Company ABC's attrition rate.
- GDP growth %: -0.89 Correlation coefficient with Company ABC's attrition rate.
- Company ABC's Glassdoor rating: -0.95 Correlation coefficient with Company ABC's attrition rate.

From the Excel Correlation analysis, Inflation % has very little impact on Company ABC's attrition rate as they have very weak correlation of between -0.20 to 0.20.

14.5) Multiple Regression Example: Predict "Company Attrition Rate" with "Glassdoor Company Ratings".

In this example, we want to use multiple regression to predict the company's attrition rate, based on changes in external factors (Unemployment %, GDP growth %, Inflation %, and Glassdoor Company ratings).

1) Install "Analysis ToolPak", an Excel add-in

"Analysis ToolPak" is an add-in for Microsoft Excel that comes with Microsoft Excel. To be able to run regression using Excel, you need to first install "Analysis ToolPak", an Excel add-in program that provides data analysis tools. To load the Analysis ToolPak add-in, follow these steps:

On the File tab, click Options.

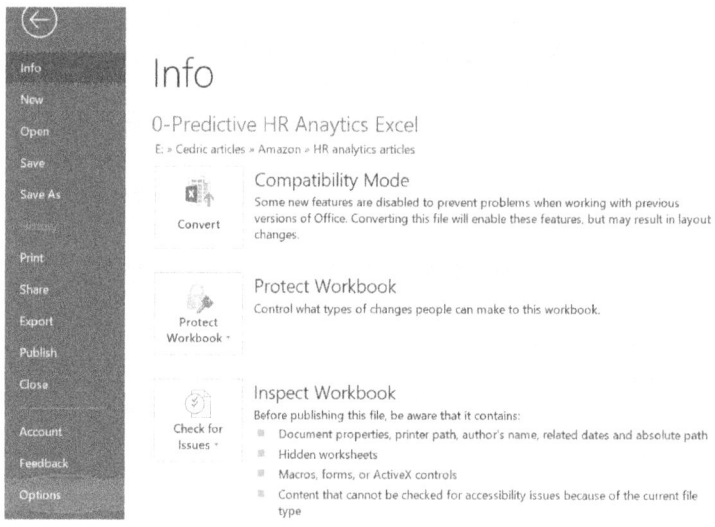

Under Add-ins, click Analysis ToolPak and click the "Go" button.

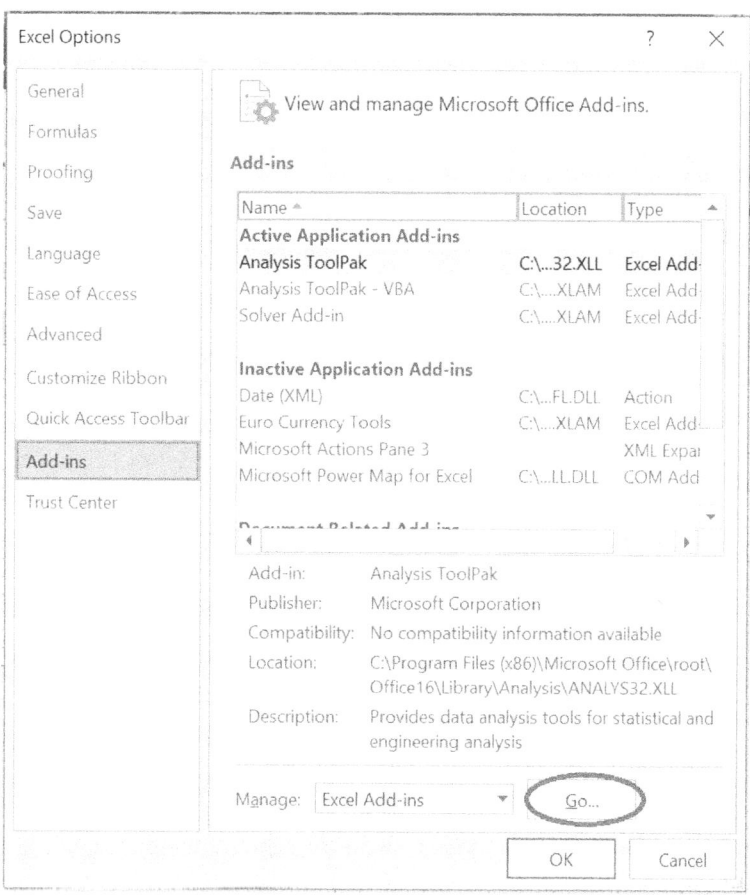

Click "Analysis ToolPak" and click on OK.

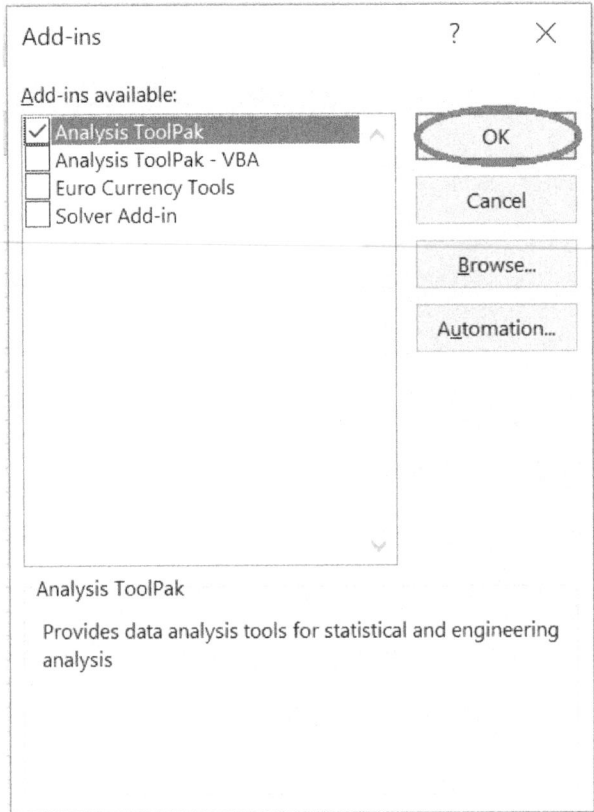

On the Data tab, in the Analysis group, you are now able to click on "Data Analysis".

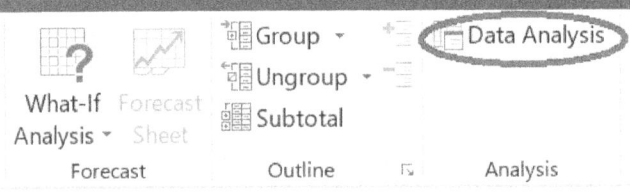

2) Copy the example data in the following table, and paste it in cell A1 of a new Excel worksheet.

	A	B	C	D	E
1	Year	Unemployment %	GDP growth %	Company ABC's Glassdoor rating	Company ABC's attrition rate
2	2009	2.3	4.1	1	20
3	2010	2.4	3.9	1	21
4	2011	2.3	4.3	2	18
5	2012	2.5	3.5	2	15
6	2013	2.6	2.8	2	17
7	2014	2.8	2.4	3	13
8	2015	3.1	2.6	3	13
9	2016	2.9	2.5	3	12
10	2017	3.2	2.1	4	12
11	2018	3.3	2.3	4	11
12	2019	3.4	2.2	4	11

3) On the Data tab, in the Analysis group, click on "Data Analysis".

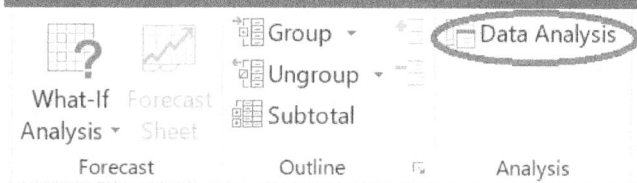

4) Select "Regression" and click "OK".

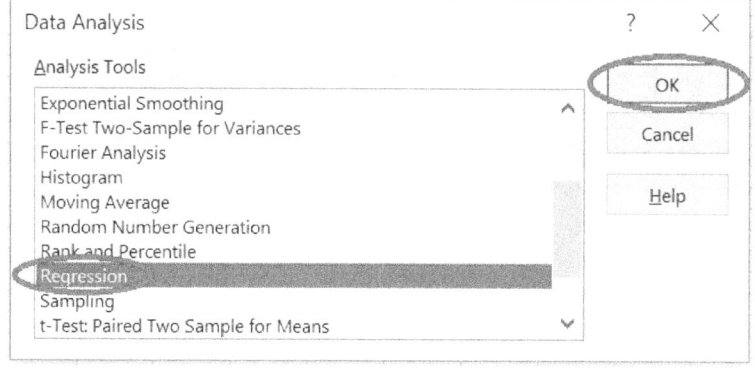

5) After you click OK in the "Data Analysis" dialog box, you will see a "Regression" dialog box.
6) For "Input Y Range", select cells (E1:E12). This is the predictor variable or dependent variable.
7) For "Input X Range", select cells (B1:D12). These are the explanatory variables or independent variables.
8) Check "Labels" box.
9) Click the "Output Range" box, and select cell A14.
10) Click "OK".

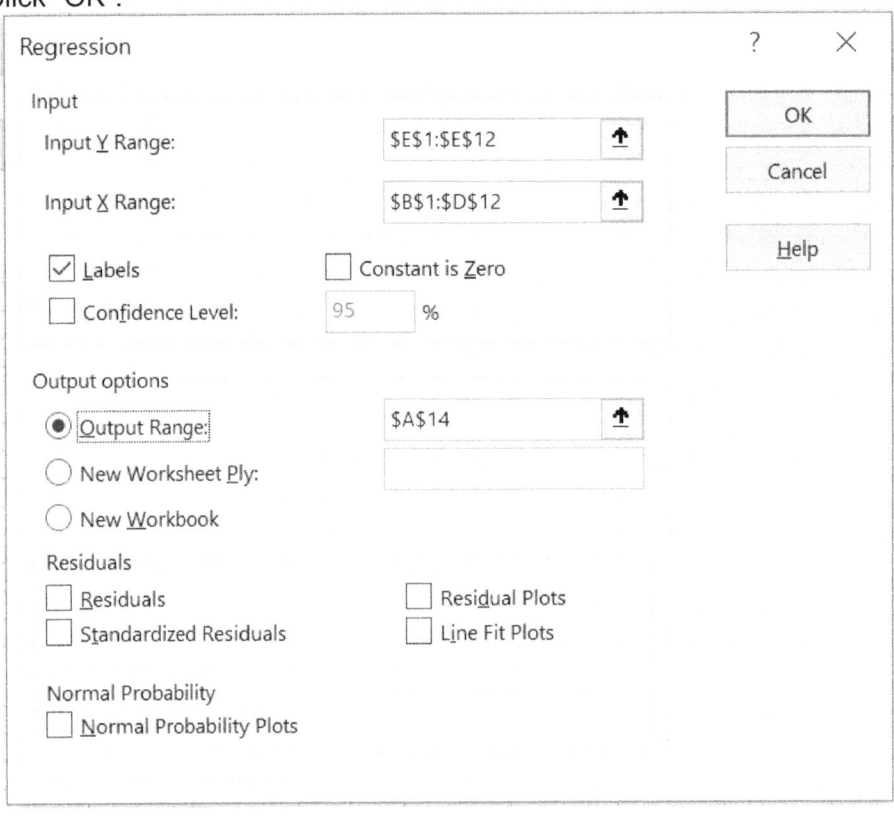

After you click "OK", Excel generates the following Summary Output. Round the numbers to 3 decimal places.

SUMMARY OUTPUT

Regression Statistics	
Multiple R	0.961
R Square	0.923
Adjusted R Square	0.890
Standard Error	1.203
Observations	11

ANOVA

	df	SS	MS	F	Significance F
Regression	3	121.498	40.499	27.963	0.000
Residual	7	10.138	1.448		
Total	10	131.636			

	Coefficients	Standard Error	t Stat	P-value	Lower 95%	Upper 95%	Lower 95.0%	Upper 95.0%
Intercept	13.708	9.814	1.397	0.205	-9.499	36.915	-9.499	36.915
Unemployment %	1.631	3.181	0.513	0.624	-5.890	9.152	-5.890	9.152
GDP growth %	1.316	1.136	1.159	0.285	-1.369	4.001	-1.369	4.001
Company ABC's Glassdoor rating	-2.795	1.036	-2.698	0.031	-5.246	-0.345	-5.246	-0.345

R Square
In the output, R Square is 0.923, which means it is a good fit. 92% of the variation in Company ABC's Attrition Rate (Output) is explained by the independent variables (Input): Unemployment %, GDP growth %, and Company ABC's Glassdoor rating. The closer R Square is to "1", the better the regression line fits the data.

Significance F and P-values
To determine if your results are statistically significant (i.e. reliable), check "Significance F". If the value of "Significance F" is less than 0.05, it is statistically significant (i.e. reliable). If "Significance F" is bigger than 0.05, don't use this set of independent variables. Delete those variables with "P-value" that is bigger than 0.05 and run the regression again until "Significance F" drops below 0.05. Most or all of your P-values should be lower than 0.05. In our example below "P-value" is 0.205, 0.624, 0.285, and 0.031 for Intercept, Unemployment %, GDP growth %, Company ABC's Glassdoor rating, respectively.

Coefficients

From the Summary Output, the regression line is:

SUMMARY OUTPUT

Regression Statistics	
Multiple R	0.961
R Square	0.923
Adjusted R Square	0.890
Standard Error	1.203
Observations	11

ANOVA

	df	SS	MS	F	Significance F
Regression	3	121.498	40.499	27.963	0.000
Residual	7	10.138	1.448		
Total	10	131.636			

	Coefficients	Standard Error	t Stat	P-value	Lower 95%	Upper 95%	Lower 95.0%	Upper 95.0%
Intercept	13.708	9.814	1.397	0.205	-9.499	36.915	-9.499	36.915
Unemployment %	1.631	3.181	0.513	0.624	-5.890	9.152	-5.890	9.152
GDP growth %	1.316	1.136	1.159	0.285	-1.369	4.001	-1.369	4.001
Company ABC's Glassdoor rating	-2.795	1.036	-2.698	0.031	-5.246	-0.345	-5.246	-0.345

Based on the above coefficients,
- For each unit increase in Unemployment %, Company ABC's Attrition Rate increase by 1.631.
- For each unit increase in GDP growth %, Company ABC's Attrition Rate increase by 1.316.
- For each unit increase in Company ABC's Glassdoor rating, Company ABC's Attrition Rate increase by -2.795.

Regression formula

Based on the Summary Output, if "Unemployment %" is 3.3, "GDP growth %," is 2.3, "Company ABC's Glassdoor rating" is 5, then predicted "Company ABC's Attrition Rate"
= 13.708 + (1.631 * Unemployment %) + (1.316 * GDP growth %) - (2.795 * Company ABC's Glassdoor rating)
= 13.708 + (1.631 * 3.3) + (1.316 * 2.3) - (2.795 * 5)
= 8.1

15) Employee Engagement Analytics

Numerous researches have shown that there is a relationship between Employee Engagement and variables such as: employee demographics, employee churn, absenteeism, employees' health, inventory shrinkage, sales, safety, profit, innovation, customer satisfaction, and total shareholder return.

Engagement's impact on Absenteeism
- Marks & Spencer found that absenteeism in stores at the top quartile of engagement scores were twenty-five percent lower compared to those in the bottom quartile. [1]
- While only 4 in 10 employees say that they had opportunities to learn and grow, Gallup suggests that improving that ratio to 8 in 10 can allow your business to achieve a 44% drop in absenteeism and a 16% jump in productivity. [2]

Engagement's impact on Employee Attrition
- Corporate Leadership Council found that engaged employees are eighty-seven per cent less likely to resign. [3]
- Standard Chartered found their branches with high employee engagement had 46% lower voluntary turnover. [4]

Employee Demographics' impact on Engagement
- Shoshana Dobrow Riza (London School of Economics and Political Science), Yoav Ganzach (Tel Aviv University), and Yihao Liu (University of Florida) found that Job satisfaction tends to improve as we get older but also tends to decrease the longer we stay at a particular job. [5]
- Eric van Duin discovered that a manager's tenure strongly correlated to its team's engagement at PostNL N.V. The shorter the manager's tenure, the higher the engagement of the team. With the findings, HR reviewed their training programme for managers with long tenures, and as a result improved their engagement with their teams. [6]

Engagement's impact on Health
- Engaged employees take an average of 2.7 sick days per year, whereas disengaged employees take 6.2 sick days per year. [7]

Engagement's impact on Innovation
- Krueger & Killham found that fifty-nine percent of engaged employees indicated that their job 'brings out their most creative ideas', compared to only three percent for disengaged employees. [8]

Engagement's impact on Inventory shrinkage
- Research showed that low employee engagement levels are linked to increased inventory shrinkage. Inventory shrinkage refers to the loss of inventory. For example, if the inventory records show that there are 5,520 units of Product ABC, but a physical count show that there are only 5,510 units, there is an inventory shrinkage of 10 units. Inventory shrinkage can be due to employee theft, shoplifting, damage, etc. The MacLeod Review cited a meta-analysis that found inventory shrinkage was fifty-one percent worse in workplaces with low engagement compared to those with engaged employees. As an example, if the number of engaged employees increase by five percent in a business with $10M inventory shrinkage, the company might save $250,000. [9]

Engagement's impact on Profitability
- Standard Chartered found that branches where employee engagement was high has sixteen percent higher profit margin growth compared to branches where employee engagement was low. [10]
- A study by ISS found that profitability is high, when employee engagement and customer advocacy are high. The average profitability in countries scoring highest on both employee engagement and customer advocacy was 7.75 percent, versus 4.52 percent for the lowest scoring groups. ISS tested this link by combining selected questions from their Employee Engagement Survey and their Customer Experience Survey with their profitability. In the diagram below, the eNPS (employee net promoter score) and cNPS (customer net promoter score) scores are depicted on the X and Y axis respectively, while the average profitability is shown in the boxes. The figure shows that if the employees and the customers report a high level of satisfaction (as measured by the NPS), then the country will have significantly higher margins than if either of the two scores are low. Implication of this study is that companies should reinvest more in customer service improvements, employee training, on-boarding processes, and service innovation. [11]

Link between margins and employee & customer net promoter scores at a country level

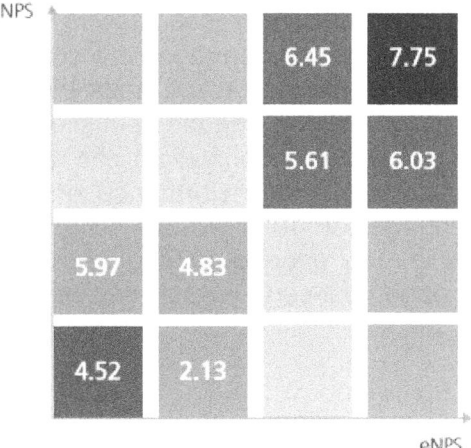

Source: ISS Prospectus, 2014

Source: Morten Kamp Andersen (proacteur), Simon Svegaard (ISS) & Peter Ankerstjerne (ISS). (2015), Linking Customer Experience with Service Employee Engagement, ISS White Paper, pp 10-11.

Engagement's impact on Quality
- DDI studied the employee engagement in two hundred organizations, and found that employees with higher engagement scores have lower Quality errors. In a Fortune 100 manufacturing company, Quality errors as measured by external and internal parts per million was 5,658 for the low-engagement group and only 52 for the high-engagement group! [12]

Engagement's impact on Safety
- MolsonCoors found that engaged employees were five times less likely than non-engaged employees to have a safety incident. [13]

Engagement's impact on Sales
Research has shown that Employee Engagement affects Sales.
- A company featured in the 2016 Engage for Success report reported that engaged salespeople sold an average of $70,000 more per year, compared to disengaged salespeople - which is around 7% extra sales revenue per engaged employee. [14]
- Standard Chartered Bank found that those branches with highly engaged employees produced twenty percent higher returns compared to branches with lower engagement scores. [15]
- Marks and Spencer found that one percent improvement in their employee engagement produced three percent increase in sales per square foot. [15]
- JCPenney found that stores with top engagement scores produced around ten percent more sales per square foot and have thirty six percent higher operating income than stores with low engagement scores. [15]
- Best Buy (a leader in HR predictive analytics) can accurately predict how employee engagement impacts the performance of their stores. A 0.1% increase in employee engagement results in an increase of over $100,000 in the store's annual income. The enormous impact of engagement prompted Best Buy to make its engagement surveys quarterly instead of annually. [16]

Engagement's impact on Service

- A study by ISS found that Employee Engagement correlates strongly with Customer Experience. The strength of this correlation was 0.55. Diagram below shows the customer net promoter score (cNPS) and the employee net promoter score (eNPS). Net promoter score (NPS) is a tool to measure the loyalty of a firm's customers and employees. The NPS was based on the question: "How likely is it that you would recommend a company, product or service to a friend or colleague?" This answer was scored based on a 0-10 scale. Promoters (loyal enthusiasts) are those who respond with a score of 9 or 10. Passives are those who respond with a score of 7 or 8. Detractors (unhappy customers) are those who respond with a score of 0 to 6. Calculate the NPS by subtracting the percentage who are detractors from the percentage who are promoters. The drivers behind customer experience are: Motivation and engagement of service staff, Amount of training and quality of service staff, and the ability to act on customer expectations by customer service staff. With the findings, ISS recommendations were to monitor metrics on service employee engagement in areas such as the eNPS (employee net promoter score), employee churn, absenteeism and training hours. [17]

Correlation between eNPS and cNPS

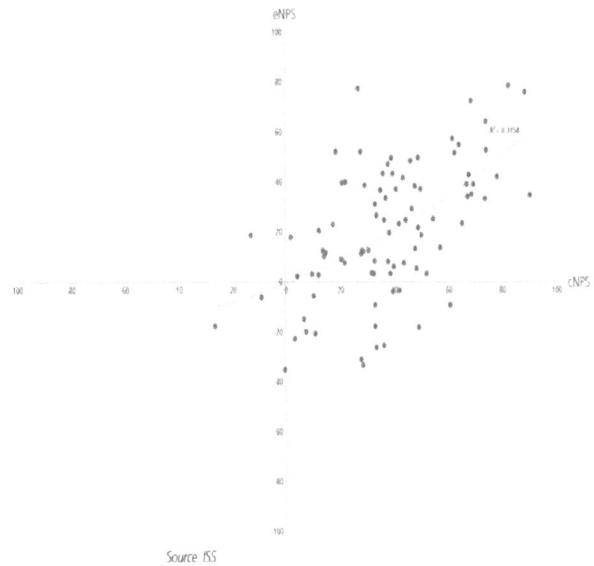

Source: Morten Kamp Andersen (proacteur), Simon Svegaard (ISS) & Peter Ankerstjerne (ISS). (2015), Linking Customer Experience with Service Employee Engagement, ISS White Paper.

Engagement's impact on Total Shareholder Returns (TSR)
- A study conducted across 39 organizations showed that organizations with highly engaged employees achieve seven times greater 5-year total shareholder return (TSR) than organizations whose employees are less engaged. In companies where 60 to 70 percent of employees were engaged, average total shareholder's return (TSR) is 24.2 percent. in companies with only 49 to 60 percent of their employees engaged, TSR fell to 9.1 percent; companies with engagement below 25 percent suffered negative TSR. [18]

Leadership's impact on Engagement
- Gallup's research in the US shows that the actions of leaders account for up to 70% variance in employee engagement scores. [19]

References:
(1) Culture Amp (2018) Why Employee Engagement Matters. https://www.cultureamp.com/resources/guides/foundation-guides/why-employee-engagement-matters.html#ref-3 (14 November 2018)
(2) Tanvir Haque (2018) Four Strategies To Boost Employee Engagement That Can Improve Your Bottom Line. https://www.entrepreneur.com/article/312697 (20 November 2018)
(3) Corporate Leadership Council, Corporate Executive Board. Driving Performance and Retention through Employee Engagement: a quantitative analysis of effective engagement strategies. 2004. Cited in MacLeod 2009, p.14.
(4) MacLeod, D., and Clarke, N. Engaging for Success: Enhancing Performance Through Employee Engagement—A Report to Government. London: Department for Business Innovation and Skills. 2009. Crown copyright.
(5) Association for Psychological Service (2018) Job Satisfaction Tends to Increase with Age https://www.psychologicalscience.org/news/minds-business/job-satisfaction-tends-to-increase-with-age.html (21 November 2018)

(6) Nigel Guenole, Jonathan Ferrar, and Sheri Feinzig (2017). The Power of People, Pearson FT Press.
(7) Harter, J.K., Schmidt, F. L., Kilham, E. A., Asplund, J.W., (2006), Gallup Q12 Meta-Analysis. Cited in MacLeod, 2009, p. 36.
(8) Krueger, J. & Killham, E 'The Innovation Equation.' Gallup Management Journal, 2007. Cited in MacLeod 2009, p. 12.
(9) Culture Amp (2018) Why Employee Engagement Matters. https://www.cultureamp.com/resources/guides/foundation-guides/why-employee-engagement-matters.html#ref-3 (14 November 2018)
(10) Culture Amp (2018) Why Employee Engagement Matters. https://www.cultureamp.com/resources/guides/foundation-guides/why-employee-engagement-matters.html#ref-3 (14 November 2018)
(11) Morten Kamp Andersen (proacteur), Simon Svegaard (ISS) & Peter Ankerstjerne (ISS). (2015), Linking Customer Experience with Service Employee Engagement, ISS White Paper, pp 10-11.
(12) Richard S. Wellins, Paul Bernthal (2015) Employee Engagement: The Key to Realizing Competitive Advantage. DDI. https://www.ddiworld.com/ddi/media/monographs/employeeengagement_mg_ddi.pdf?ext=.pdf (20 November 2018)
(13) Kevin Kruse (2012) Why Employee Engagement? (These 28 Research Studies Prove the Benefits) https://www.forbes.com/sites/kevinkruse/2012/09/04/why-employee-engagement/#36bbedf63aab (20 November 2018)
(14) Court-Smith, J., 'The Evidence: Case Study Heroes and Engagement Data Daemons', Engage for Success, April 2016, p. 23.
(15) Parent, J. D., & Lovelace, K. J. (2015). The Impact of Employee Engagement and a Positive Organizational Culture on an Individual's Ability to Adapt to Organization Change.2015 Eastern Academy of Management Proceedings: Organization Behavior and Theory Track, 1-20. (14 November 2018)
(16) Mohit Sharma, Talent Analytics: From Buzzword to Reality (2018), https://sightsinplus.com/2018/05/25/talent-analytics-from-buzzword-to-reality/ (September 2018)
(17) Morten Kamp Andersen (proacteur), Simon Svegaard (ISS) & Peter Ankerstjerne (ISS). (2015), Linking Customer Experience with Service Employee Engagement, ISS White Paper, pp 3-9.
(18) Kevin Kruse (2012) Why Employee Engagement? (These 28 Research Studies Prove the Benefits) https://www.forbes.com/sites/kevinkruse/2012/09/04/why-employee-engagement/#36bbedf63aab (20 November
(19) Tanvir Haque (2018) Four Strategies To Boost Employee Engagement That Can Improve Your Bottom Line. https://www.entrepreneur.com/article/312697 (20 November 2018)

15.1) Employee Net Promoter Score

Employee Net Promoter Score (eNPS) is an approach for companies to measure employee loyalty. Employee Net Promoter Score measures the probability your employee will recommend your company as a place to work and the products/services sold. Example of eNPS question:

Using a scale of 0 to 10, how likely will you recommend this company as a place to work?

Fred Reichheld, a researcher from Bain & Company, wrote an article in Harvard Business Review about how one question can be used to predict companies' growth, "It turned out that a single survey question can, in fact, serve as a useful predictor of growth…. In fact, in most of the industries that I studied, the percentage of customers who were enthusiastic enough to refer a friend or colleague…correlated directly with differences in growth rates among competitors". [1]

Reichheld group the answers were in three groups [2]:
- Promoters: score 9 to 10 for the question and are very likely to recommend. To get actionable insights, ask the promoters what makes them willing to recommend.
- Passively Satisfied: score 7-8 for the question. They are neutral because they are neither going to promote or talk negatively about the company.
- Detractors: score 0-6 for the question and will not recommend. To get actionable insights, ask the detractors the reason for the score they gave.

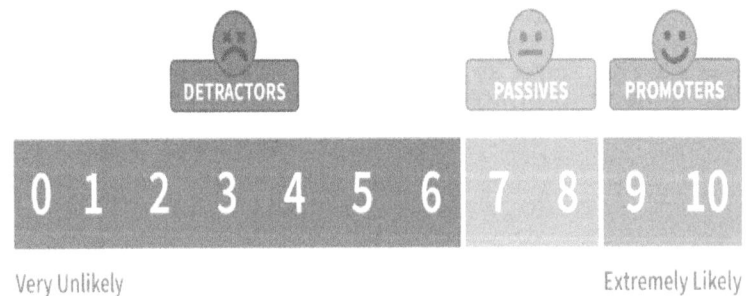

Source: Jacob Shriar, Employee Net Promoter Score, https://www.officevibe.com/employee-engagement-solution/employee-net-promoter-score (2 November 2018)

Formula to Calculate The eNPS Score:

$$eNPS = \% \text{ Promoters} - \% \text{ Detractors}$$

If the results you get from an eNPS survey is: 30 Detractors, 50 Passives, 10 Promoters,

$$eNPS = 10\%(\text{Promoters}) - 30\%(\text{Detractors}) = -20\%$$

People who scored 7 or 8 are not included in the computation because they are neutral. As the eNPS score is represented as a number, remove the percentage sign and your eNPS is -20. Keep in mind that the real value is not so much not the score, but the written feedback from employees about what you company can improve in. Here some initiatives that you can use to improve your eNPS score, and ultimately your business results:

- Communicate your results and plan – Employees will be frustrated if you conduct a survey and don't share the results of the survey and don't follow up with a plan for implementing changes.

- Analyzing the data – To be able to see trends, you can group your data in various ways: age groups, gender, length of service, department.

References:
(1) Frederick F. Reichheld, Harvard Business Review (2003), The One Number You Need to Grow (2 November 2018)
(2) Jacob Shriar, Employee Net Promoter Score, https://www.officevibe.com/employee-engagement-solution/employee-net-promoter-score (2 November 2018)

15.2) Case 1: Upwork Global Inc. – Impact Of Flexible Work Arrangements On Employee Morale

There is a lot of evidence to show the impact of flexible work arrangements on employee morale. Based on a study by PGi [1]:
- 82 percent of telecommuters showed lower stress levels
- 80 percent of employees showed higher morale when working from home
- 69 percent showed lower absenteeism

Apple hires At Home Advisors for customer support, promoting the benefits of this job with the following description: "Comfort, convenience, and a no-hassle commute are all reasons people like to work from home". Dell's "Connected Workplace" initiative gives employees several flexible work options such as flextime, job sharing, remote work, part-time work, and compressed work weeks. Also worth highlighting is the benefit that flexible work arrangements offers in terms of attraction and retention. AfterCollege's data showed that 68 percent of Millennial job seekers indicated that the option to work remotely, greatly increase their interest in the company. The Millennial Branding report also found that 45 percent of Millennials chooses workplace flexibility over salary. Flexible work arrangements are not merely about the work location or hours. It is about the potential to tap a wider global talent pool. Flexible work arrangements not only result in happier employees, it improves the organization's nimbleness, responsiveness and profitability. [1]

There are various options for Flexible work arrangements:
- Flexible schedule (e.g. working four 10-hour shifts, rather than five eight-hour days)
- Alternate daily schedules (e.g. 10:00 am to 6:00 pm working hours, or 8:00 am to 4:00 pm working hours)
- Part-time
- Unlimited vacation leave
- Work from home

References:
(1) Upwork Global Inc., https://www.upwork.com/hiring/top-companies-ditching-9-5-flexible-working-arrangements/?utm_source=facebook&utm_campaign=RTVNP&utm_medium=social-paid&vt_src=facebook&vt_cmp=RTVNP&utm_content=%5BRetarget_2-7Days_IncludesRegs%5D (17 Oct 2018)

15.3) Case 2: iNostix - Merging engagement survey outcomes with business data

In a project with a transport company, iNostix developed an 'Impact Map', linking engagement results with HR and business data. In the impact map below, low engagement for Drivers, causes the following [1]:
- more accidents.
- lower client satisfaction.
- higher insurance costs.
- more medical costs.
- higher absenteeism.
- more company doctor visits.

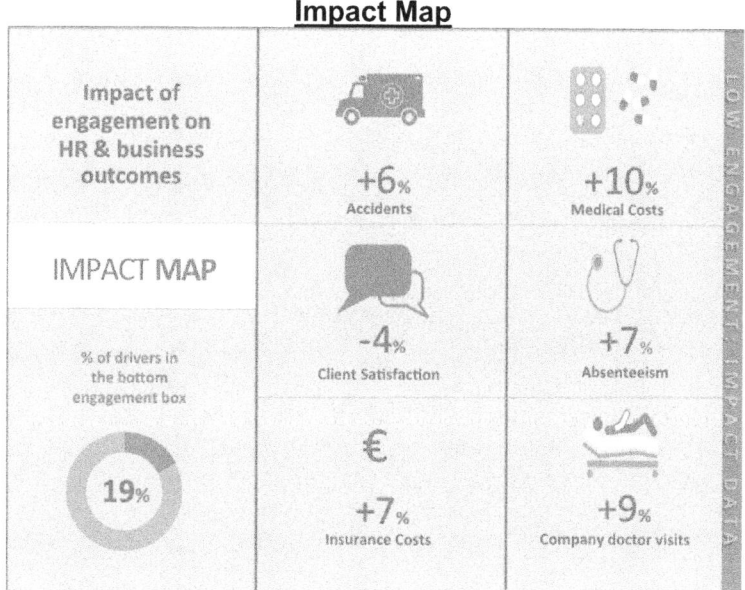

Source: Inostix (2014), Merging engagement survey outcomes with business data – HR Analytics. http://www.inostix.com/blog/en/merging-engagement-survey-outcomes-with-business-data-hr-analytics/ (2016 April 1)

Most HR professionals have a habit of overwhelming their audiences with lots of tables or figures. Communicating engagement results effectively and clearly with a good story, is the key to success for any analytics project. Inostix recommends these techniques for presentations [1]:
- enrich survey data with business data.
- develop impact maps to demonstrate the impact of engagement on the business.
- communicate actionable recommendations and evidence-based conclusions, instead of the usual decks of tables and numbers.

References:
(1) Inostix (2014), Merging engagement survey outcomes with business data – HR Analytics. http://www.inostix.com/blog/en/merging-engagement-survey-outcomes-with-business-data-hr-analytics/ (2016 February 1)

15.4) Correlation Example: Analyze Engagement

Correlation can be used to help understand what drives employee satisfaction, and how employee satisfaction affects business issues such as customer satisfaction, revenue, and profitability.

In this example, we use correlation to find out whether Staff Engagement Index is a good predictor of: Customer Satisfaction, Revenue per staff, and Operating Income per staff.

	A	B	C	D	E
1		Staff Engagement Index	Customer Satisfaction Index	Revenue per staff	Operating Income per staff
2	Australia	51%	55%	600	1.0
3	China	67%	70%	800	4.0
4	Hong Kong	39%	40%	400	0.5
5	Indonesia	54%	55%	500	1.8
6	India	71%	70%	750	4.0
7	Sri Lanka	53%	50%	600	1.0
8	Malaysia	45%	50%	550	1.0
9	New Zealand	70%	68%	700	3.0
10	Singapore	69%	65%	640	3.0
11	USA	63%	70%	720	4.0
12	Thailand	56%	60%	620	2.5

1) Install "Analysis ToolPak", an Excel add-in

"Analysis ToolPak" is an add-in for Microsoft Excel that comes with Microsoft Excel. To be able to run regression using Excel, you need to first install "Analysis ToolPak", an Excel add-in program that provides data analysis tools. To load the Analysis ToolPak add-in, follow these steps:

- On the File tab, click Options.

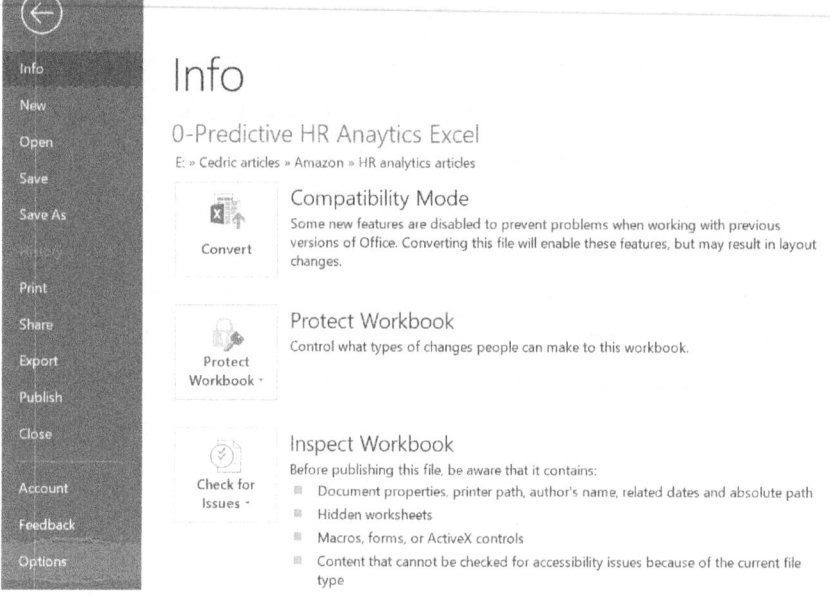

- Under Add-ins, click Analysis ToolPak and click the "Go" button.

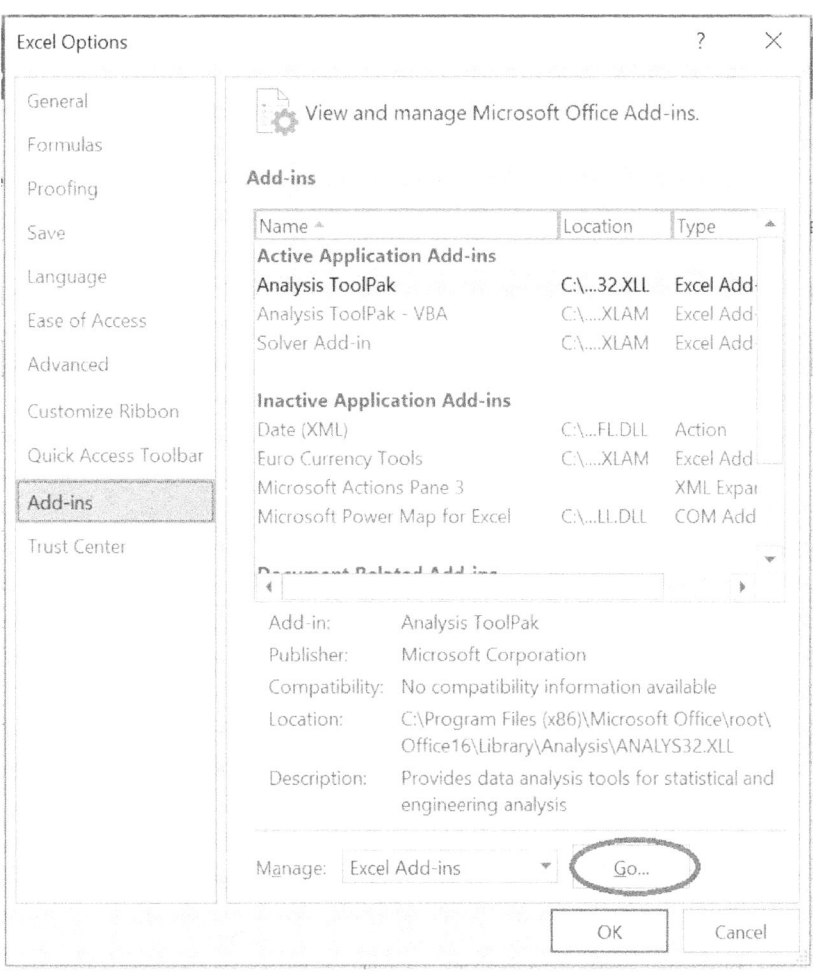

- Click "Analysis ToolPak" and click on OK.

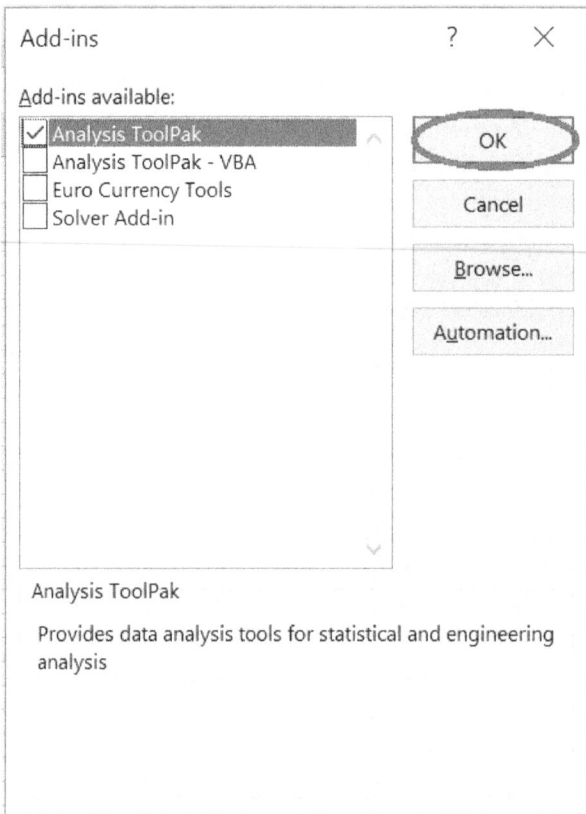

- On the Data tab, in the Analysis group, you are now able to click on "Data Analysis".

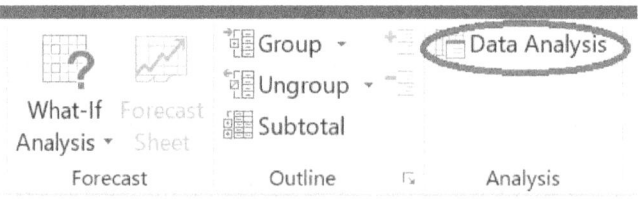

2) Copy the example data in the following table, and paste it in cell A1 of a new Excel worksheet.

	A	B	C	D	E
1		Staff Engagement Index	Customer Satisfaction Index	Revenue per staff	Operating Income per staff
2	Australia	51%	55%	600	1.0
3	China	67%	70%	800	4.0
4	Hong Kong	39%	40%	400	0.5
5	Indonesia	54%	55%	500	1.8
6	India	71%	70%	750	4.0
7	Sri Lanka	53%	50%	600	1.0
8	Malaysia	45%	50%	550	1.0
9	New Zealand	70%	68%	700	3.0
10	Singapore	69%	65%	640	3.0
11	USA	63%	70%	720	4.0
12	Thailand	56%	60%	620	2.5

3) Select "Correlation" and click "OK".

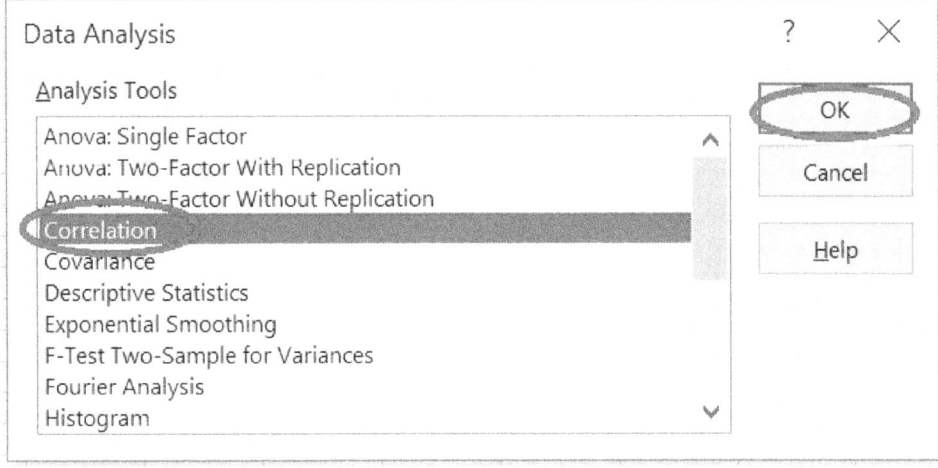

4) After you click OK in the "Data Analysis" dialog box, you will see a "Correlation" dialog box.
5) For "Input Range", select cells (B1:E12).
6) Check "Labels in first row"
7) For "Output Range", select cells (A14).
8) Click "OK"

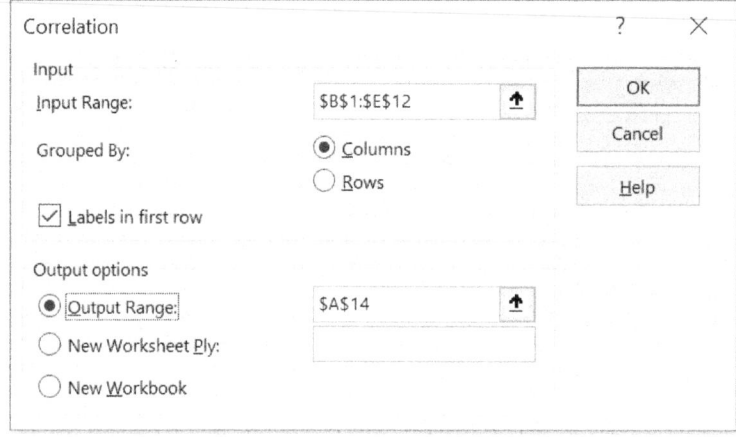

After you click "OK", Excel generates the following Correlation analysis.

	Staff Engagement Index	Customer Satisfaction Index	Revenue per staff	Operating Income per staff
Staff Engagement Index	1			
Customer Satisfaction Index	0.94	1		
Revenue per staff	0.87	0.92	1	
Operating Income per staff	0.90	0.96	0.87	1

A negative correlation coefficient means that an increase in X is associated with a decrease in Y. Similar to a positive correlation, a negative correlation shows a connection between two variables, and the relative strengths are the same. In other words, a correlation coefficient of 0.85 has the same strength as a correlation coefficient of -0.85. Correlation coefficients are always values between -1 and 1, where "-1" means that there is a perfect linear negative correlation, while "1" shows a perfect linear positive correlation. A correlation coefficient of zero, or near to zero, means that there is no meaningful relationship between variables. Correlation coefficient of 0.91 or -0.92 shows a very strong positive and negative correlation respectively. However, correlation does not mean causation.

An example of negative correlation is the amount of snowfall and the temperature. As the temperature increases, the amount of snowfall decreases. An example of positive correlation is the relationship between temperature and ice cream sales. As temperature increases, so do ice cream sales.

7) Highlight those numbers that have strong positive correlation (above 0.85) and strong negative correlation (below -0.85).

	Staff Engagement Index	Customer Satisfaction Index	Revenue per staff	Operating Income per staff
Staff Engagement Index	1			
Customer Satisfaction Index	0.94	1		
Revenue per staff	0.87	0.92	1	
Operating Income per staff	0.90	0.96	0.87	1

From the Excel Correlation analysis, Staff Engagement Index is a good predictor of Customer Satisfaction, Revenue per staff, and Operating Income per staff, as they have strong correlation of below -0.75 and above 0.75:
- **Customer Satisfaction Index**: correlation of 0.94 with Staff Engagement Index)
- **Revenue per staff:** correlation of 0.87 Staff Engagement Index)
- **Operating Income per staff:** correlation of 0.90 Staff Engagement Index)

15.5) Multiple Regression Example: Analyze Engagement

Best Buy (a leader in HR predictive analytics) can predict that a 0.1% increase in employee engagement results in an increase of over $100,000 in the store's annual income. [1]

In this section, we show you how to develop your own hypothesis to predict how employee engagement affects the results of your company's revenue.

	A	B	C	D
1	Business Unit	Advertising expense	Average Engagement Score	Revenue
2	1	96	10.0	1310
3	2	89	9.0	1260
4	3	99	9.7	1210
5	4	83	10.0	1160
6	5	95	8.6	1100
7	6	77	9.2	1050
8	7	87	8.6	1000
9	8	79	6.7	950
10	9	75	3.3	890
11	10	76	3.3	840

References:

(1) Mohit Sharma, Talent Analytics: From Buzzword to Reality (2018), https://sightsinplus.com/2018/05/25/talent-analytics-from-buzzword-to-reality/ (24 September 2018)

1) Install "Analysis ToolPak", an Excel add-in

"Analysis ToolPak" is an add-in for Microsoft Excel that comes with Microsoft Excel. To be able to run regression using Excel, you need to first install "Analysis ToolPak", an Excel add-in program that provides data analysis tools. To load the Analysis ToolPak add-in, follow these steps:

On the File tab, click Options.

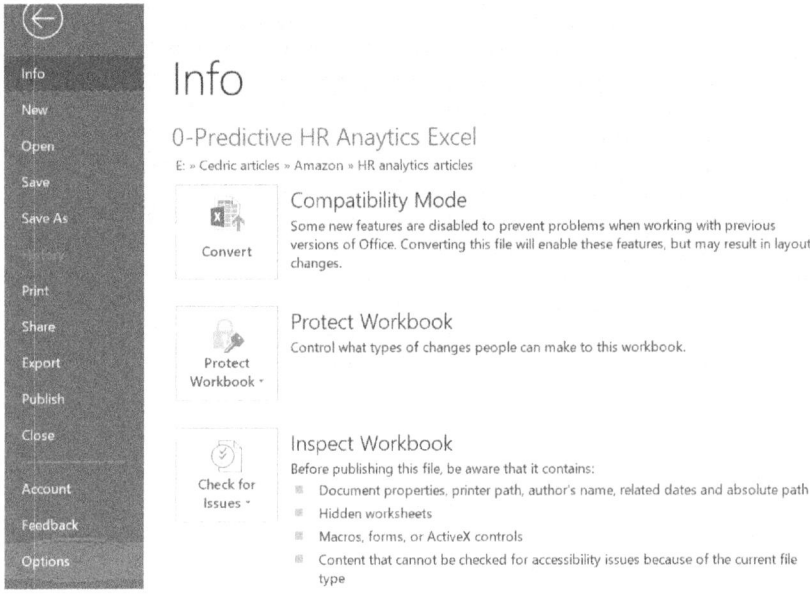

Under Add-ins, click Analysis ToolPak and click the "Go" button.

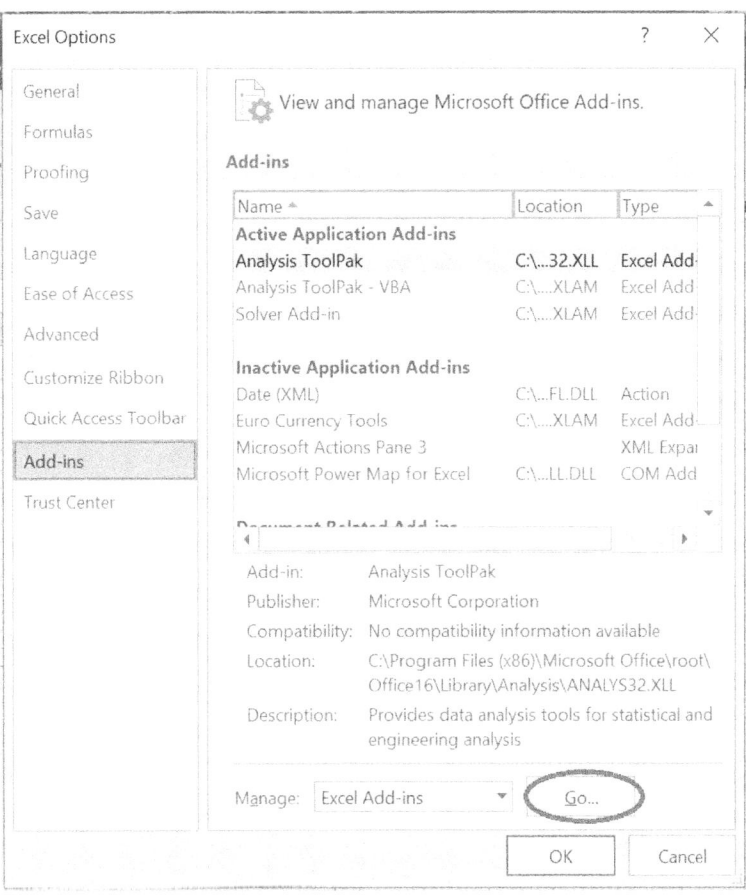

Click "Analysis ToolPak" and click on OK.

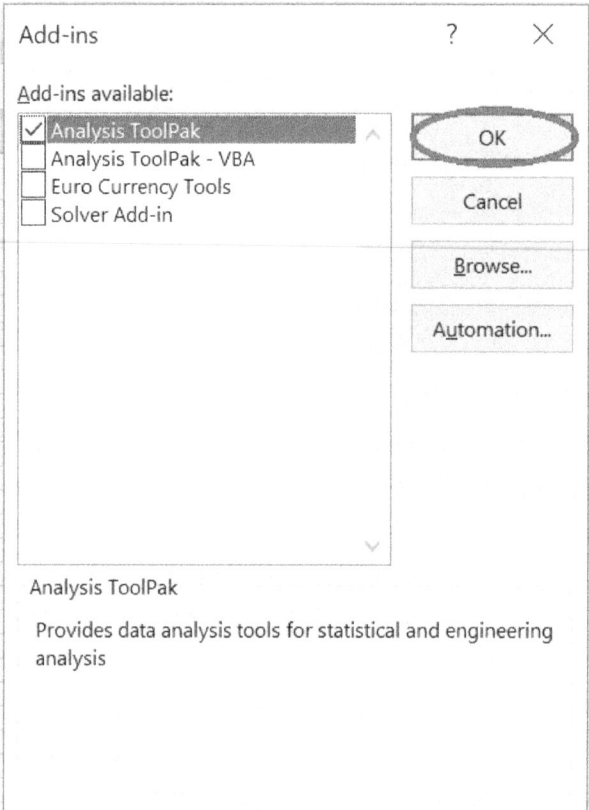

On the Data tab, in the Analysis group, you are now able to click on "Data Analysis".

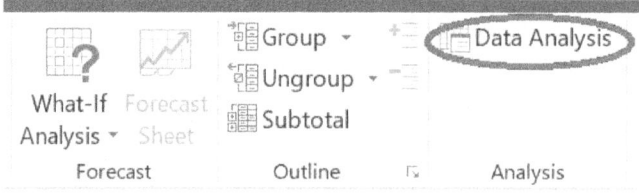

2) Copy the example data in the following table, and paste it in cell A1 of a new Excel worksheet.

	A	B	C	D
1	Business Unit	Advertising expense	Average Engagement Score	Revenue
2	1	96	10.0	1310
3	2	89	9.0	1260
4	3	99	9.7	1210
5	4	83	10.0	1160
6	5	95	8.6	1100
7	6	77	9.2	1050
8	7	87	8.6	1000
9	8	79	6.7	950
10	9	75	3.3	890
11	10	76	3.3	840

3) On the Data tab, in the Analysis group, click on "Data Analysis".

4) Select "Regression" and click "OK".

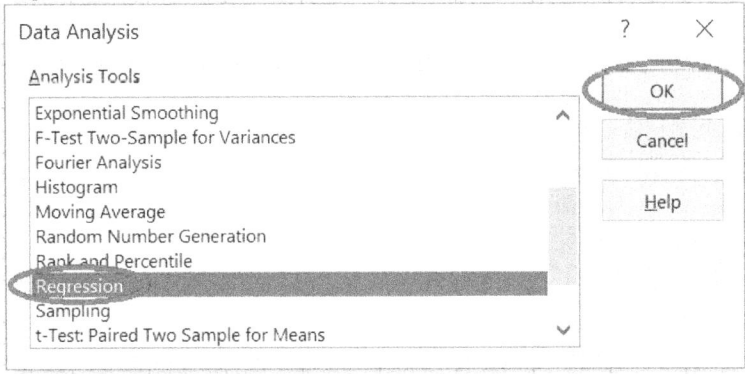

5) After you click OK in the "Data Analysis" dialog box, you will see a "Regression" dialog box.
6) For "Input Y Range", select cells (D1:D11). This is the predictor variable or dependent variable.
7) For "Input X Range", select cells (B1:C11). These are the explanatory variables or independent variables.
8) Check "Labels" box.
9) Click the "Output Range" box, and select cell A13.
10) Click "OK".

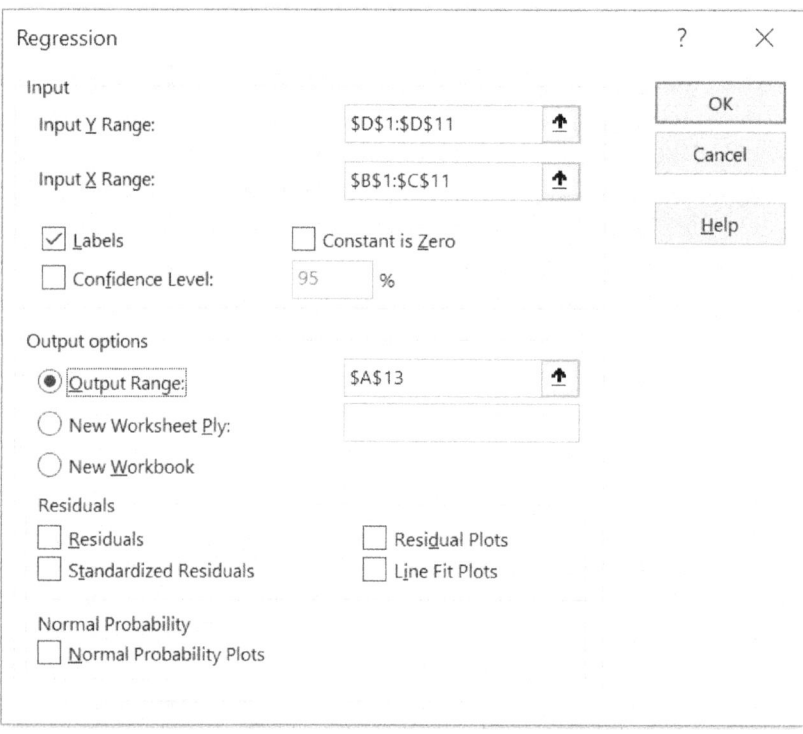

After you click "OK", Excel generates the following Summary Output. Round the numbers to 3 decimal places.

SUMMARY OUTPUT

Regression Statistics	
Multiple R	0.897
R Square	0.805
Adjusted R Square	0.749
Standard Error	79.540
Observations	10

ANOVA

	df	SS	MS	F	Significance F
Regression	2	182523.387	91261.694	14.425	0.003
Residual	7	44286.613	6326.659		
Total	9	226810.000			

	Coefficients	Standard Error	t Stat	P-value	Lower 95%	Upper 95%	Lower 95.0%	Upper 95.0%
Intercept	207.283	285.180	0.727	0.491	-467.061	881.627	-467.061	881.627
Advertising expense	6.833	4.064	1.681	0.137	-2.778	16.443	-2.778	16.443
Average Engagement Score	36.331	14.103	2.576	0.037	2.982	69.679	2.982	69.679

R Square: In the output, R Square is 0.805, which means it is a good fit. 80% of the variation in Sales (Output) is explained by the independent variables (Input), Advertising expense, and Average Engagement Score. The closer R Square is to "1", the better the regression line fits the data.

Significance F and P-values: To determine if your results are statistically significant (i.e. reliable), check "Significance F" (0.001). If the value of "Significance F" is less than 0.05, it is statistically significant (i.e. reliable). If "Significance F" is bigger than 0.05, don't use this set of independent variables. Delete those variables with "P-value" that is bigger than 0.05 and run the regression again until "Significance F" drops below 0.05. Most or all of your P-values should be lower than 0.05. In our example below "P-value" is 0.491, 0.137 and 0.037, for Intercept, Advertising expense and Average Engagement Score respectively.

Coefficients

From the Summary Output, the regression line is:

SUMMARY OUTPUT

Regression Statistics	
Multiple R	0.897
R Square	0.805
Adjusted R Square	0.749
Standard Error	79.540
Observations	10

ANOVA

	df	SS	MS	F	Significance F
Regression	2	182523.387	91261.694	14.425	0.003
Residual	7	44286.613	6326.659		
Total	9	226810.000			

	Coefficients	Standard Error	t Stat	P-value	Lower 95%	Upper 95%	Lower 95.0%	Upper 95.0%
Intercept	207.283	285.180	0.727	0.491	-467.061	881.627	-467.061	881.627
Advertising expense	6.833	4.064	1.681	0.137	-2.778	16.443	-2.778	16.443
Average Engagement Score	36.331	14.103	2.576	0.037	2.982	69.679	2.982	69.679

Sales (Output)
= 207.283 + 6.833 * Advertising expense + 36.331 * Average Engagement Score

Based on the above regression formula,
- For each unit increase in Advertising expense, Sales increase by 6.833.
- For each unit increase in Average Engagement Score, Sales increase by 36.331.

Coefficients can also be used for forecasting. For example, if "Advertising expense" is 90, "Average Engagement Score" is 8.5, then **predicted "Sales"**
= 207.283 + 6.833 * Advertising expense + 36.331 * Average Engagement Score
= 207.283 + (6.833 * 90) + 36.331 * (8.5)
= $1131

16) Social Media Analytics

Measuring your posts to identify trends and see what is working, is an important part of social media. With some many different metrics, it can be difficult to know which measurement is more important. To ascertain what analytics is important, you need to define your social media objectives: [1]

- **Brand awareness**: if your objective is to raise brand awareness, "impression" is the measurement to look at.

- **Community**: If your objective is to build a community, "engagement" is the measurement to look at.

- **Drive Users**: If your objective is to drive users to your website, traffic and conversions is the measurement to look at.

Types of social media measurements for social media analytics

Facebook Insights is free for all business pages, and it lets you to view your page profile analytics, page views, page likes, reach, and post engagements of your posts. Regardless of your social media strategy, the enormous amount of information available can be overwhelming. Here are various types of measurements for social media analytics:

- **Number of page "likes":** When someone "likes" your page on Facebook, they automatically follow your page as well. This means that your page will be listed in their "liked" directory and your posts will be seen in their feed. [2]
- **Number of page "followers":** Facebook users also have the option to "follow" a page without "liking" it. "Followers" will still see your posts in their newsfeed, but they won't be considered a like on your page. This option was set up for people who didn't want to befriend someone on Facebook but still wanted to see their posts. For page owners, the most important metric is page followers as these are the people who have opted to see your content in their newsfeed. [2]

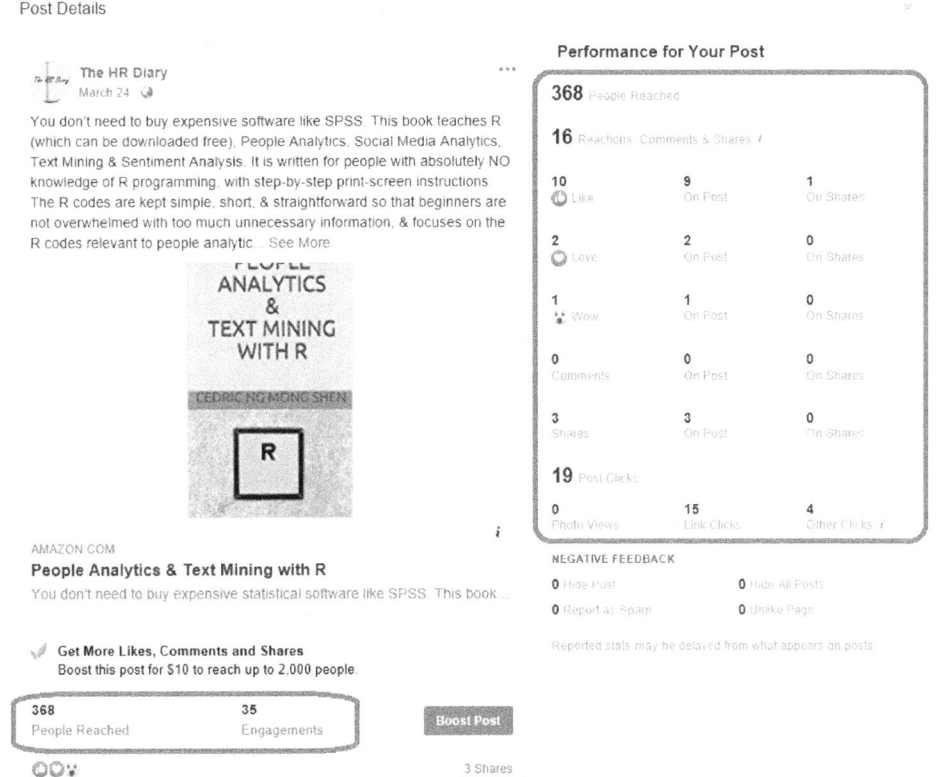

- **Post reach:** The number of people who had any posts from your page on their screens, broken down by total, organic and promotions. This is one of the most important analytics.
- **Engagement:** Total number of times someone interacted or engaged with a post. Engagement is an important metrics because it shows if people are interacting with your posts.

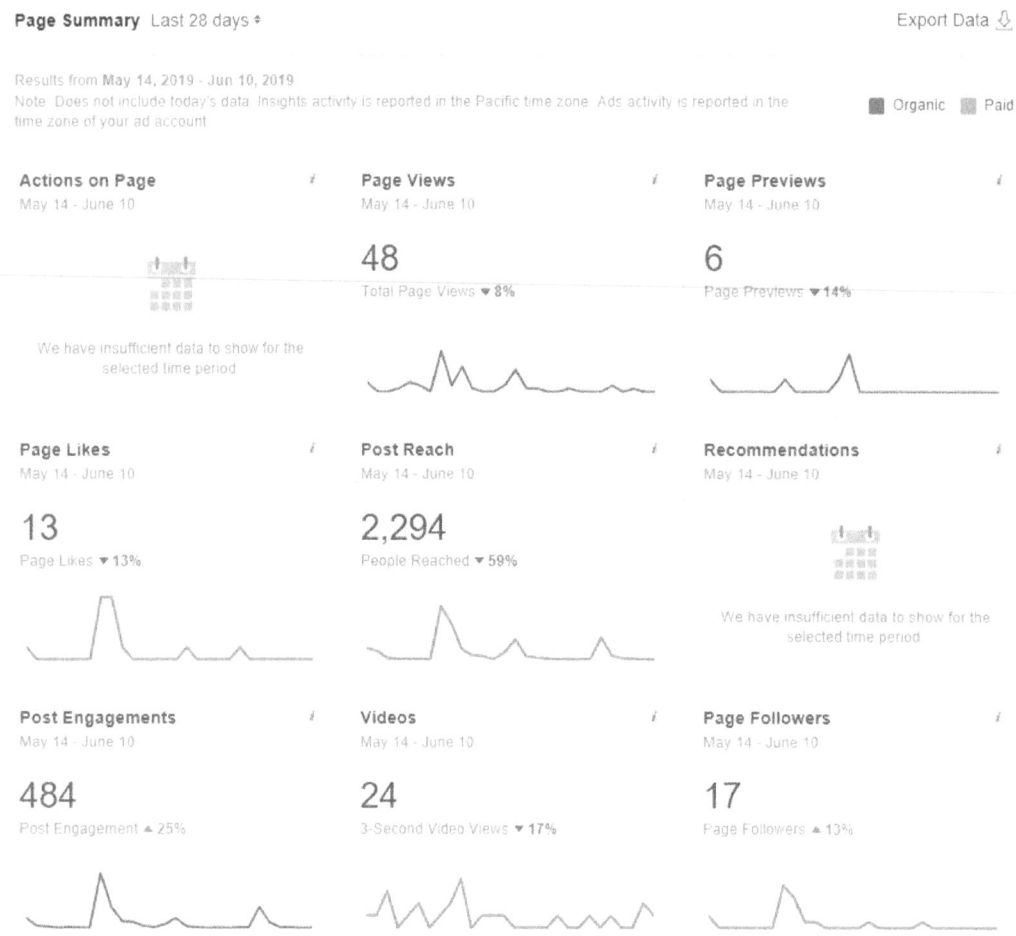

- **Impressions:** The number of users who saw the post.
- **Page actions:** The number of clicks on your business's contact information and call-to-action button.
- **Page views:** The number of times your profile page has been viewed.
- **Page previews:** The number of times people hovered over your page's name or profile picture to preview your content on Facebook.
- **Engagement rate:** The number of engagements divided by the number of impressions.
- **Post clicks:** Any clicks on the entire post.

References:
(1) Saige Driver (2018) Guide to Social Media Analytics. https://www.businessnewsdaily.com/10694-understanding-social-media-analytics.html *(11 June 2019)*
(2) Daragh Walsh (2018) In Facebook what's the difference between "like" and "follow" a page? https://www.quora.com/In-Facebook-whats-the-difference-between-like-and-follow-a-page *(11 June 2019)*

17) Social Network Analysis (SNA)

Social Network Analysis (SNA) is a hot topic and a powerful technique to improve organization effectiveness. Any social organization from the smallest team to the largest companies carries with it a social network. In today's world, social network collaboration is critical to organization effectiveness. SNA focus on relationships and their structure rather than solely on individual attributes.

SNA is based on the community behavior concept. Social network analysis can be used to understand your customers and their communities. In marketing campaign, if you want to sell your products to your customers, convince their friends (their influencers).

Sociogram

In Social Network Analysis, employees are represented as nodes and the relationships between them as links (ties) are illustrated in a sociagram. Dr Moreno, a social scientist developed a sociogram (a diagram of points and lines to represent people relations) to spot social leaders and isolates, and uncover reciprocity and asymmetry in friendship. One of the configurations he noticed was the sociometric star – everybody's friend. [1]

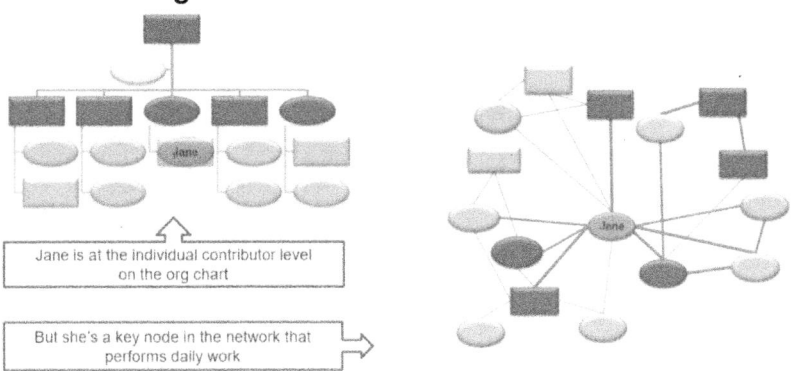

Source: Dr. Michael Moo (2016) A Network Approach to People Analytics: How to Use Social Network Analysis in HR. https://static1.squarespace.com/static/55007c24e4b001deff386756/t/57dbfff9d2b8574909f34203/1474035707614/Moon%2C+Michael.pdf (12 June 2019)

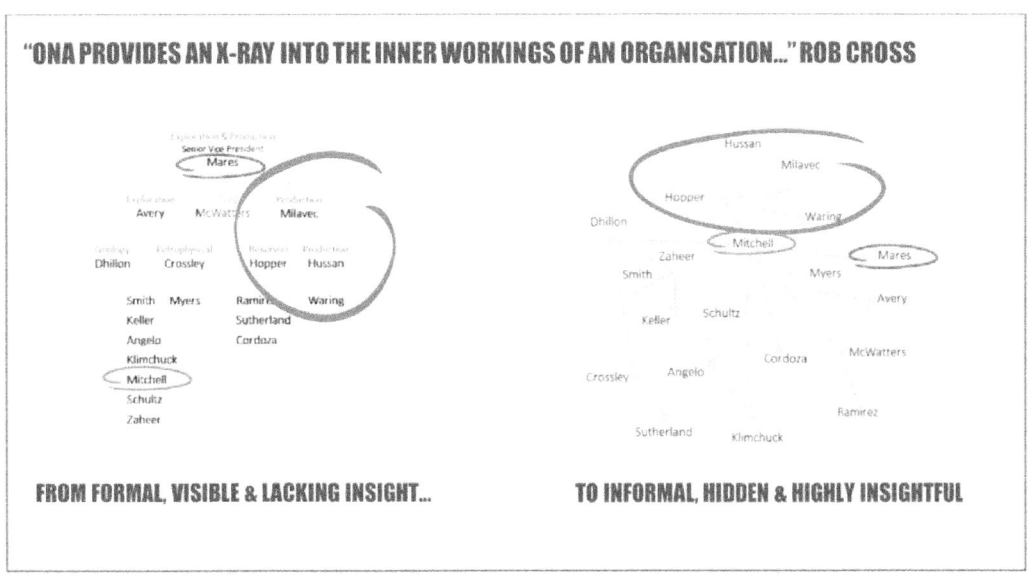

Source: David Green (2018) The role of Organisational Network Analysis in People Analytics. https://www.linkedin.com/pulse/role-organisational-network-analysis-people-analytics-david-green (12 June 2019)

Measures of SNA

Social Network Analysis (SNA) works on the measure of centrality and there are many approaches to calculate centrality. Centrality measures lets us know who is the most influential person in the network. Influencers are deemed to be the most connected customers in the network. The influence of each node in the network is represented by the score, and the most influential nodes typically have the highest in score. [2]

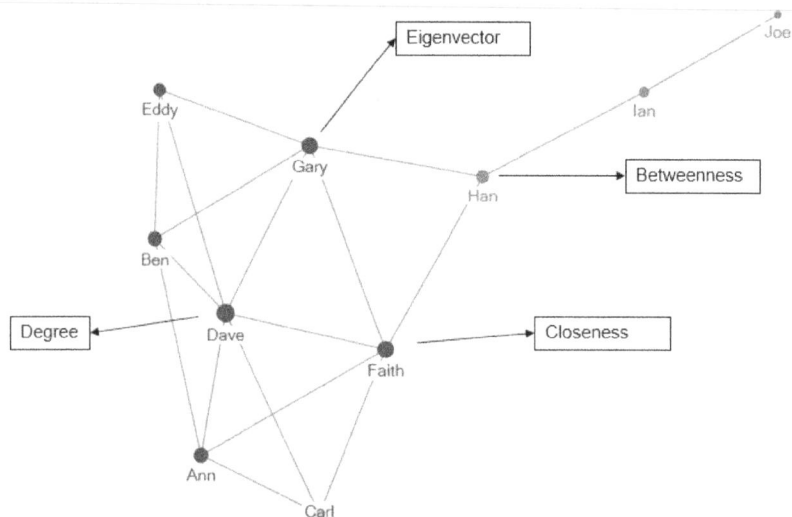

	Graph Metrics			
Vertex	Degree	Betweenness Centrality	Closeness Centrality	Eigenvector Centrality
Dave	6	3.667	0.067	0.171
Faith	5	8.333	0.071	0.142
Gary	5	8.333	0.071	0.142
Ann	4	0.833	0.059	0.125
Ben	4	0.833	0.059	0.125
Carl	3	0.000	0.056	0.102
Eddy	3	0.000	0.056	0.102
Han	3	14.000	0.067	0.070
Ian	2	8.000	0.048	0.017
Joe	1	0.000	0.034	0.004

Typical measures of centrality are:

- **Degree centrality:** How many people can this person reach directly? Dave has a Degree of 6 because he is directly connected to 6 other individuals.

- **Betweenness centrality:** How likely is this person to be the most direct route between two people in the network? Though Dave has more direct connections compared to Han, it is not everything. Han has the highest "Betweenness Centrality" of 14 as he is a **bridge between two important constituencies**.

- **Closeness centrality:** How fast can this person reach everyone in the network? Though Faith and Gary have fewer connections than Dave, the pattern of their direct and indirect ties enable them to access all the nodes in the network faster than anyone else. They have the shortest paths to everyone else. **Unlike other centrality metrics, a lower Closeness Centrality score indicates a more central** (i.e. important) position in the network.

- **Eigenvector centrality:** How well is this person connected to other well-connected people? Joe has a low Eigenvector Centrality metric as he is connected to Ian who is not popular. Gary and Faith has high Eigenvector Centrality metric as they are connected to Dave, the most popular person in the network.

References:
(1) Ken Thompson (2016) Social Network Analysis: an introduction. http://www.bioteams.com/2006/03/28/social_network_analysis.html#more *(6 June 2019)*
(2) Dr. Archana Kumari and Ashutosh Srivastava (2018) Enhancing Churn Prediction Models Using Social Network Analysis In Telecom Industry. https://www.analyticsindiamag.com/enhancing-churn-prediction-models-using-social-network-analysis-in-telecom-industry/ (6 June 2019)

17.1) Organizational Network Analysis (ONA)

When social network analytics is applied to organizations, it is called Organizational Network Analysis (ONA). ONA provides insights that can unlock innovation, drive productivity and improve performance while enhancing employee experience and wellness.

ONA terminologies.

Organizational Network Analysis (ONA) is a structured way to visualize how communications flow through an organization. Organizational networks consist of nodes and ties, which helps you to understand how information in your organization flow, can flow, and should flow. Every organization has people (nodes) who serve as critical conduits for exchange of information. A connection delivers value when needed information is exchanged. Visualizing the relationships between nodes and ties lets us identify important connections and potential obstacles to collaboration and information flow. By identifying and managing central nodes, change can be adopted more successfully. [1]

- **Central node:** Central Nodes are the people who seem to know everyone. They can be anywhere in an organization hierarchy, are able to quickly influence groups, and are highly engaged in company developments.

- **Knowledge broker:** Knowledge brokers create bridges between groups, facilitating information sharing across groups.

- **Peripheral:** Peripherals are often overlooked and unconnected to the rest of the organization. Peripherals are flight risks, and if they are talented, it becomes a risk for the organization.

- **Ties:** Ties are the formal and informal relationships between nodes.

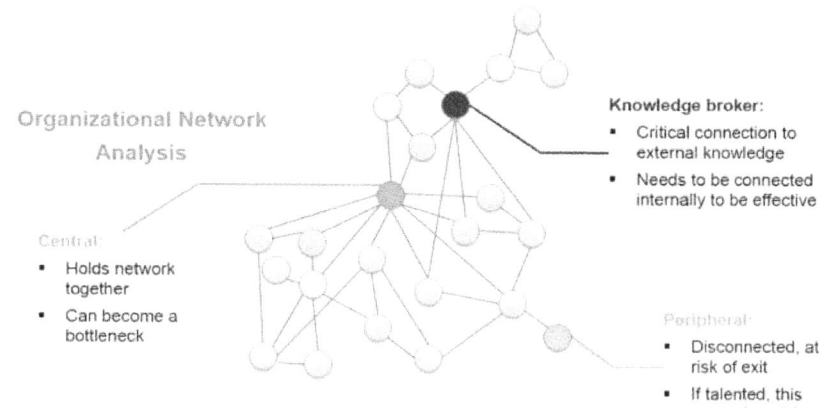

Source: Deloitte (2019) Organizational network analysis - Gain insight, drive smart https://www2.deloitte.com/us/en/pages/human-capital/articles/organizational-network-analysis.html (6 June 2019)

Active vs Passive ONA Data Sources

ACTIVE VS. PASSIVE ONA	ACTIVE	PASSIVE
EXAMPLE DATA SOURCES:	Self-reported e.g. Surveys	Email, calendar & phone metadata, social media (in/external), collaboration platforms e.g. Slack, wearables and sensors e.g. digital badges
KEY CHARACTERISTICS:	• Point-in-time, • Can be labour intensive, • Response rates critical, • Enables a deep-dive on a specific topic • Understand how employees feel	• Real-time and continuous, • Provides scale, • Privacy is a key consideration, • Clear communication on the 'why' and benefit to employees required • Understand what employees do

Source: David Green (2018) The role of Organisational Network Analysis in People Analytics. https://www.linkedin.com/pulse/role-organisational-network-analysis-people-analytics-david-green (12 June 2019)

There are two main types of ONA data sources (Active and Passive): [2]
- **Active ONA:**
Survey based Active ONA can produce great insights, but it may take significant time, and is a point-in-time.
- **Passive ONA:**
Passive ONA runs continuously and provides an objective, unbiased view of how people are actually working and collaborating.

Privacy

As with any people analytics project – when using passive data sources such as email, there needs to be clear communication with employees on what data do you want to use, for what purpose, what is the benefit to employees, establishing data collection governance, opt in/out, etc. [2]

References:
(1) Deloitte (2019) Organizational network analysis - Gain insight, drive smart https://www2.deloitte.com/us/en/pages/human-capital/articles/organizational-network-analysis.html (6 June 2019)
(2) David Green (2018) The role of Organisational Network Analysis in People Analytics. https://www.linkedin.com/pulse/role-organisational-network-analysis-people-analytics-david-green (12 June 2019)

17.2) Real world examples of ONA

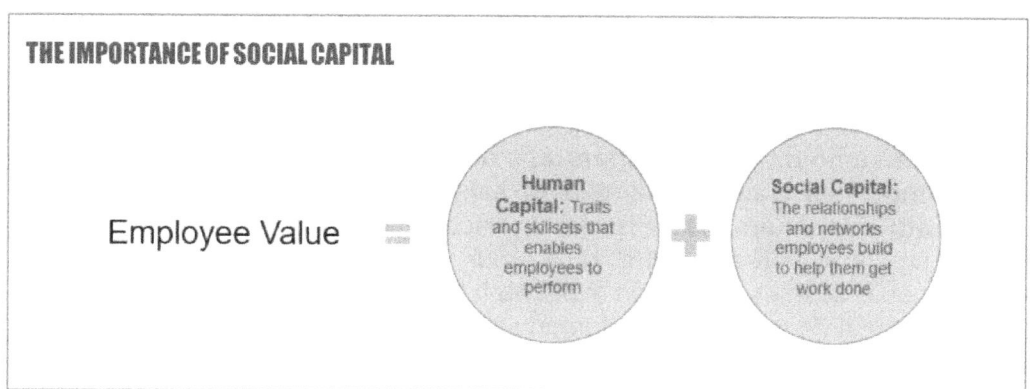

Source: Greg Newman (2017) Retain or let go? The data that you need to react correctly when an employee resigns.. https://www.linkedin.com/pulse/retain-release-data-you-need-react-correctly-when-employee-newman/ (12 June 2019)

According to Greg Newman, ONA adds the dynamic of 'Social Capital' to the static 'Human Capital' data. Social capital includes networks and relationships that employees build to help them get work done, and has a big influence on performance. [1]

- **Human Capital**.

Human Capital data focuses on a recording a static data, such as academic qualifications, tenure, skills, performance review results and experience.

- **Social Capital**.

Social Capital looks at the network that an employee has developed, the group of people the employee leverages for working, innovating, expertise, mentorship, and leadership.

Several researches have shown that a person's social network affects their performance.:
- Genpact, a global professional services firm, found that the **best performing engineers have more meetings and relationships** than others. (2)
- Most people think that the time salespeople spend with customers is the most important factor affecting their sales. Interestingly, **VoloMetrix found that the size and amount of a salesperson's network inside their company is a more important leading indicator of sales, than the time salespeople spend with customers**. Their research discovered that regardless of what you are sell, who you are sell it to, or your work location anywhere in the world, success in sales correlates highly with three actionable metrics. Merely increasing the time your underperforming salesperson spend with customers will not increase sales much. The second and third factor are more important. Corporate buyers want a salesperson who is credible, who understands their needs, and who is able to address their concerns quickly and competently. Doing this well requires a salesperson is able to get access to management when needed, and have a comprehensive understanding of what their company can offer to the buyer above and beyond the current sales. (3)
  - Spending sufficient time with customers.
  - Having a big network in your own organization.
  - Spending time with your manager and other senior people in your organization.
- Conventional wisdom says **the more connections you have in your professional network, the more successful you will be**. An employee with more connections, is more likely to receive information and more likely to be promoted. **New research by Brooks Holtom, found that who you are connected to matters most**. The degree to which an individual is connected to people with strong, positive reputations means it is less likely for them to leave their current job, because It is hard to replicate that social network when you move to a new firm. Holtom analyzed the social networks of employees in two different large firms for 18 months to see who stayed and was successful, and who quit. He found that **employees who were more connected to people with strong reputations were more likely to stay**. But, when well-connected people leave, the losses can be extensive, because their network often follow them. Companies should identify well-connected employees for retention. (4)

References:

(1) Greg Newman (2017) Retain or let go? The data that you need to react correctly when an employee resigns.. https://www.linkedin.com/pulse/retain-release-data-you-need-react-correctly-when-employee-newman/ (12 June 2019)

(2) Josh Bersin (2018). What Emails Reveal About Your Performance At Work. https://joshbersin.com/2018/10/what-emails-reveal-about-your-performance-at-work/# (12 November 2018)

(3) Ryan Fuller (2015) What Makes Great Salespeople. Harvard Business Review. https://hbr.org/2015/07/what-makes-great-salespeople (13 November 2018)

(4) Georgetown University (2016) Employees' Internal Social Networks Can Predict Turnover https://msb.georgetown.edu/newsroom/news/employees'-internal-social-networks-can-predict-turnover# (12 June 2019)

17.3) Applications of ONA in HR

Organizational Network Analysis (ONA) has many HR Applications. It can be used for: Communications Gap Analysis, Diversity and Inclusion Analysis, Flight Risk Analysis, Organizational Development, Organizational Design, Talent Management, Mergers and Acquisitions, Workspace Configuration, Onboarding, Promotion, Knowledge Management, Identifying Under & High-performing teams, Identifying Hubs Connecting Departments, Identifying High-Potentials, Change Management, and Leadership Development.

- **Communications Gap Analysis**: identify gaps in communication within or between departments, and identify opportunities to improve connections across silos.

- **Diversity and Inclusion Analysis**.
  Social network analysis can be used to evaluate how diverse a person's network is.

- **Organizational Development**.
  Identify key players influencers, formulate targeted training interventions, review network changes over time.

- **Organizational Design**.
  Formal organization charts often show little similarity to the network of people who in reality, carry out the work. Organizational Network Analysis can offer precious insights during organizational design, making it easier to setup effectiveness organizational structures that is structured to increase collaboration and that puts people where they can deliver the most results. [1]

- **Talent Management.**
    Organizational Network Analysis can reveal which positions and units are interacting to get work done. ONA can help identify redundant roles in a network, freeing talents to fill more value-added activities. (1)

- **Mergers and Acquisitions.**
    Cigna uses social network analysis to understand acquired companies' informal networks and identify their go-to people. It then uses the information to support integration efforts. (2)

- **Workspace Configuration**.
    A large company division analyzed its organizational network to assess its employee's interdependencies, and to identify informal leaders. As a result of the analysis, the workspace was reconfigured such that people who needed to work together closely were place on the same floor. Person-to-person connectivity has improved significantly after the workspace configuration. (2)

- **Onboarding**.
    Some companies assign mentors to new hires based on the mentor's network. If a new hire is replacing someone, the company gives the new hire a map of the departing employee's informal network and make introductions. (2)

- **Promotion.**
    Through social network analysis, some companies identified hires who had developed the most extensive informal networks and were hence deemed candidates for promotion. (2)

- **Flight Risk Analysis.**
Peripherals who are unconnected to the rest of the organization are flight risks, and if they are talented, it becomes a risk for the organization. Conventional wisdom says the more connections you have in your professional network, the more successful you will be, as you are more likely to receive information and more likely to be promoted. New research by Brooks Holtom, finds that who you are connected to matters most. The degree to which an individual is connected to people with strong, positive reputations means it is less likely for them to leave their current job, because it is hard to replicate that social network when they move to a new firm. Holtom analyzed the social networks of employees in two different large firms for 18 months to see who stayed and was successful. He found that employees who were more connected to people with strong reputations were more likely to stay, and those who were not as well connected were more likely to quit. When well-connected people leave, the losses can be extensive, because others often follow them. Thus, companies should identify well-connected employees for retention. [3]

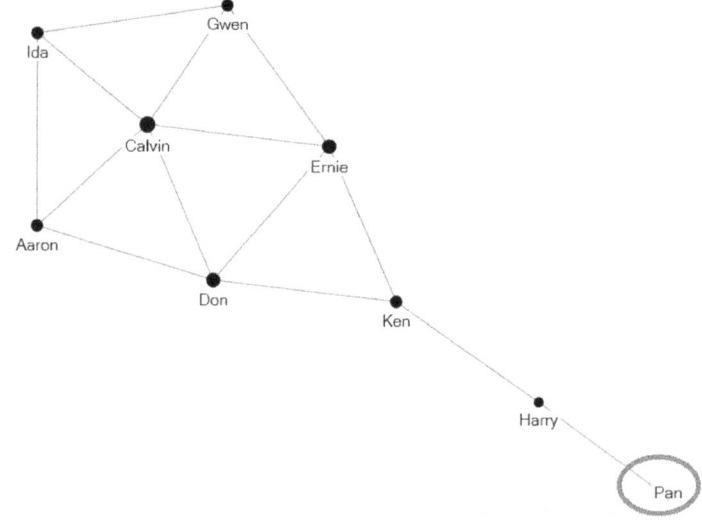

- **Knowledge Management.**

Organizational Network Analysis (ONA) can be used in knowledge management to identify groups with no direct access to knowledge center. There are two major sources of network data [4]:
- **Passive Data:** Analyze for patterns of interaction among groups or individuals. E.g. Email logs, Twitter, Facebook, Flickr, etc.
- **Survey Data:** Electronic web or paper-based surveys. Survey participants can include:
    - **Bounded network** – Participant must identify level of interaction with a specified group (e.g. within a company).
    - **Unbounded network** – list of key contacts is open ended and defined by the participant (e.g. not limited to company).

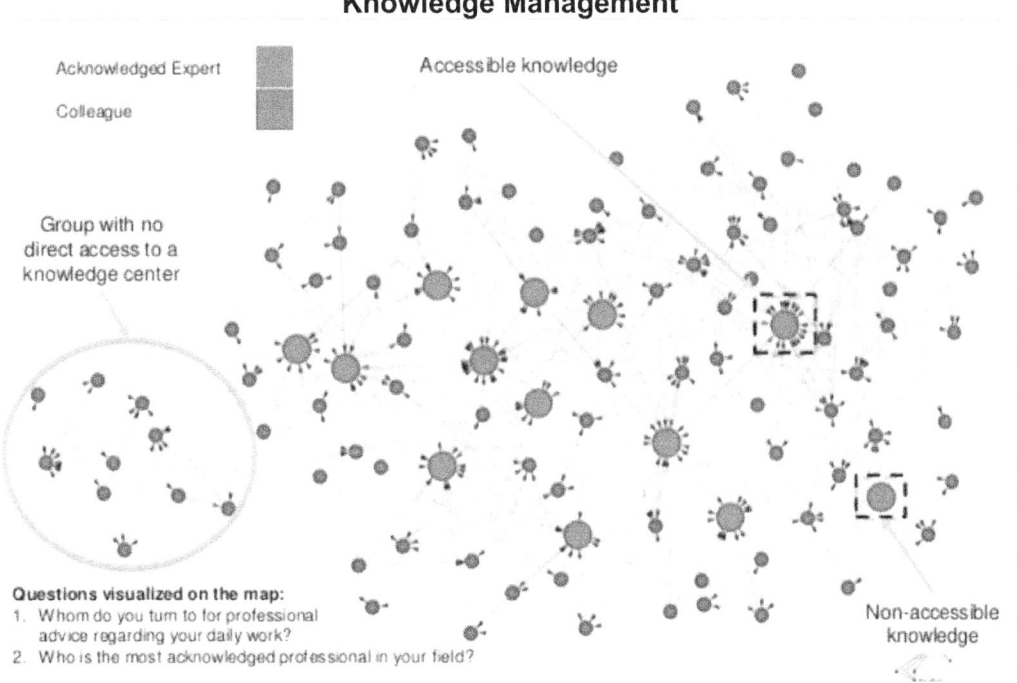

Source: Patti Anklam (2015) ONA and the Tools Landscape. https://www.slideshare.net/panklam/ona-and-the-tools-landscape (6 June 2019)

- **Identifying Under & High-performing teams**

Social Network Analysis (SNA) can be used to identify Under-performing and High-performing teams. However, we have to be cautious of assuming that more connections are better, as excessive networking can drain productivity. Your network can be energy sappers, for whom a problem shared is a problem doubled - such people de-energizes others. Cross and Parker shared an example of a software company where, although it was the most connected group, no work could be done. In general, rather than pursuing number, quality of connections is better. (5)

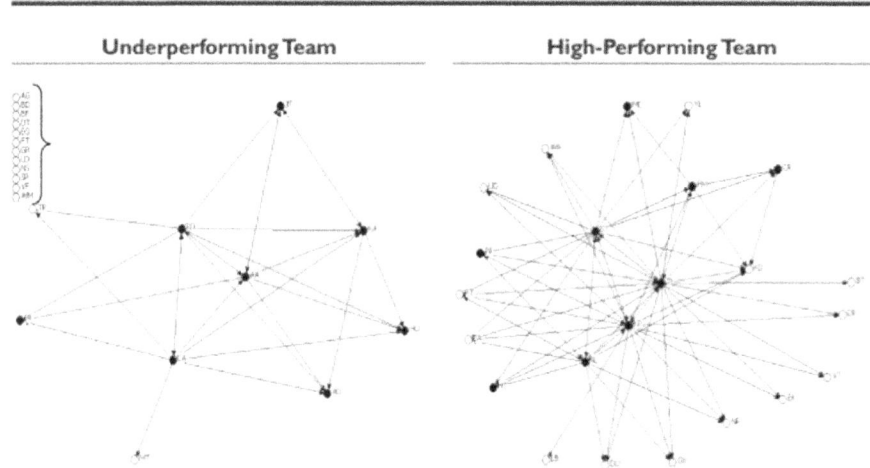

Source: Dr. Michael Moo (2016) A Network Approach to People Analytics: How to Use Social Network Analysis in HR. https://static1.squarespace.com/static/55007c24e4b001deff386756/t/57dbfff9d2b8574909f34203/1474035707614/Moon%2C+Michael.pdf (6 June 2019)

- **Identifying Hubs Connecting Departments**

Organizations can benefit from SNA by identifying gaps in communication within or between departments, and identifying opportunities to improve connections across silos. (6)

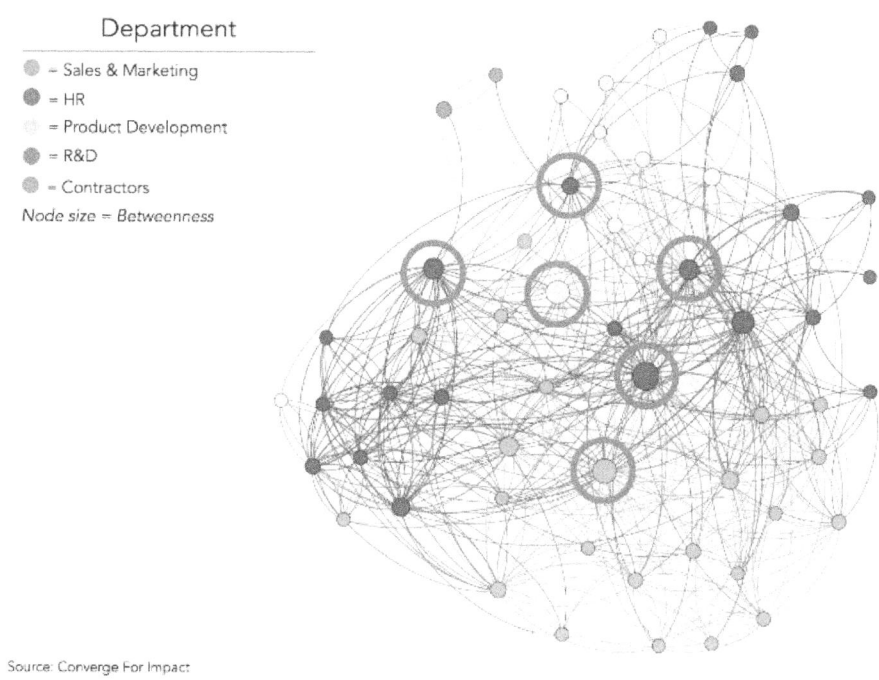

Source: Converge For Impact

Source: David Ehrlichman *(2018)* Asking the Right Questions: Collecting Meaningful Data About Your Network.   *https://blog.kumu.io/asking-the-right-questions-collecting-meaningful-data-about-your-network-dcb4b5f9383c (6 June 2019)*

- **Identifying High-Potentials**

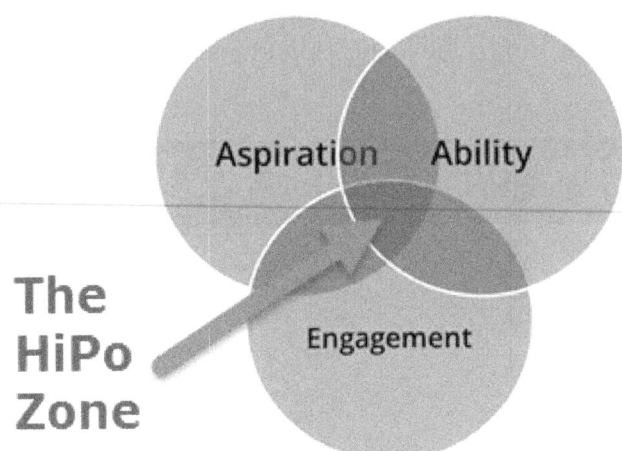

Source: Greg Newman (2017) How can we better identify HiPos using network data? https://www.trustsphere.com/identify-hipos-using-network-data/ (6 June 2019)

Broadly, Hi-Potential (HiPos) employees have a mix of aspiration, ability, and engagement. Specifically, these are the characteristics of HiPos [7]:

• **Collaboration**: HiPos know what they want and can work with others to get it.

• **Energy**: HiPos are constantly motivated to improve themselves.

• **Courage**: HiPos are not afraid of facing difficult challenges.

• **Productivity**: HiPos produce more work in shorter time.

• **Influence**: HiPos can talk to other people in a way that makes people like them, leading to better networking skills.

Most company's process for identifying Hi-Potential employees are limited to asking managers whether an employee is a candidate for the High Potential program? Organizational Network Analysis ("ONA") is a quantifiable method of measuring an employee's potential. High performers usually have more immersed, broader and stronger networks than their peers. ONA provides a structured way to visualize how communications, information, and decisions flow through an organization.[7]

Network Reach – measures how and where the employee's network reach, how are their relationships with more senior employees, with different teams, departments and across different physical locations. [7]

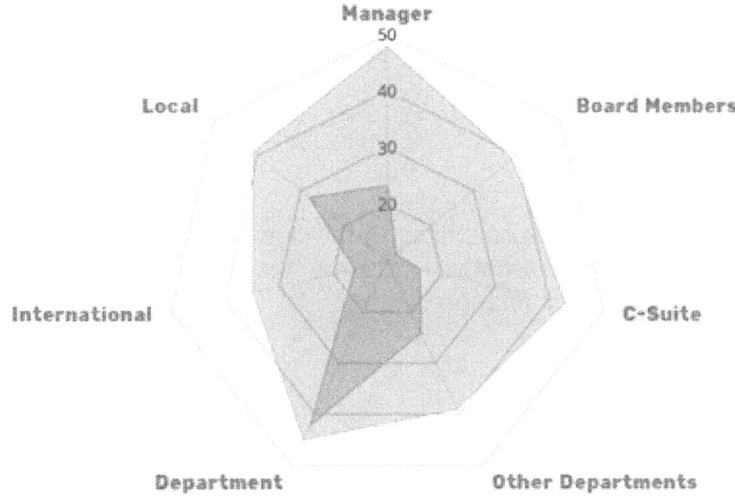

Source: Greg Newman (2017) How can we better identify HiPos using network data? https://www.trustsphere.com/identify-hipos-using-network-data/ (6 June 2019)

Network immersion and strength – how strong are the relationships that the employee has with colleagues? The stronger those relationships, the better value is created. [7]

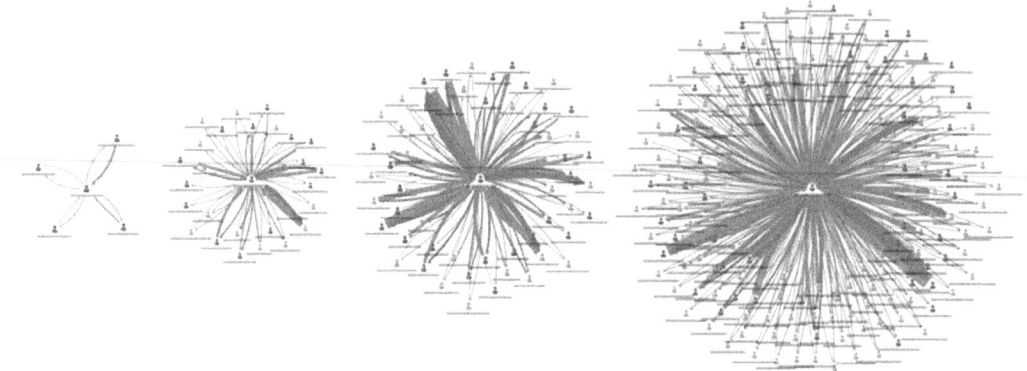

Source: Greg Newman (2017) How can we better identify HiPos using network data? https://www.trustsphere.com/identify-hipos-using-network-data/ (6 June 2019)

Ability to influence: HiPos that are critical to the flow of information are the key connectors that hold networks together, and are the key employees who can drive and support organizational change. [7]

Source: Greg Newman (2017) How can we better identify HiPos using network data? https://www.trustsphere.com/identify-hipos-using-network-data/ (6 June 2019)

- **Change Management**

Often, leaders in organization charts are not the real hubs of information flow or go-to people. Organizational Network Analysis can be used in change management to identify individuals that can be leverage in change programs. [8]

Change Management

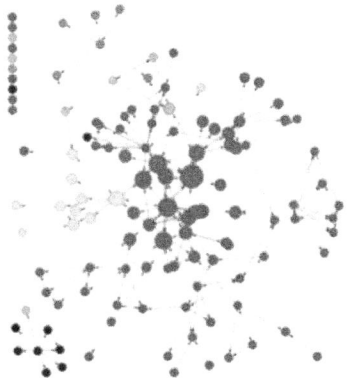

- Network analysis can be used to identify key information & opinion conduits
- These individuals can be leveraged in change programs
- Goal: find the smallest number of people who can reach the largest number of employees in the smallest number of steps

Source: Patti Anklam (2015) ONA and the Tools Landscape. https://www.slideshare.net/panklam/ona-and-the-tools-landscape (6 June 2019)

- **Leadership Development**

Social network analysis can be used to evaluate a person's network connectivity before and after a Leadership Development Program. The leadership program can look into the participant's development in areas such as, their network strength and reach, their leadership style and how they and their teams communicate. [9]

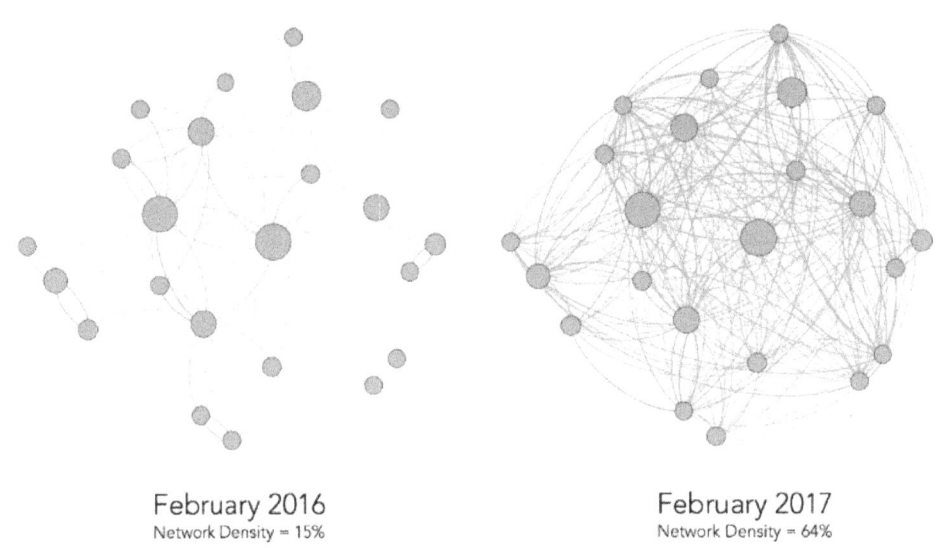

Source: David Ehrlichman (2018) Asking the Right Questions: Collecting Meaningful Data About Your Network.   https://blog.kumu.io/asking-the-right-questions-collecting-meaningful-data-about-your-network-dcb4b5f9383c (6 June 2019)

References:
(1) Deloitte (2019) Organizational network analysis - Gain insight, drive smart https://www2.deloitte.com/us/en/pages/human-capital/articles/organizational-network-analysis.html (6 June 2019)
(2) Bill Roberts (2014) The value of Social Network Analysis https://www.shrm.org/hr-today/news/hr-magazine/pages/0114-social-network-analysis.aspx (6 June 2019)
(3) Brooks Holtom (2016) Employees' Internal Social Networks Can Predict Turnover https://msb.georgetown.edu/newsroom/news/employees'-internal-social-networks-can-predict-turnover# (6 June 2019)
(4) Dr. Michael Moo (2016) A Network Approach to People Analytics: How to Use Social Network Analysis in HR. https://static1.squarespace.com/static/55007c24e4b001deff386756/t/57dbfff9d2b8574909f34203/1474035707614/Moon%2C+Michael.pdf (6 June 2019)
(5) Ken Thompson (2006) Social Network Analysis in practice http://www.bioteams.com/2006/04/04/social_network_analysis.html (6 June 2019)
(6) David Ehrlichman (2018) Asking the Right Questions: Collecting Meaningful Data About Your Network. https://blog.kumu.io/asking-the-right-questions-collecting-meaningful-data-about-your-network-dcb4b5f9383c (6 June 2019)
(7) Greg Newman (2017) How can we better identify HiPos using network data? https://www.trustsphere.com/identify-hipos-using-network-data/ (6 June 2019)
(8) Patti Anklam (2015) ONA and the Tools Landscape. https://www.slideshare.net/panklam/ona-and-the-tools-landscape (6 June 2019)
(9) David Ehrlichman (2018) Asking the Right Questions: Collecting Meaningful Data About Your Network. https://blog.kumu.io/asking-the-right-questions-collecting-meaningful-data-about-your-network-dcb4b5f9383c (6 June 2019)

17.4) Tools to build an ONA Survey

There are two main steps to build an Organizational Network Analysis (ONA) Survey:

(i) Identify target population: Define the purpose for carrying out ONA in your organization and identify who will take the survey. The larger the target population, the longer the survey takes, and the more complicated the analysis becomes.

(ii) Create Survey: Start the survey with an opening statement on the survey objective and confidentiality. The survey should not take any longer than 10-15 minutes of a respondent's time. Here's a good sample online ONA survey: https://www.surveymonkey.com/r/63R8XMT?sm=JbeN1eq4AWhuxJjDhIgY7yWG9l5N4D%2fZ8bOLGMyAMcw%3d

There are several Social Network Analysis (SNA) survey tools which have a free version:
- **Qualtrics**: Qualtrics is a general purpose survey tool for gathering bounded network information. [1]
  http://www.qualtrics.com/
- **ONA Surveys**: Specialist survey tool for collecting bounded network data. [1]
  https://www.s2.onasurveys.com/
- **Survey Monkey**: Survey Monkey is a free general purpose online survey tool.

References:
(1) Ken Thompson (2008) A great free Social Network Analysis Tool. http://www.bioteams.com/2008/02/08/a_great_free.html *(6 June 2019)*

17.5) Tools to visualize your ONA survey results

There are several Social Network Analysis tools which have a free version:

- **Gephi**: Gephi is a free open source visualization and exploration software for graphs and networks.
  https://gephi.org/

- **Agna**: Agna is a platform-independent neat tool build for social network analysis, sociometry and sequential analysis. [1]
  https://mac.softpedia.com/get/Network-Admin/AGNA.shtml#download

- **Socilyzer**: Socilyzer makes social network analyses easy as it integrates questionnaire design, data collection and data visualization. [1]
  www.socilyzer.com

- **SocNetV**: SocNetV has a nice graphical user interface and is quick to calculate centralities and distances. [1]
  https://socnetv.org/

- **NodeXL**: NodeXL is a free, open-source template for Microsoft Excel that is easy to build social network graphs. You can enter a network edge list in the Excel worksheet, click a button and see your social network graph in Excel. NodeXL is ideal for those who like spreadsheets. [2]
  https://archive.codeplex.com/?p=nodexl

References:
(1) Dr. Michael Moo (2016) A Network Approach to People Analytics: How to Use Social Network Analysis in HR. (6 June 2019)
https://static1.squarespace.com/static/55007c24e4b001deff386756/t/57dbfff9d2b8574909f34203/1474035707614/Moon%2C+Michael.pdf
(2) Ken Thompson (2008) A great free Social Network Analysis Tool. http://www.bioteams.com/2008/02/08/a_great_free.html *(6 June 2019)*

17.6) Visualize ONA with Excel NodeXL

NodeXL Basic is a free open-source network analysis tool for Microsoft Excel that can help you to visualize and analyze formal and informal relationships in your organization. With NodeXL, you can enter a network edge list in Excel, click a button and it will generate a social network graph in an Excel window. [1]

1) Install NodeXL Basic:
Go to https://www.nodexlgraphgallery.org/Pages/Registration.aspx to download and install NodeXL Basic, the free version.

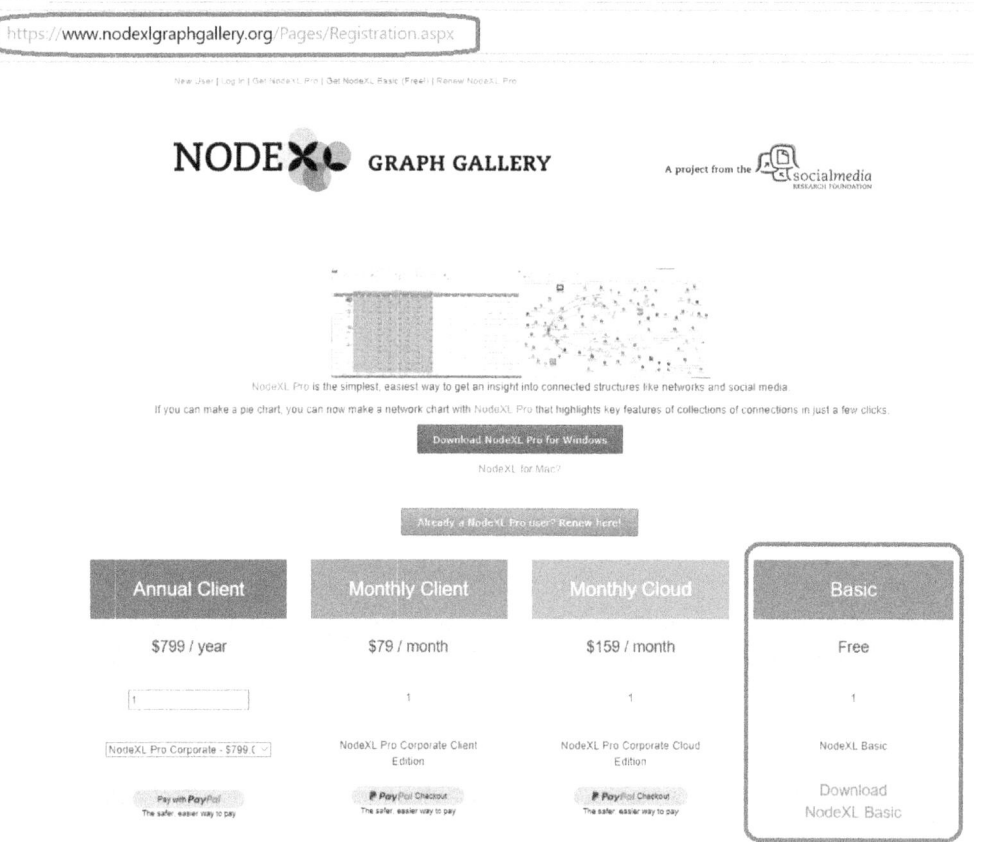

2) After you have installed NodeXL, go to Windows search function and type "**nodexl excel template**". Once found it, click on it to launch it.

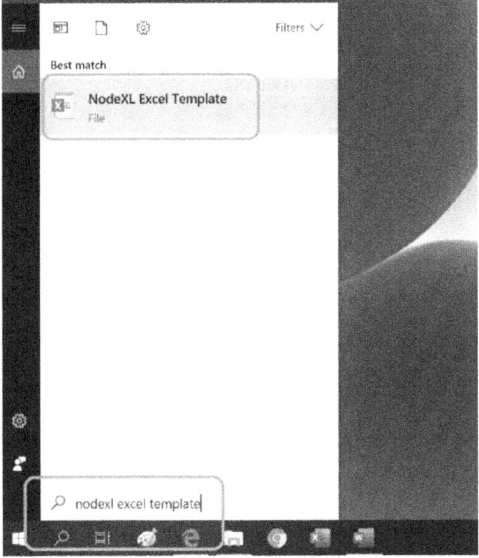

3) You will see two windows here. The window on the right will contain your network pictures. The window on the left has a number of tabs where you can put network information in.

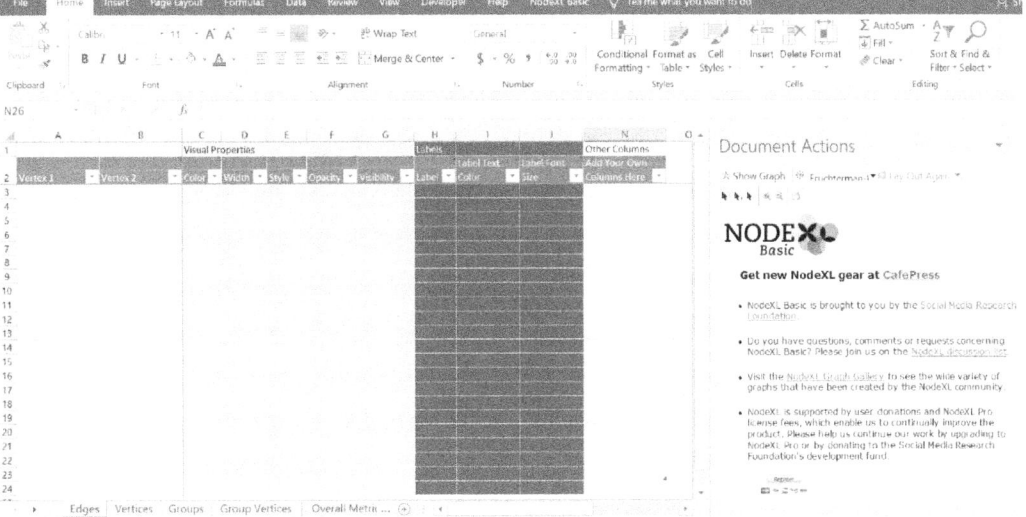

4) Copy and enter following information in "Vertex 1" and "Vertex 2" of the "Edges" tab, select "Harel-Koren Fast Multiscale" then click "Show Graph". Other than "Harel-Koren Fast Multiscale", NodeXL has other automatic layout types (e.g. Fruchterman-Reingold, Circle Spiral) that can be selected. Trying different layout type can reveal useful patterns in the data set being analyzed.

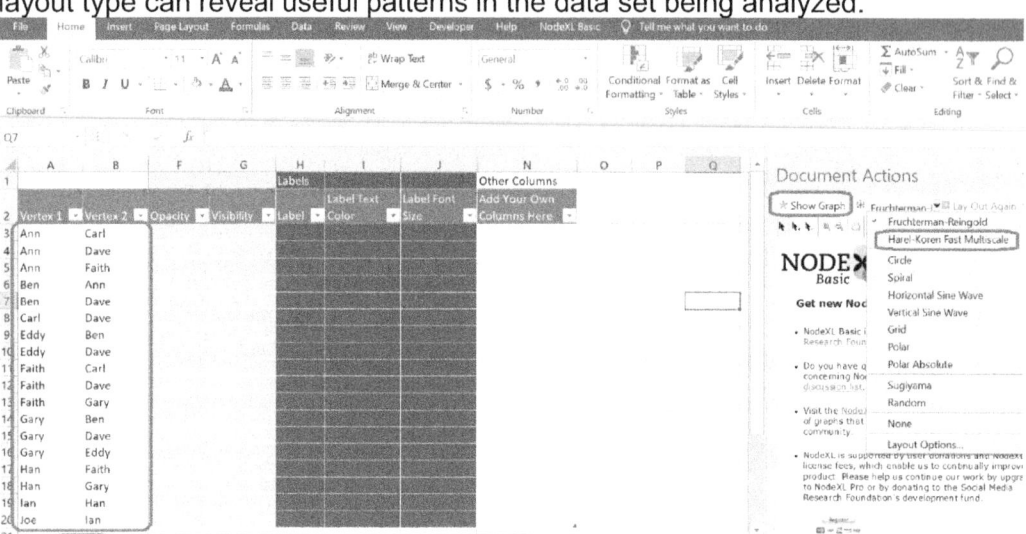

5) Your network graph will appear in the right window. Whenever you click "Refresh Graph", the layout will change.

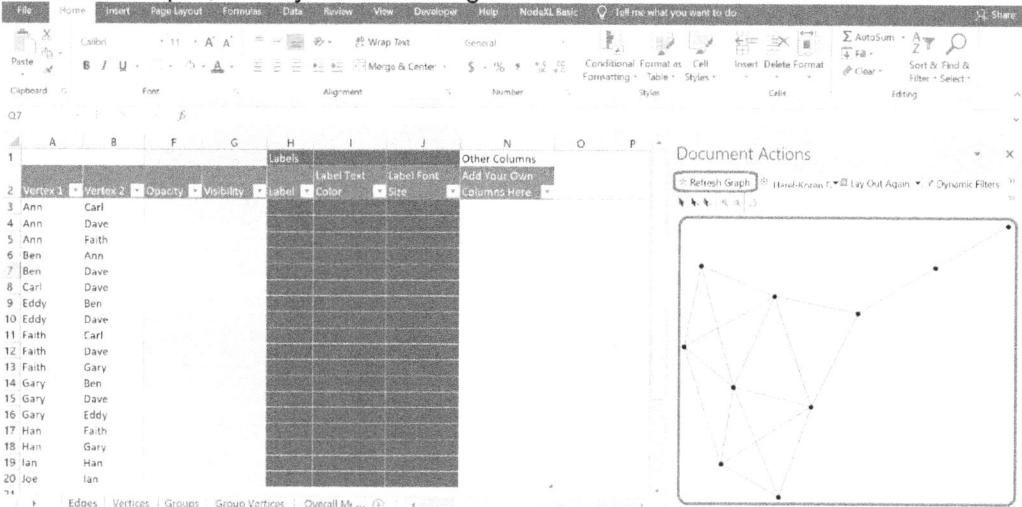

6) Preserving manual layout: To preserve the layout, chose "None" in the layout selection menu.

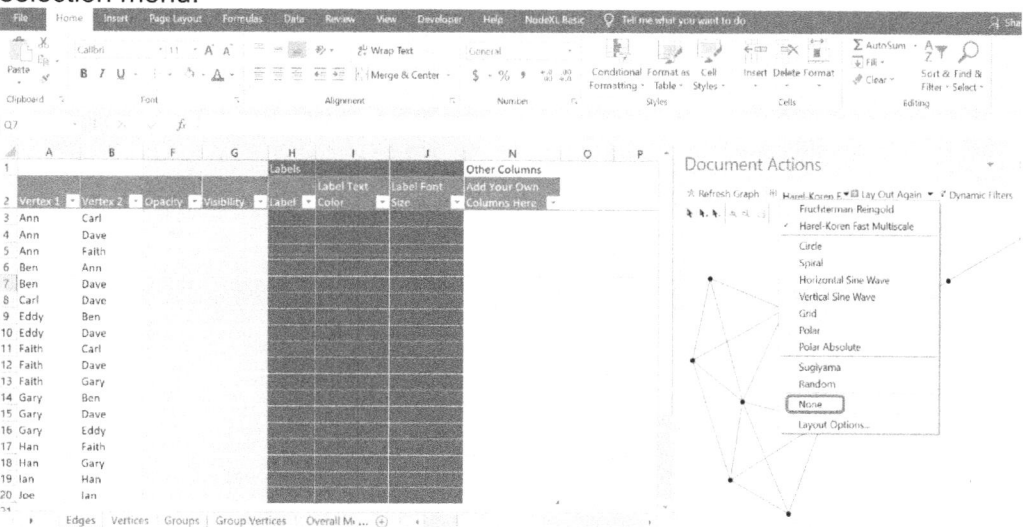

7) Add Labels: To add labels, go to the "Vertices" tab, copy the text in the "Vertex 1" column, and paste in the "Label" column. Then click "Refresh Graph". After you click "Refresh", you will see the labels in the graph.

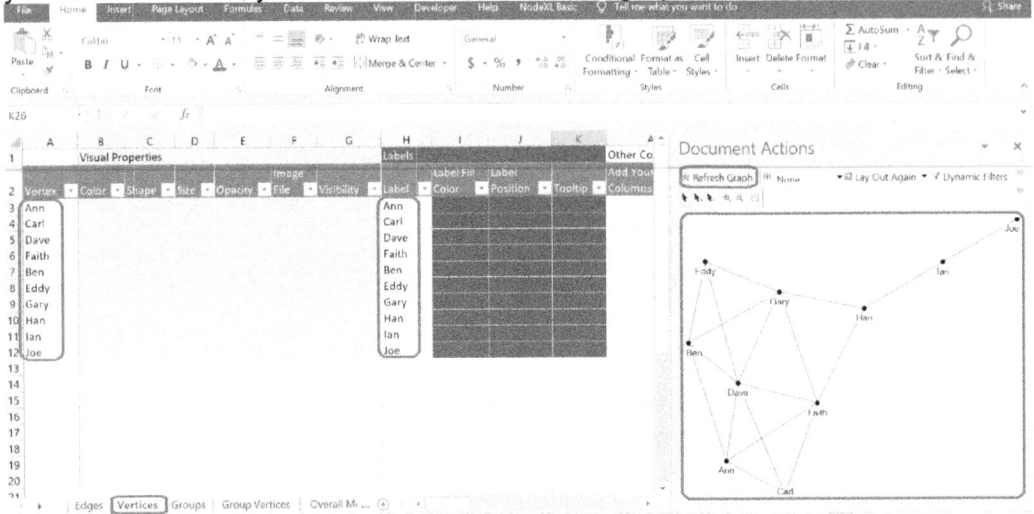

8) Changing Vertex Colors: You might want to color vertices that represent a department with blue, and another department with red. To change the color, copy the text in the color columns, then click refresh. After you click "Refresh", you will see the new vertices colors in the graph.

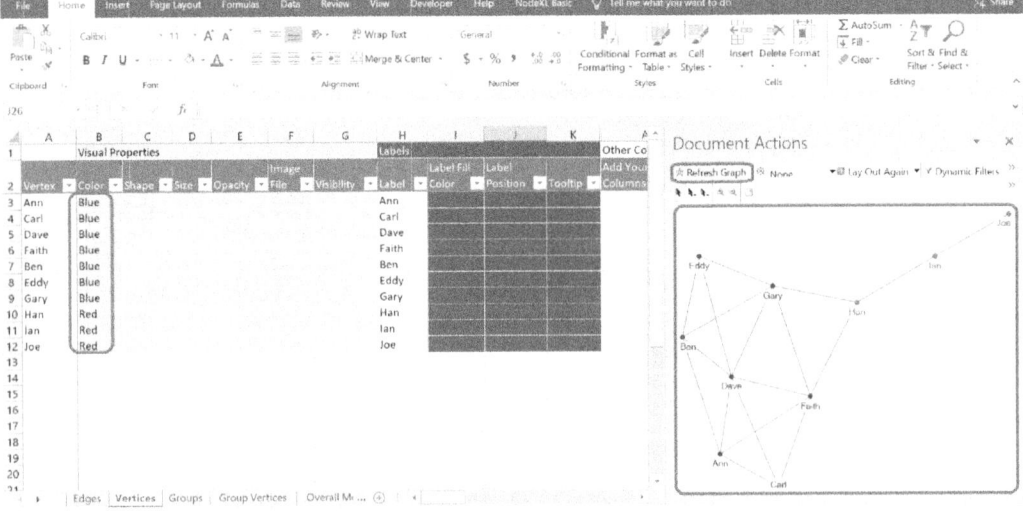

9) Changing Vertex Size: To change the vertex size, copy the text in the size columns, then click refresh.

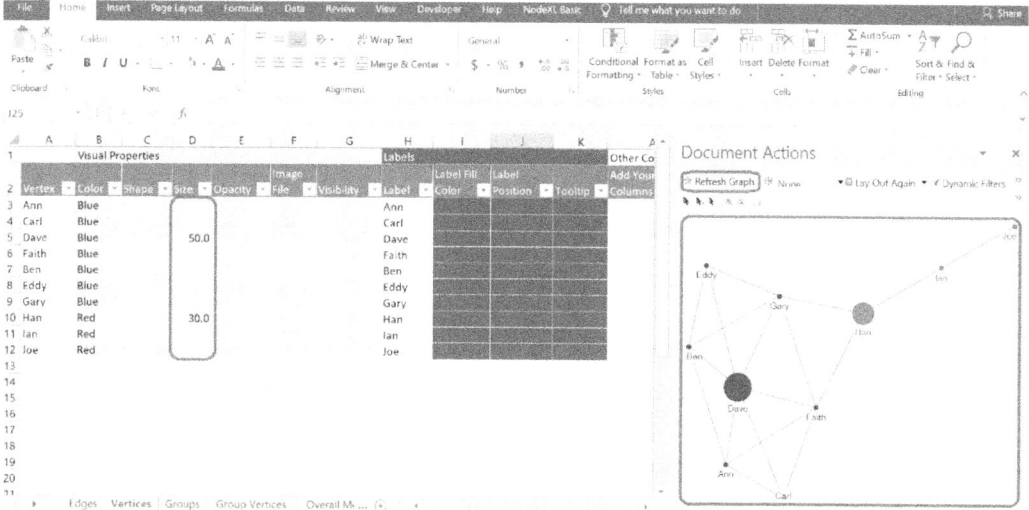

10) Calculating Graph Metrics: Network visualization is useful for identifying important vertices, locating subgroups, and for seeing how interconnected a network. However, graph metrics is useful for providing quantitative measures of a network. To calculate graph metrics, click on the "Graph Metrics" button on the "NodeXL Basic" Ribbon.

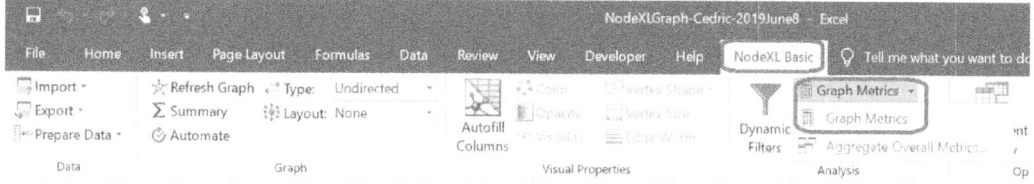

11) At the Graph Metrics window, click "Select All", then click "Calculate Metrics".

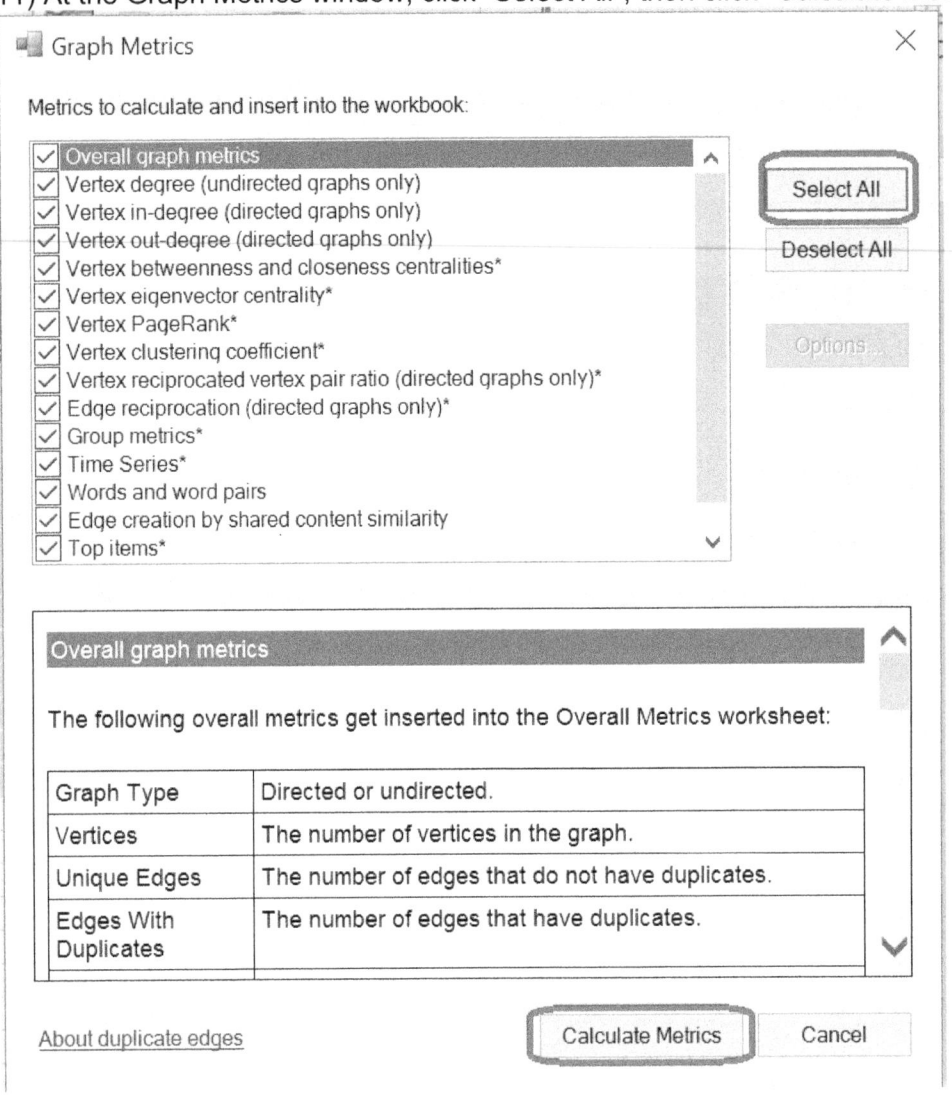

12) **Vertex Metrics:** To see the vertex-specific metrics, go to the "Vertices" worksheet, and you will see the calculated "Graph Metrics" columns.

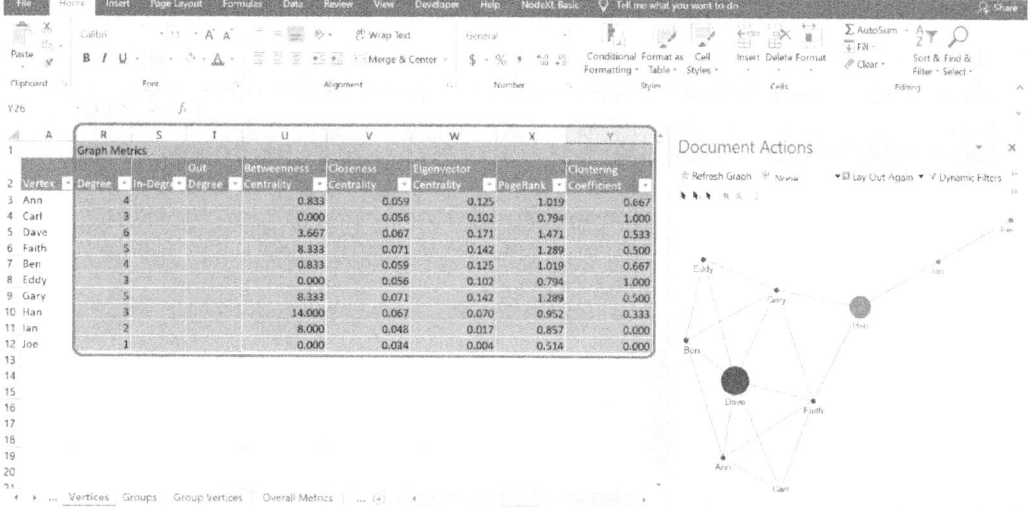

- **Degree**

The **Degree of a vertex (also called Degree Centrality) is a count of the number of direct connections a node has**. Dave has a Degree of 6 because he is directly connected to 6 other individuals. Dave has the most direct connections in the network, making him the most active node in the network. He is a 'connector' or 'hub' in this network. Joe has a Degree of only 1 because he is connected to only 1 other person. **Dave is the most popular person while Joe is the least popular.** Generally, the more connections, the better. But Dave has connections only to others in his immediate cluster -- his clique. To visualize these, copy the numbers in "Degree", and paste it in "Size", then click "Refresh", and you will see the size of the vertices change.

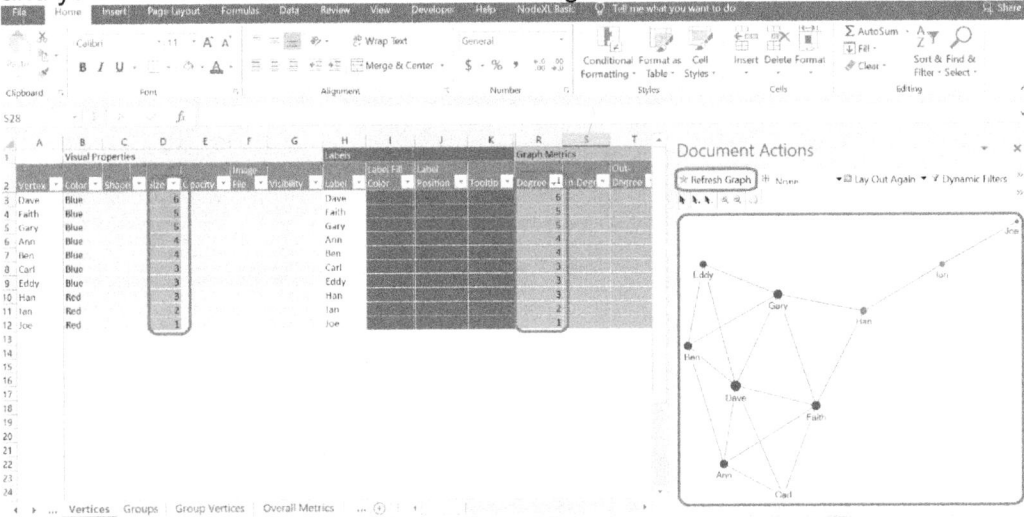

- **Betweenness Centrality**

Although Dave is more popular (he has more direct connections) compared to Han, it is not everything. Han has the highest "Betweenness Centrality" of 14 as he is a bridge between two important constituencies. **A node with high betweenness has great influence over what flows, and what does not flow in the network**. Han plays a powerful **'broker' role in the network**, but the danger to the organization is that he is potentially a single point of failure. Without Han, Ian and Joe would be cut off from information and knowledge in Dave's cluster. In contrast, Joe has a Betweenness Centrality of 0, because if he were removed from the network everyone would still be connected to everyone else and their shortest communication paths will not be change.

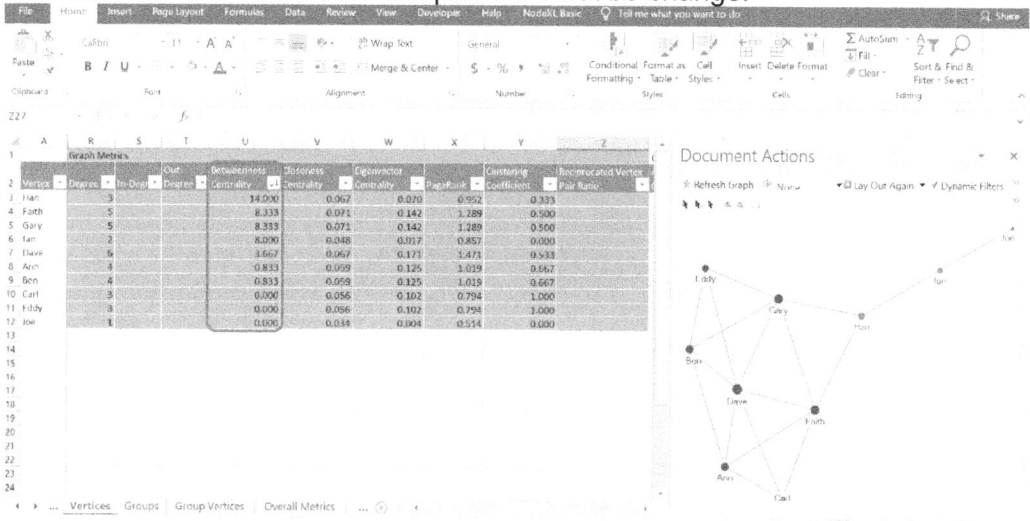

- **Closeness Centrality**

Unlike other centrality metrics, a lower Closeness Centrality score indicates a more central (i.e. important) position in the network. Faith and Gary have the lowest closeness centrality of 0.071. Though Faith and Gary have fewer connections than Dave, the pattern of their direct and indirect ties enable them to access all the nodes in the network faster than anyone else. **They have the shortest paths to everyone else.** They have the best visibility into what is happening in the network.

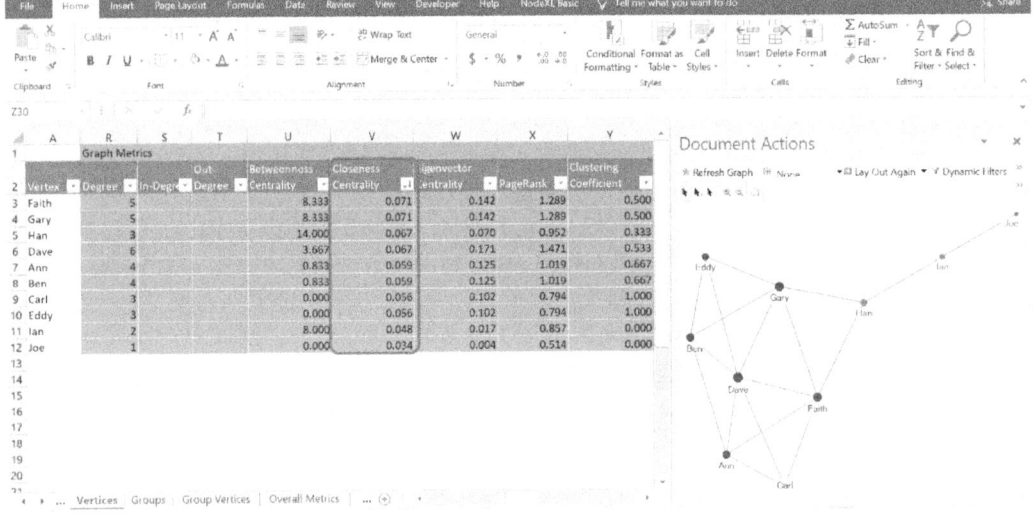

- **Eigenvector Centrality:**

A **connection to a popular individual is more important than a connection to a loner.** The Eigenvector Centrality metric considers not just how many connections a vertex has (i.e. its Degree), but also the Degree of the vertices that it connects to. Joe has a low Eigenvector Centrality metric as he is connected to Ian who is not popular. Gary and Faith has high Eigenvector Centrality metric as they are connected to Dave, the most popular person in the network,

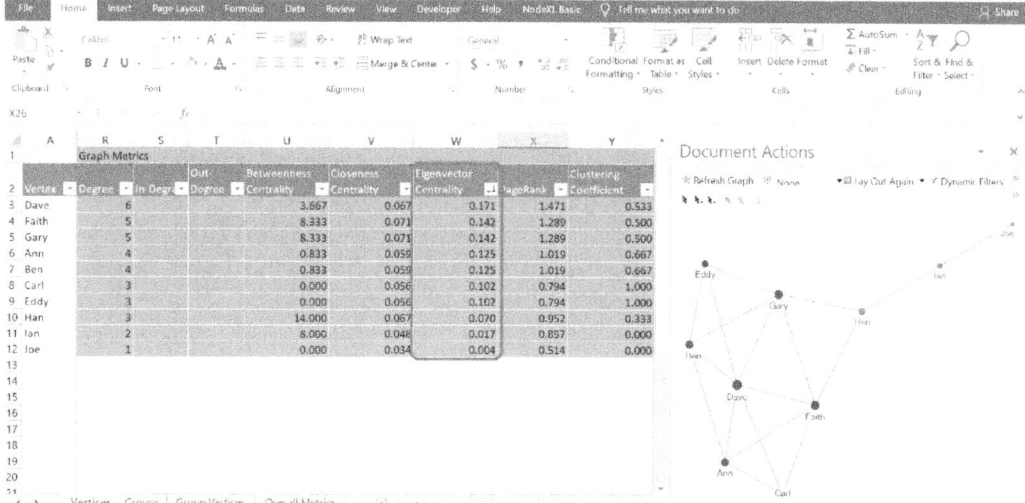

References:
(1) Derek Hansen and Ben Shneiderman (2009) Analyzing Social Media Networks: Learning by Doing with NodeXL. http://www.urbanlab.org/articles/NodeXL_tutorial_draft%5B1%5D.pdf (6 June 2019)

17.7) ONA Interventions

Organizational network interventions:

There are various ways to change organizational network patterns: [1]

- **Setup more connections**: Make introductions through events and offsite meetings.

- **Increase knowledge flow:** Setup internal company instant messaging, make existing knowledge bases more accessible.

- **Decentralize**: Setup knowledge warehouse in SharePoint to shift knowledge to the edge.

- **Connect disconnected clusters**: Setup knowledge brokering roles, expand communication avenues.

- **Facilitate trusted relationships**: Assign employees to work on projects together.

- **Leverage connectors**: Educate employees on the importance of their place in a network.

- **Increase Diversity**: Add nodes, connect and create networks.

Personal Social Network interventions:

Here's fours steps to analyze and enhance your personal social network.

(i) List your social network:
List who you turn to for information or help with your work, and rate your frequency of contact with them.

Personal social network listing template

Name	Department	Contact frequency (1) 1-2 times a month (2) 3-4 times a month (3) 5-6 times a month (4) more than 6 times a month

(ii) Visualize your organizational network using network diagrams.

(iii) Analyze your network.
- What expertise (nodes) do you need to do your job effectively?
- Where you are over dependent on certain individuals (nodes) for information? Bottlenecks are central nodes that provide the only connection between different groups.
- Identify people (nodes) that you have not leverage sufficiently.

(iv) Action plan to enhance your network.
List a couple of actions to enhance your network.

References:
(1) Patti Anklam (2015) ONA and the Tools Landscape. https://www.slideshare.net/panklam/ona-and-the-tools-landscape (6 June 2019)

17.8) Correlations Example: Predict employee churn with ONA graph metrics

In this example, we use correlation to find out whether ONA centrality metrics affects Status (whether an employee left or stay).:
- **Degree:** How many people can this person reach directly?
- **Betweenness:** Is the person a bridge between different groups?
- **Closeness:** Which person has the shortest paths to everyone. Unlike other centrality metrics, a lower score indicates a more central position.
- **Eigenvector:** How well is this person connected to other well-connected people?

Vertex	Degree Centrality	Betweenness Centrality	Closeness Centrality	Eigenvector Centrality	Status (0 = Left; 1 = Stay)
Dave	6	3.667	0.067	0.171	1
Faith	5	8.333	0.071	0.142	1
Gary	5	8.333	0.071	0.142	1
Ann	4	0.833	0.059	0.125	1
Ben	4	0.833	0.059	0.125	1
Carl	3	0.000	0.056	0.102	0
Eddy	3	0.000	0.056	0.102	0
Han	3	14.000	0.067	0.070	1
Ian	2	8.000	0.048	0.017	0
Joe	1	0.000	0.034	0.004	0

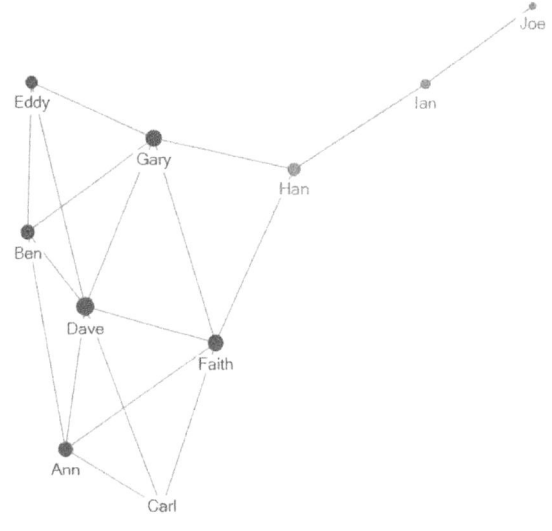

1) Install "Analysis ToolPak", an Excel add-in

"Analysis ToolPak" is an add-in for Microsoft Excel that comes with Microsoft Excel. To be able to run regression using Excel, you need to first install "Analysis ToolPak", an Excel add-in program that provides data analysis tools. To load the Analysis ToolPak add-in, follow these steps:

- On the File tab, click Options.

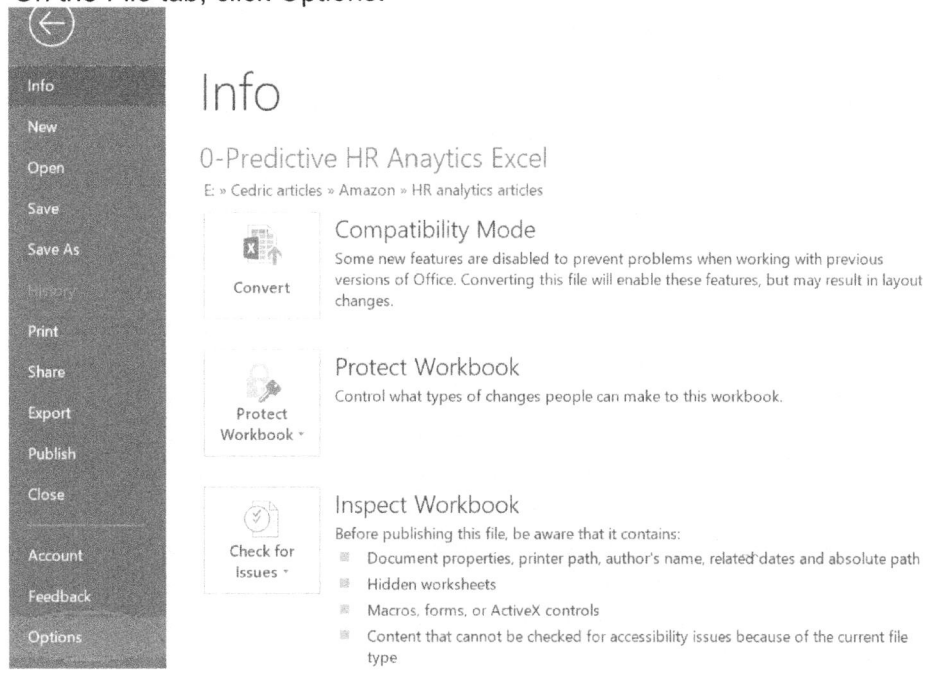

- Under Add-ins, click Analysis ToolPak and click the "Go" button.

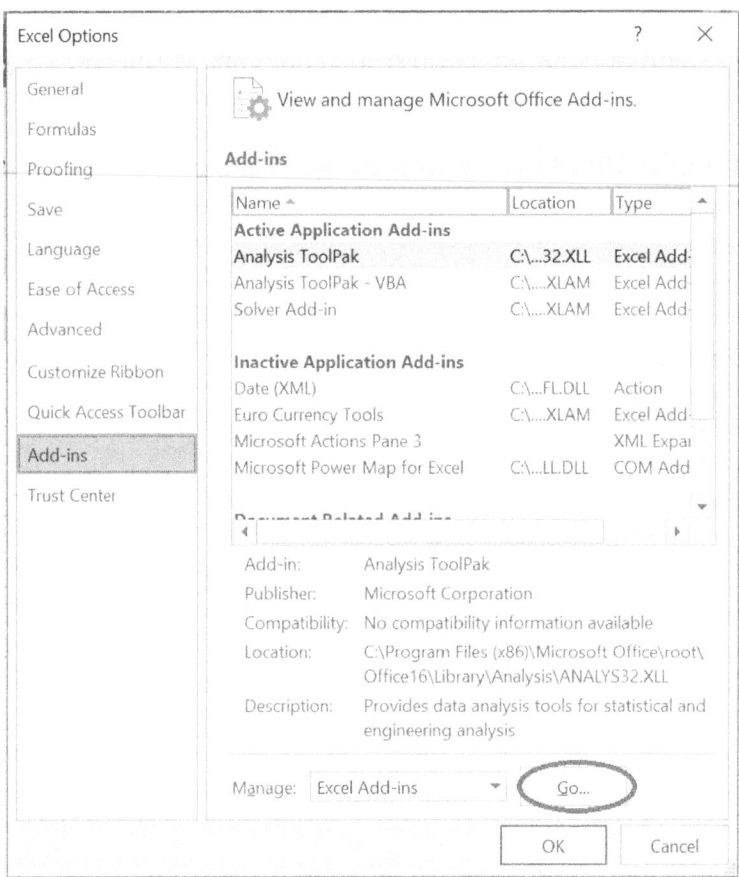

- Click "Analysis ToolPak" and click on OK.

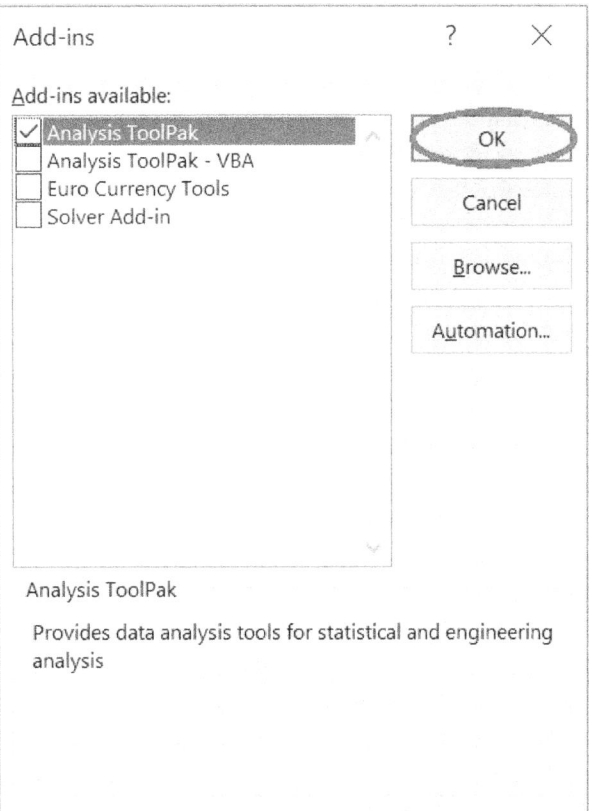

- On the Data tab, in the Analysis group, you are now able to click on "Data Analysis".

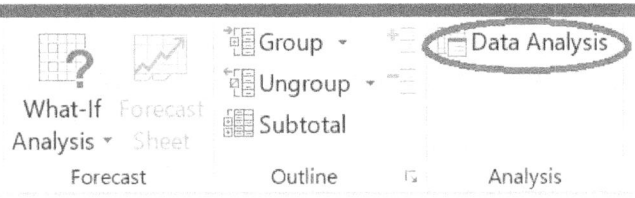

2) Copy the example data in the following table, and paste it in cell A1 of a new Excel worksheet.

	A	B	C	D	E	F
1	Vertex	Degree Centrality	Betweenness Centrality	Closeness Centrality	Eigenvector Centrality	Status (0 = Left; 1 = Stay)
2	Dave	6	3.667	0.067	0.171	1
3	Faith	5	8.333	0.071	0.142	1
4	Gary	5	8.333	0.071	0.142	1
5	Ann	4	0.833	0.059	0.125	1
6	Ben	4	0.833	0.059	0.125	1
7	Carl	3	0.000	0.056	0.102	0
8	Eddy	3	0.000	0.056	0.102	0
9	Han	3	14.000	0.067	0.070	1
10	Ian	2	8.000	0.048	0.017	0
11	Joe	1	0.000	0.034	0.004	0

3) Select "Correlation" and click "OK".

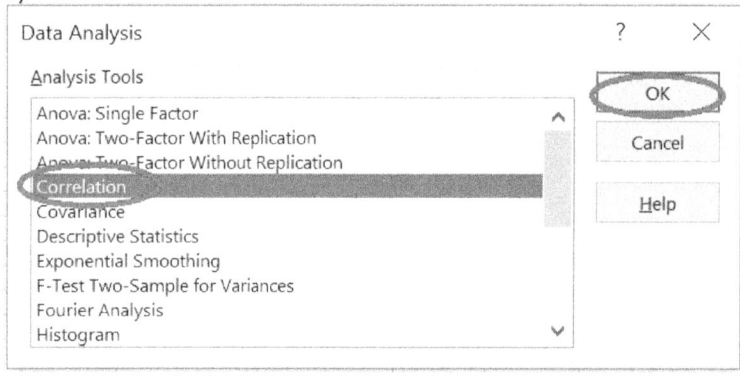

4) After you click OK in the "Data Analysis" dialog box, you will see a "Correlation" dialog box.
5) For "Input Range", select cells (B1:F11).
6) Check "Labels in first row".
7) For "Output Range", select cells (A13).
8) Click "OK".

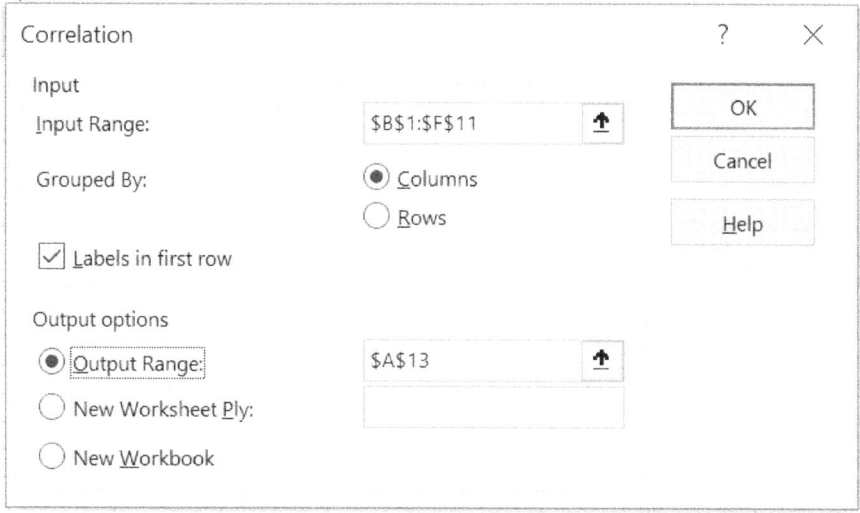

After you click "OK", Excel generates the following Correlation analysis.

	Degree Centrality	Betweenness Centrality	Closeness Centrality	Eigenvector Centrality	Status (0 = Left; 1 = Stay)
Degree Centrality	1				
Betweenness Centrality	0.17	1			
Closeness Centrality	0.87	0.52	1		
Eigenvector Centrality	0.95	-0.04	0.82	1	
Status (0 = Left; 1 = Stay)	0.77	0.42	0.78	0.69	1

A negative correlation coefficient means that an increase in X is associated with a decrease in Y. Similar to a positive correlation, a negative correlation shows a connection between two variables, and the relative strengths are the same. In other words, a correlation coefficient of 0.85 has the same strength as a correlation coefficient of -0.85. Correlation coefficients are always values between -1 and 1, where "-1" means that there is a perfect linear negative correlation, while "1" shows a perfect linear positive correlation. A correlation coefficient of zero, or near to zero, means that there is no meaningful relationship between variables. Correlation coefficient of 0.91 or -0.92 shows a very strong positive and negative correlation respectively. However, correlation does not mean causation. An example of negative correlation is the amount of snowfall and the temperature. As the temperature increases, the amount of snowfall decreases. An example of positive correlation is the relationship between temperature and ice cream sales. As temperature increases, so do ice cream sales.

9) Observations from the above Excel Correlation analysis:

	Degree Centrality	Betweenness Centrality	Closeness Centrality	Eigenvector Centrality	Status (0 = Left; 1 = Stay)
Degree Centrality	1				
Betweenness Centrality	0.17	1			
Closeness Centrality	0.87	0.52	1		
Eigenvector Centrality	0.95	-0.04	0.82	1	
Status (0 = Left; 1 = Stay)	0.77	0.42	0.78	0.69	1

From the Excel Correlation analysis, these variables are good predictors of Status (whether an employee resigns) as they have strong correlation of below -0.75 and above 0.75:
- Degree Centrality: 0.77 Correlation coefficient with Status.
- Closeness Centrality: 0.78 Correlation coefficient with Status.
- Eigenvector Centrality: 0.69 Correlation coefficient with Status.

From the Excel Correlation analysis, Betweenness Centrality has very little impact on Status (whether an employee resigns) as they have very weak correlation of between -0.20 to 0.20.

17.9) Logistics Regression Example: Predict employee churn based on their Organizational Network

Research have shown that a person's Organizational Network, Age, and Commute Time, affects employee retention:

- **Organizational Network:**

    Research by Brooks Holtom, found that the degree to which an individual is connected to people with strong, positive reputations means it is less likely for them to leave their current job, because It is hard to replicate that organizational network when you move to a new firm. [1]

- **Age:**

    The median number of years U.S. workers stay with their employers is 4.6, according to the BLS. For ages 25 to 34 that median number is only three years, while those age 55 to 64 stay at the same job for about 10 years. [2]

- **Commute time:**

    Research by consultant Jeff Parks found that at one manufacturer, a 30 to 45-minute commute, the probability of quitting jumped to more than 92 percent. [3]

The data below represents events that have already happened. We want to use this data (Degree Centrality, Age, Commute Time) to predict employee Resignation (i.e. "Resign" or "Stay") using Excel Logistic regression.
- **Resign column**: As the variables can only take 1 or 0, for the "Resign" column: Stay = 0, Resign = 1.
- **Degree centrality:** refers to how many people can this person reach directly in his organizational network? Dave has a Degree of 6 because he is directly connected to 6 other individuals.
- **Age:** refers to how old the person is.
- **Commute time**: tells you the employee's commute time to work in minutes.

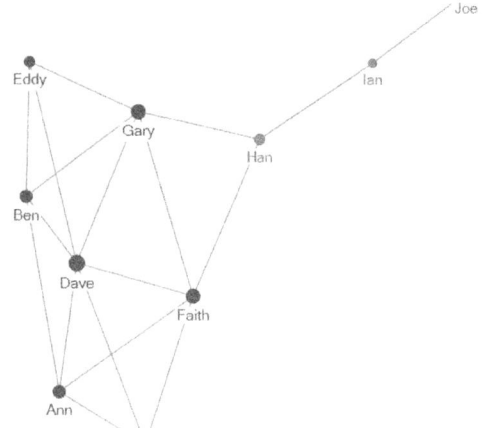

Name	Resign	Degree Centrality	Age	Commute time
Dave	0	6	50	20
Faith	0	5	62	20
Gary	0	5	50	60
Ann	0	4	34	50
Ben	1	4	39	40
Carl	0	3	55	30
Eddy	0	3	40	30
Han	0	3	55	25
Ian	1	2	21	50
Joe	1	1	34	60

1) Our objective is to create an equation with coefficients b_0 to b_3 and then enter values for Degree Centrality, Age, and Commute time to predict Staff Resignation (Resign or Stay). We have 4 coefficients (Constant, Degree Centrality, Age, Commute time). The logistic equation is:

$$\text{Logit(Resign)} = b_0 + b_1 * \text{Degree Centrality} + b_2 * \text{Age} + b_3 * \text{Commute time}$$

2) Logit is a function that takes a probability of an event as input and returns the logarithm of the odds of that event as output. We need to first assign an arbitrary value (e.g. 0.000) for these coefficients (b_0, b_1, b_2, b_3) - Later you will be shown how to use the Excel Solver to replace these starting arbitrary coefficients (e.g. 0.000) with optimized coefficients to create an equation to predict probability.

	A	B	C	D	E	F	G	H	I
1								b_0 (Intercept)=	0.000
2								b_1=	0.000
3								b_2=	0.000
4								b_3=	0.000
5									
6									
7		Name	Resign	Degree Centrality	Age	Commute time	Logit (L)	e^L	P (X)	Log-Likelihood (LL)
8	Dave	0	6	50	20	0.000	1.000	0.500	-0.693
9	Faith	0	5	62	20	0.000	1.000	0.500	-0.693
10	Gary	0	5	50	60	0.000	1.000	0.500	-0.693
11	Ann	0	4	34	50	0.000	1.000	0.500	-0.693
12	Ben	1	4	39	40	0.000	1.000	0.500	-0.693
13	Carl	0	3	55	30	0.000	1.000	0.500	-0.693
14	Eddy	0	3	40	30	0.000	1.000	0.500	-0.693
15	Han	0	3	55	25	0.000	1.000	0.500	-0.693
16	Ian	1	2	21	50	0.000	1.000	0.500	-0.693
17	Joe	1	1	34	60	0.000	1.000	0.500	-0.693
18								sum of LL:	-6.931

3) Here, you need to calculate a Logit for each record. Enter the Logit formula below for all the data records, in your Excel spreadsheet:

$$\text{Logit(Resign)} = b_0 + b_1*\text{Degree Centrality} + b_2*\text{Age} + b_3* \text{Commute time}$$

F8   fx   =$I$1+$I$2*C8+$I$3*D8+$I$4*E8

	A	B	C	D	E	F	G	H	I
1								b_0 (Intercept)=	0.000
2								b_1=	0.000
3								b_2=	0.000
4								b_3=	0.000
5									
6									
7	Name	Resign	Degree Centrality	Age	Commute time	Logit (L)			
8	Dave	0	6	50	20	0.000			
9	Faith	0	5	62	20	0.000			
10	Gary	0	5	50	60	0.000			
11	Ann	0	4	34	50	0.000			
12	Ben	1	4	39	40	0.000			
13	Carl	0	3	55	30	0.000			
14	Eddy	0	3	40	30	0.000			
15	Han	0	3	55	25	0.000			
16	Ian	1	2	21	50	0.000			
17	Joe	1	1	34	60	0.000			

4) Here, you need to calculate e^L for each record. The number e is the base of the natural logarithm. It is approximately equal to 2.71828163. e^L must be calculated for each record. Enter the Exponential formula below, in your Excel spreadsheet:

G8					f_x	=EXP(F8)		

	A	B	C	D	E	F	G	H	I
1								b_0 (Intercept)=	0.000
2								$b_1=$	0.000
3								$b_2=$	0.000
4								$b_3=$	0.000
5									
6									
7	Name	Resign	Degree Centrality	Age	Commute time	Logit (L)	e^L		
8	Dave	0	6	50	20	0.000	1.000		
9	Faith	0	5	62	20	0.000	1.000		
10	Gary	0	5	50	60	0.000	1.000		
11	Ann	0	4	34	50	0.000	1.000		
12	Ben	1	4	39	40	0.000	1.000		
13	Carl	0	3	55	30	0.000	1.000		
14	Eddy	0	3	40	30	0.000	1.000		
15	Han	0	3	55	25	0.000	1.000		
16	Ian	1	2	21	50	0.000	1.000		
17	Joe	1	1	34	60	0.000	1.000		

5) Here, you need to calculate P(X) for each record. P(X) is the probability of event X occurring. Enter the formula for probability of the event (i.e. Resign) in your Excel spreadsheet, using the formula below:

$$P(X) = e^L / (1 + e^L)$$

H8   fx =G8/(1+G8)

	A	B	C	D	E	F	G	H	I
1								b_0 (Intercept)=	0.000
2								b_1=	0.000
3								b_2=	0.000
4								b_3=	0.000
5									
6									
7	Name	Resign	Degree Centrality	Age	Commute time	Logit (L)	e^L	P (X)	
8	Dave	0	6	50	20	0.000	1.000	0.500	
9	Faith	0	5	62	20	0.000	1.000	0.500	
10	Gary	0	5	50	60	0.000	1.000	0.500	
11	Ann	0	4	34	50	0.000	1.000	0.500	
12	Ben	1	4	39	40	0.000	1.000	0.500	
13	Carl	0	3	55	30	0.000	1.000	0.500	
14	Eddy	0	3	40	30	0.000	1.000	0.500	
15	Han	0	3	55	25	0.000	1.000	0.500	
16	Ian	1	2	21	50	0.000	1.000	0.500	
17	Joe	1	1	34	60	0.000	1.000	0.500	

6) Here, you need to calculate LL, the Log-Likelihood Function. The log-likelihood function computes a probability based on the input variables values. Enter the log-likelihood formula below, in your Excel spreadsheet:

I8 | f_x =B8*LN(H8)+(1-B8)*(LN(1-H8))

	A	B	C	D	E	F	G	H	I
1							b_0 (Intercept)=		0.000
2							b_1=		0.000
3							b_2=		0.000
4							b_3=		0.000
5									
6									
7	Name	Resign	Degree Centrality	Age	Commute time	Logit (L)	e^L	P (X)	Log-Likelihood (LL)
8	Dave	0	6	50	20	0.000	1.000	0.500	-0.693
9	Faith	0	5	62	20	0.000	1.000	0.500	-0.693
10	Gary	0	5	50	60	0.000	1.000	0.500	-0.693
11	Ann	0	4	34	50	0.000	1.000	0.500	-0.693
12	Ben	1	4	39	40	0.000	1.000	0.500	-0.693
13	Carl	0	3	55	30	0.000	1.000	0.500	-0.693
14	Eddy	0	3	40	30	0.000	1.000	0.500	-0.693
15	Han	0	3	55	25	0.000	1.000	0.500	-0.693
16	Ian	1	2	21	50	0.000	1.000	0.500	-0.693
17	Joe	1	1	34	60	0.000	1.000	0.500	-0.693
18								sum of LL:	-6.931

7) The sum that we wish to maximize is the total of log-likelihood (LL):

	A	B	C	D	E	F	G	H	I
1								b_0 (Intercept)=	0.000
2								$b_1=$	0.000
3								$b_2=$	0.000
4								$b_3=$	0.000
5									
6									
7		Name	Resign	Degree Centrality	Age	Commute time	Logit (L)	e^L	P (X)	Log-Likelihood (LL)
8	Dave	0	6	50	20	0.000	1.000	0.500	-0.693
9	Faith	0	5	62	20	0.000	1.000	0.500	-0.693
10	Gary	0	5	50	60	0.000	1.000	0.500	-0.693
11	Ann	0	4	34	50	0.000	1.000	0.500	-0.693
12	Ben	1	4	39	40	0.000	1.000	0.500	-0.693
13	Carl	0	3	55	30	0.000	1.000	0.500	-0.693
14	Eddy	0	3	40	30	0.000	1.000	0.500	-0.693
15	Han	0	3	55	25	0.000	1.000	0.500	-0.693
16	Ian	1	2	21	50	0.000	1.000	0.500	-0.693
17	Joe	1	1	34	60	0.000	1.000	0.500	-0.693
18								sum of LL:	-6.931

Cell I18 formula: =SUM(I8:I17)

The objective of Logistic Regression is find the coefficients of the Logit (b_0, b_1, b_2 + ...+ b_k) that maximize LL, the Log-Likelihood Function in cell I18, to produce the Maximum Log-Likelihood (MLL) Function. The only values we can change are the guesses for the coefficient b_0 through b_3, which we have assigned an arbitrary value of 0.000. We don't have to optimize them ourselves, as we can use Solver, an Excel add-in, that adjusts the coefficient to maximize or minimize the value in the cell.

8) The Excel Solver is an add-in that is included in Excel. But it must be manually activated by you before it can be utilized for the first time. To use Solver, on the File tab, click Options.

9) Under Add-ins, click Excel Add-ins and click the "Go" button.

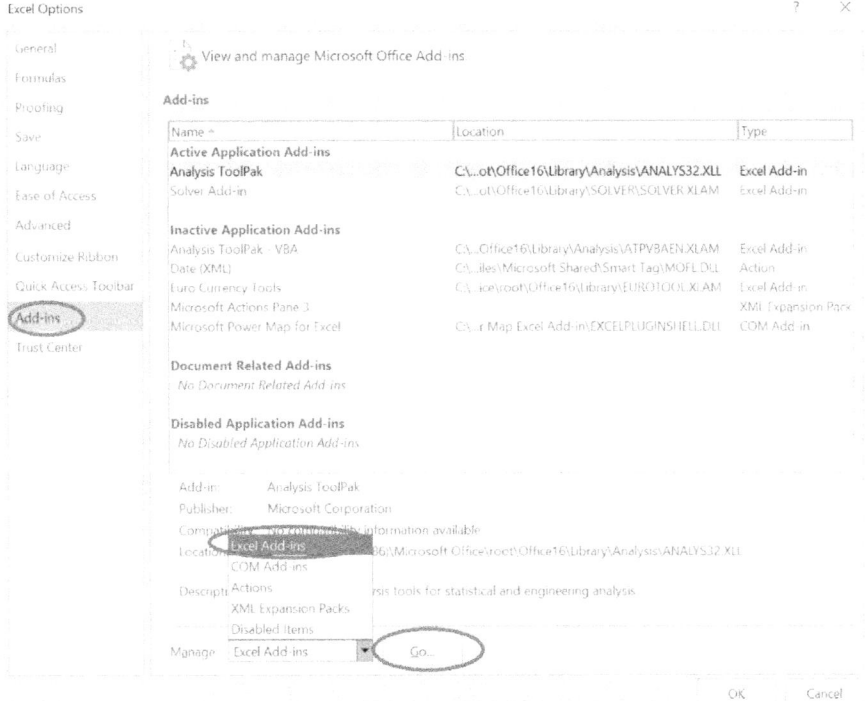

10) Click "Solver Add-In" and click on OK.

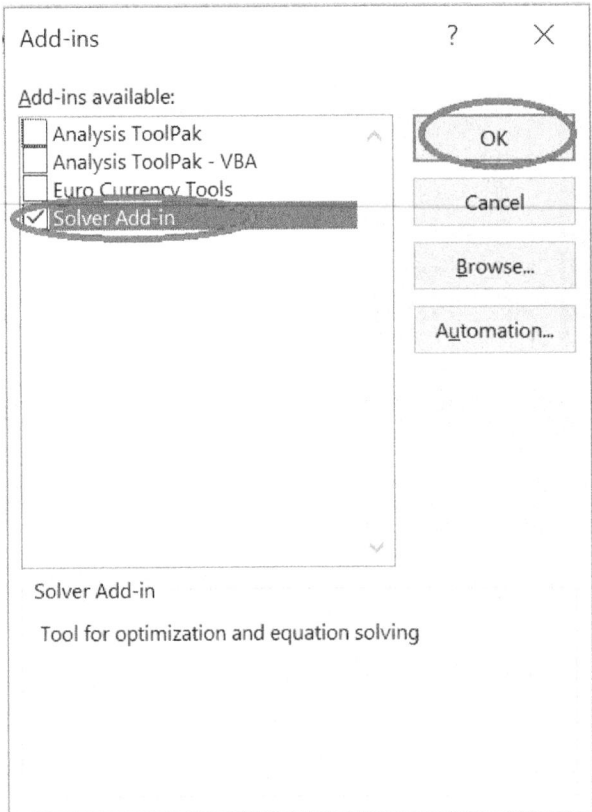

11) On the Data tab, in the Analysis group, you are now able to click on "Solver". Click the Solver button.

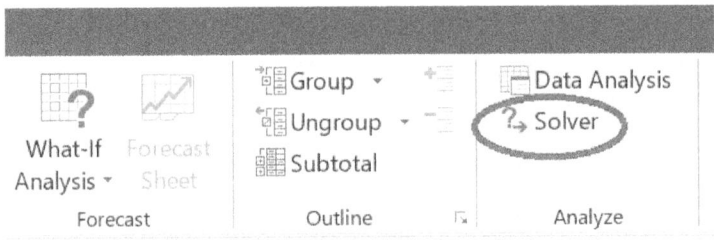

12) The objective is to maximize the sum of the log-likelihood column (LL), by changing the values in I1:I4, representing coefficients b_0-b_3.
- For "Set Objective", select cell (I18), the sum of the log-likelihood column (LL).
- Uncheck the box labeled "Make Unconstrained Variables Non-Negative".
- For "Select a Solving Method", select "GRG Nonlinear" because we are not performing a linear optimization.
- Click "Solve"

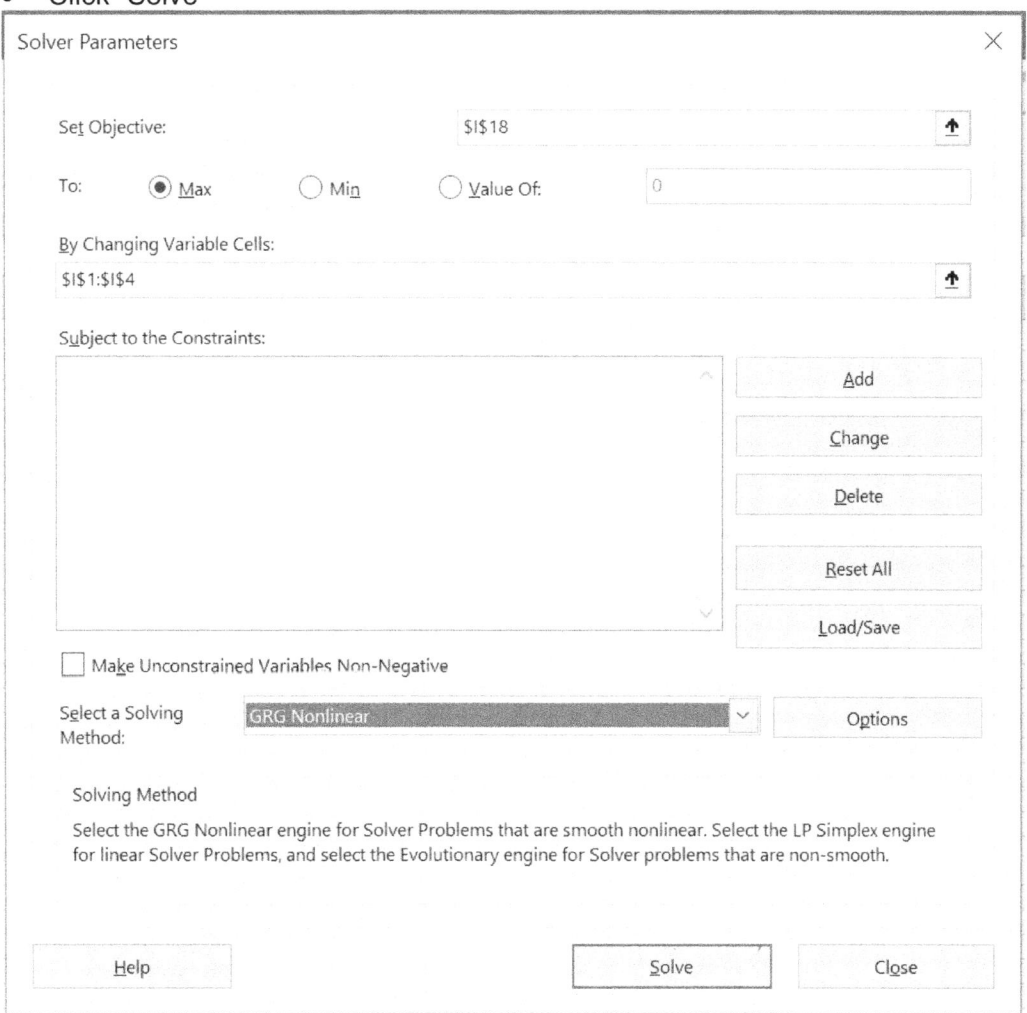

13) After clicking "Solve", you will see the screen "Solver Results". Check "Keep Solver Solution".

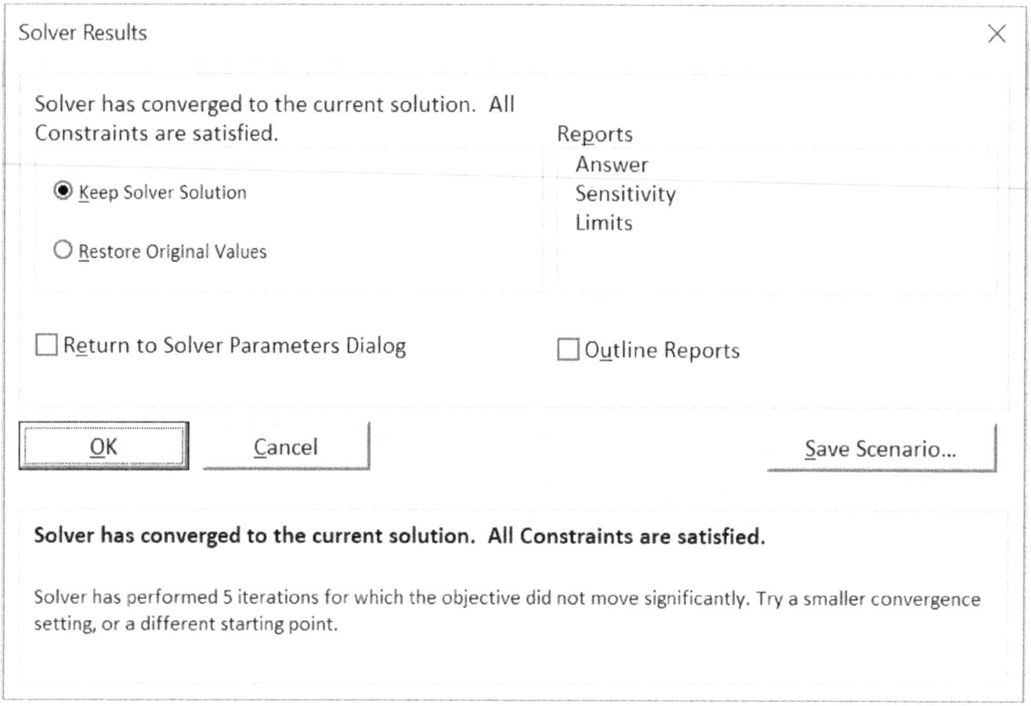

Solver Results

Solver has converged to the current solution. All Constraints are satisfied.

Reports
Answer
Sensitivity
Limits

◉ Keep Solver Solution

○ Restore Original Values

☐ Return to Solver Parameters Dialog   ☐ Outline Reports

OK   Cancel   Save Scenario...

Solver has converged to the current solution. All Constraints are satisfied.

Solver has performed 5 iterations for which the objective did not move significantly. Try a smaller convergence setting, or a different starting point.

14) After clicking "Solve" you get new values in I1:I4, representing coefficients b_0-b_3.

	A	B	C	D	E	F	G	H	I
1								b_0 (Intercept)=	7.578
2								b_1=	-0.731
3								b_2=	-0.177
4								b_3=	0.026
5									
6									
7	Name	Resign	Degree Centrality	Age	Commute time	Logit (L)	e^L	P (X)	Log-Likelihood (LL)
8	Dave	0	6	50	20	-5.120	0.006	0.006	-0.006
9	Faith	0	5	62	20	-6.510	0.001	0.001	-0.001
10	Gary	0	5	50	60	-3.340	0.035	0.034	-0.035
11	Ann	0	4	34	50	-0.044	0.957	0.489	-0.672
12	Ben	1	4	39	40	-1.190	0.304	0.233	-1.455
13	Carl	0	3	55	30	-3.549	0.029	0.028	-0.028
14	Eddy	0	3	40	30	-0.898	0.407	0.289	-0.342
15	Han	0	3	55	25	-3.680	0.025	0.025	-0.025
16	Ian	1	2	21	50	3.716	41.087	0.976	-0.024
17	Joe	1	1	34	60	2.411	11.146	0.918	-0.086
18								sum of LL:	-2.674

15) To predict whether a Staff whose Degree Centrality is 2, Age is 30, and Commute time is 30, will resign, copy those figures that are circled in your spreadsheet.

F21 fx =$I$1+$I$2*C21+$I$3*D21+$I$4*E21

	A	B	C	D	E	F	G	H	I	J
1								b_0 (Intercept)=	7.578	
2								b_1=	-0.731	
3								b_2=	-0.177	
4								b_3=	0.026	
5										
6										
7	Name	Resign	Degree Centrality	Age	Commute time	Logit (L)	e^L	P (X)	Log-Likelihood (LL)	
8	Dave	0	6	50	20	-5.120	0.006	0.006	-0.006	
9	Faith	0	5	62	20	-6.510	0.001	0.001	-0.001	
10	Gary	0	5	50	60	-3.340	0.035	0.034	-0.035	
11	Ann	0	4	34	50	-0.044	0.957	0.489	-0.672	
12	Ben	1	4	39	40	-1.190	0.304	0.233	-1.455	
13	Carl	0	3	55	30	-3.549	0.029	0.028	-0.028	
14	Eddy	0	3	40	30	-0.898	0.407	0.289	-0.342	
15	Han	0	3	55	25	-3.680	0.025	0.025	-0.025	
16	Ian	1	2	21	50	3.716	41.087	0.976	-0.024	
17	Joe	1	1	34	60	2.411	11.146	0.918	-0.086	
18								sum of LL:	-2.674	
19										
20			Degree Centrality	Age	Commute time	Logit (L)				
21			2	30	30	1.600				
22										

You will notice that the formula is cell E18 is:

Logit(L)
=$I$1+$I$2*C21+$I$3*D21+$I$4*E21
= b_0 + b_1*Degree Centrality + b_2*Age + b_3*Commute time
= 7.578 – 0.731*Degree Centrality – 0.177*Age + 0.026*Commute time

Just like a multiple linear regression, you need to enter the new b_0, b_1, b_2 and b_3 coefficient values into your logistic regression equation to predict a value. But unlike a linear regression that predicts values like sales amount, the logistic regression equation predicts probabilities.

16) Next, you need to calculate e^L. The number e is the base of the natural logarithm. It is approximately equal to 2.71828163. Enter the Exponential formula below, in your Excel spreadsheet:

G21   fx   =EXP(F21)

	A	B	C	D	E	F	G	H	I
1								b_0 (Intercept)=	7.578
2								b_1=	-0.731
3								b_2=	-0.177
4								b_3=	0.026
5									
6									
7	Name	Resign	Degree Centrality	Age	Commute time	Logit (L)	e^L	P (X)	Log-Likelihood (LL)
8	Dave	0	6	50	20	-5.120	0.006	0.006	-0.006
9	Faith	0	5	62	20	-6.510	0.001	0.001	-0.001
10	Gary	0	5	50	60	-3.340	0.035	0.034	-0.035
11	Ann	0	4	34	50	-0.044	0.957	0.489	-0.672
12	Ben	1	4	39	40	-1.190	0.304	0.233	-1.455
13	Carl	0	3	55	30	-3.549	0.029	0.028	-0.028
14	Eddy	0	3	40	30	-0.898	0.407	0.289	-0.342
15	Han	0	3	55	25	-3.680	0.025	0.025	-0.025
16	Ian	1	2	21	50	3.716	41.087	0.976	-0.024
17	Joe	1	1	34	60	2.411	11.146	0.918	-0.086
18								sum of LL:	-2.674
19									
20			Degree Centrality	Age	Commute time	Logit (L)	e^L		
21			2	30	30	1.600	4.955		

17) Calculate P(X). P(X) is the probability of event X occurring. Enter the formula for probability of the event (i.e. Buy) in your spreadsheet, using this formula:

$$P(X) = e^L / (1 + e^L)$$

H21   fx   =G21/(1+G21)

	A	B	C	D	E	F	G	H	I
1								b_0 (Intercept)=	7.578
2								b_1=	-0.731
3								b_2=	-0.177
4								b_3=	0.026
5									
6									
7	Name	Resign	Degree Centrality	Age	Commute time	Logit (L)	e^L	P (X)	Log-Likelihood (LL)
8	Dave	0	6	50	20	-5.120	0.006	0.006	-0.006
9	Faith	0	5	62	20	-6.510	0.001	0.001	-0.001
10	Gary	0	5	50	60	-3.340	0.035	0.034	-0.035
11	Ann	0	4	34	50	-0.044	0.957	0.489	-0.672
12	Ben	1	4	39	40	-1.190	0.304	0.233	-1.455
13	Carl	0	3	55	30	-3.549	0.029	0.028	-0.028
14	Eddy	0	3	40	30	-0.898	0.407	0.289	-0.342
15	Han	0	3	55	25	-3.680	0.025	0.025	-0.025
16	Ian	1	2	21	50	3.716	41.087	0.976	-0.024
17	Joe	1	1	34	60	2.411	11.146	0.918	-0.086
18								sum of LL:	-2.674
19									
20			Degree Centrality	Age	Commute time	Logit (L)	e^L	P (X)	
21			2	30	30	1.600	4.955	0.832	

Thus, if a Staff's Degree Centrality is 2, Age is 30, and Commute time is 30, his probability of resigning is 0.832, which is closer to 1 than to 0. Closer to 1 means probably "Resign", while closer to 0 means probably "Stay".

References:
(1)   Georgetown University (2016) Employees' Internal Social Networks Can Predict Turnover https://msb.georgetown.edu/newsroom/news/employees'-internal-social-networks-can-predict-turnover# (12 June 2019)
(2) Caroline Zaayer Kaufman (2018) Have you had a 'normal' number of jobs? https://www.monster.com/career-advice/article/have-had-normal-number-jobs (22 November 2018)
(3) Dr. John Sullivan (2015) How Commute Issues Can Dramatically Impact Employee Retention https://www.tlnt.com/how-commute-issues-can-dramatically-impact-employee-retention/   (21 November 2018)

17.10) Multiple Regression Example: Predict Employee's Sales based on their Organizational Network

VoloMetrix found that the size and amount of a "salesperson's network inside their company" is a more important leading indicator of sales, than the "time salespeople spend with customers". Their research discovered that regardless of what you are sell, who you are sell it to, or your work location anywhere in the world, success in sales correlates highly with three actionable metrics. Merely increasing the time your underperforming salesperson spend with customers will not increase sales much. The second and third factor are more important. Corporate buyers want a salesperson who is credible, who understands their needs, and who is able to address their concerns quickly and competently. Doing this well requires a salesperson is able to get access to management when needed, and have a comprehensive understanding of what their company can offer to the buyer above and beyond the current sales. [1]

- Spending sufficient time with customers.
- Having a big network in your own organization.
- Spending time with your manager and other senior people in your organization.

The data below represents events that have already happened. **We want to use the Salesperson's Organizational Network (Degree Centrality) and Time with Customers (hours) to predict a Salesperson's Sales** using Excel Multiple Regression. "Degree Centrality" refers to how many people can this person reach directly in his organizational network? In this example, Dave has a Degree of 6 because he is directly connected to 6 other individuals.

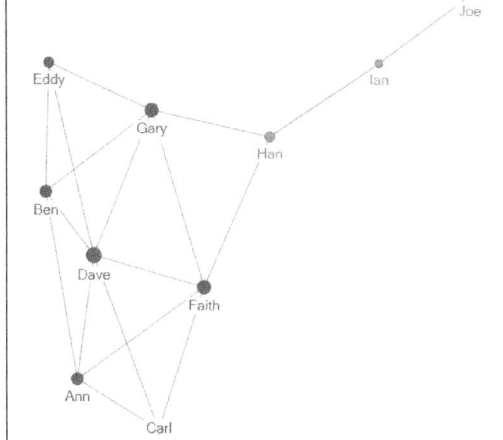

Salesperson	Degree Centrality	Time with Customers (hours)	Sales
Dave	6	200	1900
Faith	5	160	1850
Gary	5	170	1800
Ann	4	160	1750
Ben	4	150	1700
Carl	3	140	1680
Eddy	3	130	1650
Han	3	120	1600
Ian	2	90	1500
Joe	1	50	1400

1) Install "Analysis ToolPak", an Excel add-in

"Analysis ToolPak" is an add-in for Microsoft Excel that comes with Microsoft Excel. To be able to run regression using Excel, you need to first install "Analysis ToolPak", an Excel add-in program that provides data analysis tools. To load the Analysis ToolPak add-in, follow these steps:

On the File tab, click Options.

Under Add-ins, click Analysis ToolPak and click the "Go" button.

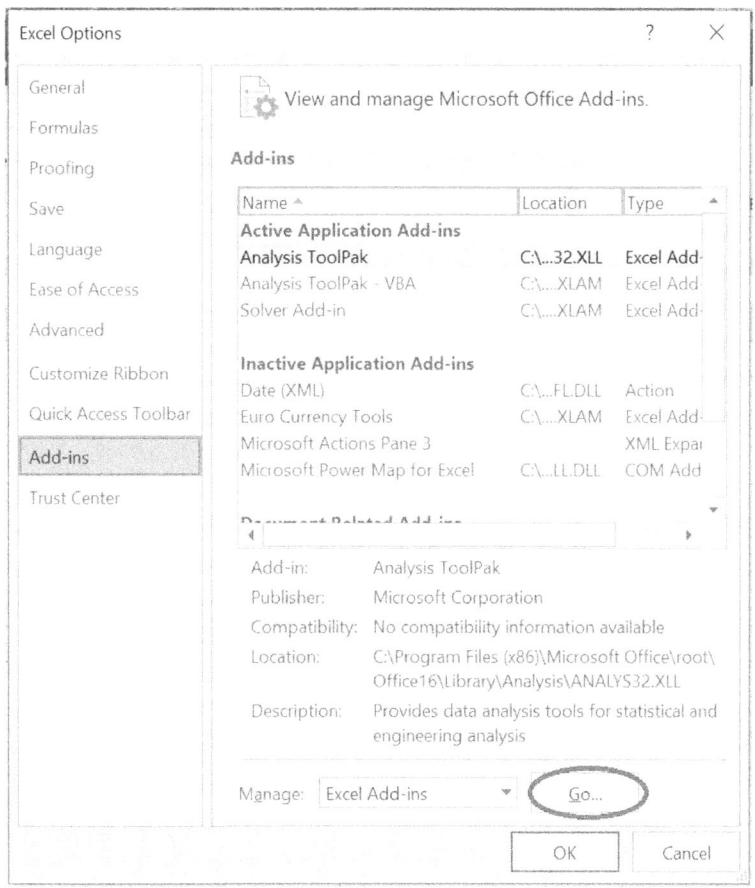

Click "Analysis ToolPak" and click on OK.

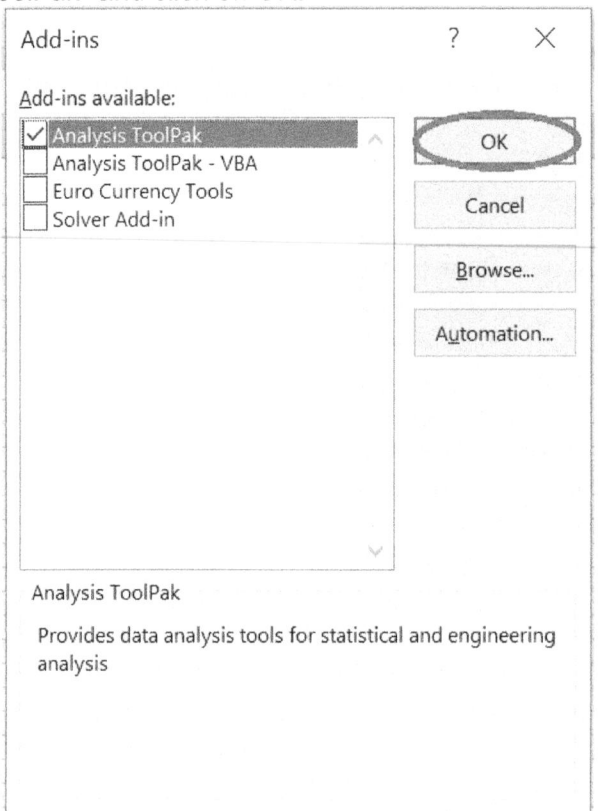

On the Data tab, in the Analysis group, you are now able to click on "Data Analysis".

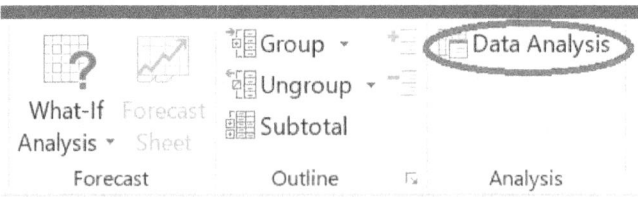

2) Copy the example data in the following table, and paste it in cell A1 of a new Excel worksheet.

	A	B	C	D
1	Salesperson	Degree Centrality	Time with Customers (hours)	Sales
2	Dave	6	200	1900
3	Faith	5	160	1850
4	Gary	5	170	1800
5	Ann	4	160	1750
6	Ben	4	150	1700
7	Carl	3	140	1680
8	Eddy	3	130	1650
9	Han	3	120	1600
10	Ian	2	90	1500
11	Joe	1	50	1400

3) On the Data tab, in the Analysis group, click on "Data Analysis".

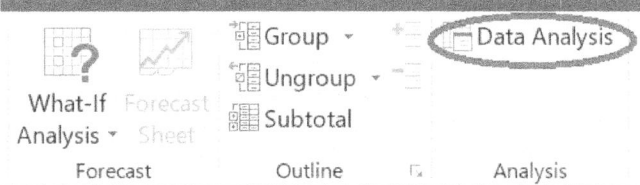

4) Select "Regression" and click "OK".

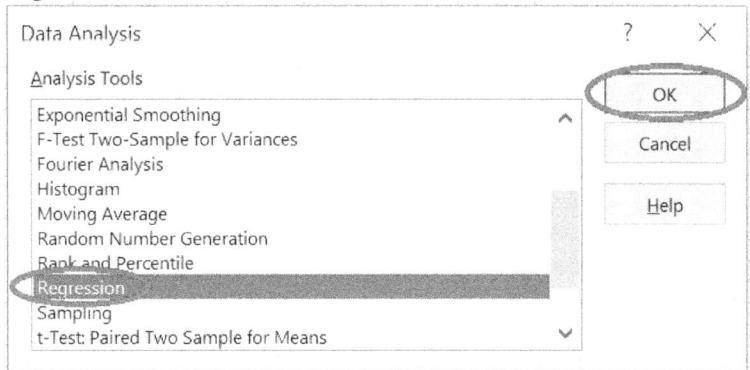

5) After you click OK in the "Data Analysis" dialog box, you will see a "Regression" dialog box.
6) For "Input Y Range", select cells (D1:D11). This is the predictor variable or dependent variable.
7) For "Input X Range", select cells (B1:C11). These are the explanatory variables or independent variables.
8) Check "Labels" box.
9) Click the "Output Range" box, and select cell A13.
10) Click "OK".

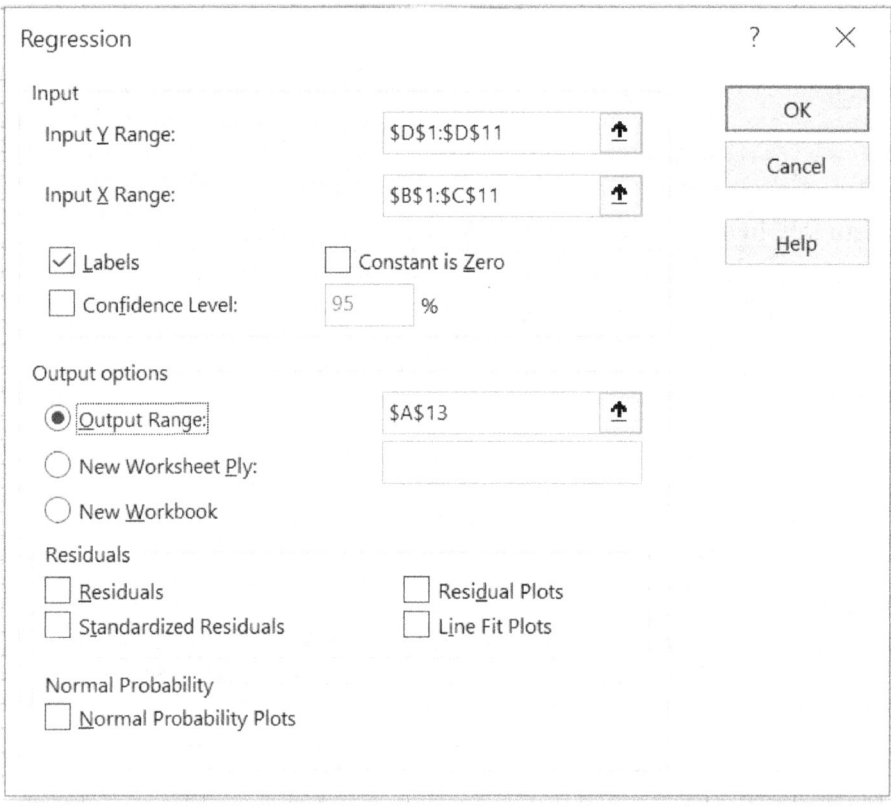

After you click "OK", Excel generates the following Summary Output. Round the numbers to 3 decimal places.

	A	B	C	D	E	F	G	H	I
1	Salesperson	Degree Centrality	Time with Customers (hours)	Sales					
2	Dave	6	200	1900					
3	Faith	5	160	1850					
4	Gary	5	170	1800					
5	Ann	4	160	1750					
6	Ben	4	150	1700					
7	Carl	3	140	1680					
8	Eddy	3	130	1650					
9	Han	3	120	1600					
10	Ian	2	90	1500					
11	Joe	1	50	1400					
12									
13	SUMMARY OUTPUT								
14									
15	Regression Statistics								
16	Multiple R	0.987							
17	R Square	0.974							
18	Adjusted R Square	0.967							
19	Standard Error	28.237							
20	Observations	10							
21									
22	ANOVA								
23		df	SS	MS	F	Significance F			
24	Regression	2	209428.846	104714.423	131.335	0.000			
25	Residual	7	5581.154	797.308					
26	Total	9	215010.000						
27									
28		Coefficients	Standard Error	t Stat	P-value	Lower 95%	Upper 95%	Lower 95.0%	Upper 95.0%
29	Intercept	1258.346	40.226	31.282	0.000	1163.227	1353.465	1163.227	1353.465
30	Degree Centrality	56.192	23.646	2.376	0.049	0.278	112.106	0.278	112.106
31	Time with Customers (hours)	1.623	0.834	1.947	0.093	-0.348	3.595	-0.348	3.595

R Square - In the output, "R Square" is 0.974, which means it is a good fit. 97% of the variation in Sales (Output) is explained by the independent variables (Input), Degree Centrality, and Time with Customers. The closer R Square is to "1", the better the regression line fits the data.

Significance F and P-values - To determine if your results are statistically significant (i.e. reliable), check "Significance F" (0.001). If the value of "Significance F" is less than 0.05, it is statistically significant (i.e. reliable). If "Significance F" is bigger than 0.05, don't use this set of independent variables. Delete those variables with "P-value" that is bigger than 0.05 and run the regression again until "Significance F" drops below 0.05. Most or all of your P-values should be lower than 0.05. In our example below "P-value" is 0.000, 0.049 and 0.093, for Intercept, Degree Centrality and Time with Customers respectively.

Coefficients

From the Summary Output, the regression line is:

SUMMARY OUTPUT

Regression Statistics	
Multiple R	0.987
R Square	0.974
Adjusted R Square	0.967
Standard Error	28.237
Observations	10

ANOVA

	df	SS	MS	F	Significance F
Regression	2	209428.846	104714.423	131.335	0.000
Residual	7	5581.154	797.308		
Total	9	215010.000			

	Coefficients	Standard Error	t Stat	P-value	Lower 95%	Upper 95%	Lower 95.0%	Upper 95.0%
Intercept	1258.346	40.226	31.282	0.000	1163.227	1353.465	1163.227	1353.465
Degree Centrality	56.192	23.646	2.376	0.049	0.278	112.106	0.278	112.106
Time with Customers (hours)	1.623	0.834	1.947	0.093	-0.348	3.595	-0.348	3.595

Y
= Sales
= 1258.346 + 56.192 * Degree Centrality + 1.623 * Time with Customers

Based on the above regression formula,
- For each unit increase in Degree Centrality, Sales increase by 56.192.
- For each unit increase in Time with Customers, Sales increase by 1.623.

Coefficients can also be used for forecasting. For example, if a Salesperson's "Degree Centrality" is 3, "Time with Customers" is 150, then **predicted "Sales"**
= 1258.346 + 56.192 * Degree Centrality + 1.623 * Time with Customers
= 1258.346 + (56.192 * 3) + (1.623 * 150)
= 1670

References:
(1) Ryan Fuller (2015) What Makes Great Salespeople. Harvard Business Review. https://hbr.org/2015/07/what-makes-great-salespeople (13 November 2018)

18) Diversity and Inclusion Analytics

Joshbersin, (2018) Diversity and Inclusion Is A Business Strategy, Not An HR Program https://joshbersin.com/2018/08/diversity-and-inclusion-is-a-business-strategy-not-an-hr-program/ (9 November 2018)

For years, Schneider Electric repatriated French nationals into jobs in various countries. Then, CEO Jean-Pascal Tricoire and CHRO Olivier Blum, realized that this was no longer viable. Today, as Olivier explains, Schneider is a "multi-hub business model," where "We want everyone everywhere in the company to have the same chance of success, irrespective of their nationality or location." Diversity and inclusion are now core to Schneider's strategy, and Schneider grew by 6.2% in Q1 2018. [1]

"Diversity" and "Inclusion" are different:
- **Diversity** is the who: who are in that department, who are recruited, who are promoted, who we are we monitoring. Workplace diversity is about hiring and retaining employees of different characteristics and walks of life in terms of: ethnicity, age, gender, education, socioeconomic background, sexual orientation, and religious beliefs.
- **Inclusion**, is the how. Inclusion means how does the company actually doing something to include different people in their team, or organization. Inclusion are the behaviors that welcome and embrace diversity. Inclusion goes beyond recruiting employees with different backgrounds, gender, ethnicities, etc. It is a mindset in where companies actively provide each employee with equal access to career opportunities in the company.

Research have shown that there is a relationship between "Diversity and Inclusion" and factors such as: Market Share, Revenue, Earnings Before Interest and Taxes (EBIT), and Absenteeism.

Diversity & Inclusion impact on Absenteeism
- Deloitte found that diversity and inclusion correlates with actual absenteeism. Data from one organization showed that if 10 percent more employees feel included, the company will increase work attendance by about one day per year per employee. [2]

Diversity & Inclusion impact on Earnings Before Interest and Taxes (EBIT)
McKinsey's report, Diversity Matters, studied 366 public companies across various industries in Canada, Latin America, United Kingdom, and United States. The research found that [3]:
- Companies in the top quartile for racial and ethnic diversity are 35 percent more likely to have financial returns above their national industry medians.
- Companies in the top quartile for gender diversity are 15 percent more likely to have financial returns above their respective national industry medians.
- In the United States, there is a linear relationship between racial and ethnic diversity and better financial performance: for every 10 percent increase in racial and ethnic diversity on the senior-executive team, earnings before interest and taxes (EBIT) rise 0.8 percent.
- In the United Kingdom, greater gender diversity on the senior-executive team corresponded to the highest performance uplift in our data set: for every 10 percent increase in gender diversity, EBIT rose by 3.5 percent.

Diversity & Inclusion impact on Market Share
- A study by the Harvard Business Review (HBR) found that a team with a member who shares a client's ethnicity is 152 percent more likely to understand that client than another team. In this research, two kinds of diversity were scrutinized: inherent and acquired. Inherent diversity involves traits you are born with such as gender, ethnicity, and sexual orientation. Acquired diversity involves traits you gain from experience: Working in another country can help you appreciate cultural differences, while selling to female consumers can give you gender smarts. HBR found that companies whose leaders exhibit at least three inherent and three acquired diversity traits, out-innovate and out-perform others. Employees at these companies are 45 percent likelier to report that their firm's market share grew over the previous year and 70 percent likelier to report that the firm captured a new

market. Managers tend to have their own personal biases and preferences that may impact their judgment when recruiting employees. To increase the Sales Team Diversity in your company, having a wide range of team members involved in the recruiting process can help minimize against individual biases holding too much weight. [4]

Diversity & Inclusion impact on Sales
- According to research published in the American Sociological Review, workplace diversity is one of the most important predictors of a business' sales revenue, customer numbers and profitability. In the research, Cedric Herring found that companies with the highest racial diversity has 15 times more sales revenue on average than those with the lowest racial diversity. **For every percentage increase in the rate of racial or gender diversity, there was an increase in sales revenues of 9 and 3 percent, respectively.** Companies with a more diverse workforce reported higher customer numbers than those organizations with less diversity. The difference is even larger for gender diversity rates. [5]
- For Cristian Renella, CEO of elMejorTrato.com, hiring for diversity has led to creating new products. "All of us were under 35 when we founded the company nine years ago and our demographics hadn't changed," Rennella says. So, they hired team members aged 55 and older. The shift brought a key insight. "We realized that our service, with some modifications, could also be useful for a market segment older than 50 years, something that we hadn't realized before because nobody on our team was in that segment," Rennella says. Sales jumped 24.4 percent the following year. "Now we are focusing on [hiring] mothers. No woman who works in our company is a mother and we have realized [this segment may] be very important for us" Rennella says. The lesson? Hire the customer you want to attract and let them teach you how they'd like to be sold to. [6]

References:
(1) Joshbersin, (2018) Diversity and Inclusion Is A Business Strategy, Not An HR Program https://joshbersin.com/2018/08/diversity-and-inclusion-is-a-business-strategy-not-an-hr-program/ (9 November 2018)
(2) Deloitte Australia (Deloitte) and the Victorian Equal Opportunity and Human Rights Commission (2013) A new recipe to improve business performance, https://www2.deloitte.com/content/dam/Deloitte/au/Documents/human-capital/deloitte-au-hc-diversity-inclusion-soup-0513.pdf
(3) Vivian Hunt, Dennis Layton, and Sara Prince (2015), Why diversity matters, https://www.mckinsey.com/business-functions/organization/our-insights/why-diversity-matters (24 September 2018)
(4) Sylvia Ann Hewlett, Melinda Marshall, Laura Sherbin, How Diversity Can Drive Innovation, (2013) https://hbr.org/2013/12/how-diversity-can-drive-innovation (24 September 2018)
(5) American Sociological Association (2009), Research links diversity with increased sales revenue and profits, more customers, https://www.eurekalert.org/pub_releases/2009-03/asa-rld033009.php (24 September 2018)
(6) Tim Beyers (2018), Diversity at Work: 3 Best Practices For Small Business Growth, https://www.capitalone.com/small-business/sparkiq/article/diversity-at-work-3-best-practices-for-small-business-growth/ (24 September 2018)

18.1) How to Convert Diversity into an Index

Most companies measure diversity (e.g. ethnicity) using traditional breakdowns (70% Chinese : 20% Korean : 10% Japanese). By monitoring a diversity index as a supplement to traditional diversity breakdowns, it is easier to track shifts in diversity representation across your company, and for hypothesis testing.

The Simpson's Diversity Index (SDI), is a better way to quantify diversity into a trackable metric. Simpson's Diversity Index started out as a tool for measuring species diversity – but here, we will use it to measure workplace diversity.

The formula to compute the Simpson's Diversity Index is:

$$D = 1 - \frac{\Sigma n(n-1)}{N(N-1)}$$

- n = number of individuals of each ethnicity
- N = total number of individuals of all ethnicity

The range is from 0 to 1, where:
- High scores (close to 1) indicate high diversity.
- Low scores (close to 0) indicate low diversity.

The Simpson's Diversity Index captures two elements of diversity (Richness and Evenness):
- Richness refers to the number of different groups represented (e.g. number of ethnic group). The more ethnic group present in a sample, the 'richer' the sample. It does not consider the number of individuals of each ethnic group.
- Evenness refers to the spread across those groups (e.g. are the ethnic group spread evenly). It considers the number of individuals of each ethnic group.

In the Simpson Index, both richness and evenness are important, and are incorporated in a single, clean snapshot. Thus, your company is not diverse if only 2 ethnic group are represented compared with 4 ethnic groups (i.e. low in richness), and if you have 97 members in 1 ethnic group and 1 member in each of 3 other ethnic groups (i.e. low in evenness).

Assuming there are two companies. Company A has 330 Japanese, 400 Koreans and 330 Chinese. Company B has 20 Japanese, 30 Koreans and 950 Chinese. Both companies have the same richness (3 ethnic groups) and the same total number of employees (1000). However, Company A has more evenness than Company B. This is because the total number of employees in Company A is quite evenly distributed between the three ethnic groups. In Company B, most of the employees are Chinese, with only a few Japanese and Koreans. Company B is therefore considered to be less diverse than Company A.

	Company A	Company B
Japanese	330	20
Korean	340	30
Chinese	330	950
Total	1000	1000
Diversity Index	= 1- (((330*329)+(340*339)+(330*329))/(1000*999))	= 1- (((20*19)+(30*29)+(950*949))/(1000*999))
	0.67	0.10

18.2) Multiple Regression Example: Predict Ethnic & Gender Diversity's Impact On EBIT

Research have shown that there is a relationship between "Diversity and Inclusion" and "Earnings Before Interest and Taxes" (EBIT):
- McKinsey's report found that in the United States, there is a linear relationship between ethnic diversity and better financial performance: for every 10 percent increase in ethnic diversity on the senior-executive team, earnings before interest and taxes (EBIT) rise 0.8 percent. [1]
- McKinsey's report found that in the United Kingdom, greater gender diversity on the senior-executive team corresponded to the highest performance uplift in our data set: for every 10 percent increase in gender diversity, EBIT rose by 3.5 percent. [1]

Consider the table below. **In this example, we want to predict the effect of "Advertising expense", "Ethnic Diversity", and "Gender Diversity" on "EBIT".**

	A	B	C	D	E	F	G	H	I	J
1			\multicolumn{4}{c}{Ethnicity Diversity}		\multicolumn{3}{c}{Gender Diversity}	
2	Year	Advertising expense	Chinese	Korean	Japanese	Total	Female	Male	Total	EBIT
3	2019	95	34	33	33	100	50	50	100	1200
4	2018	90	30	30	40	100	45	55	100	1200
5	2017	95	25	30	45	100	40	60	100	1200
6	2016	80	20	25	55	100	35	65	100	1160
7	2015	95	15	25	60	100	30	70	100	1100
8	2014	75	10	25	65	100	25	75	100	1050
9	2013	85	5	20	75	100	20	80	100	1000
10	2012	80	0	20	80	100	10	90	100	950
11	2011	75	0	10	90	100	5	95	100	850
12	2010	75	0	0	100	100	0	100	100	800

The Regression data analysis tool cannot analyze Diversity data in numbers of each Diversity (5 Chinese : 20 Korean : 75 Japanese), or in percentages of each Diversity (5% Chinese : 20% Korean : 75% Japanese). The "Diversity numbers" needs to be converted into a "Diversity Index", using the Simpson's Diversity Index formula.

1) Add a Diversity Index column for Ethnicity and Gender in the table as shown below.

	A	B	C	D	E	F	G	H	I	J	K	L
1				Ethnicity Diversity				Gender Diversity				
2	Year	Advertising expense	Chinese	Korean	Japanese	Total	Ethnicity Diversity index	Female	Male	Total	Gender Diversity index	EBIT
3	2019	95	34	33	33	100	0.7	50	50	100	0.5	1200
4	2018	90	30	30	40	100	0.7	45	55	100	0.5	1200
5	2017	95	25	30	45	100	0.7	40	60	100	0.5	1200
6	2016	80	20	25	55	100	0.6	35	65	100	0.5	1160
7	2015	95	15	25	60	100	0.6	30	70	100	0.4	1100
8	2014	75	10	25	65	100	0.5	25	75	100	0.4	1050
9	2013	85	5	20	75	100	0.4	20	80	100	0.3	1000
10	2012	80	0	20	80	100	0.3	10	90	100	0.2	950
11	2011	75	0	10	90	100	0.2	5	95	100	0.1	850
12	2010	75	0	0	100	100	0.0	0	100	100	0.0	800

The Diversity Index is derived using this formula:

$$D = 1 - \frac{\Sigma n(n-1)}{N(N-1)}$$

As an example, "Ethnicity Diversity Index" for cell G5
= 1-(((C5*(C5-1))+(D5*(D5-1))+(E5*(E5-1)))/(F5*(F5-1)))
= 1-(((25*(25-1))+(30*(30-1))+(45*(45-1)))/(100*(100-1)))
= 0.7

As an example, "Gender Diversity Index" for cell K5
= 1-((H5*(H5-1))+(I5*(I5-1)))/(J5*(J5-1))
= 1-((40*(40-1))+(60*(60-1)))/(100*(100-1))
= 0.5

2) Now that you have the "Ethnicity Diversity Index" and "Gender Diversity Index", delete the columns with the Diversity numbers.

	A	B	C	D	E	F	G	H	I	J	K	L
1	Year	Advertising expense	Ethnicity Diversity				Ethnicity Diversity index	Gender Diversity			Gender Diversity index	EBIT
			Chinese	Korean	Japanese	Total		Female	Male	Total		
3	2019	95	34	33	33	100	0.7	50	50	100	0.5	1200
4	2018	90	30	30	40	100	0.7	45	55	100	0.5	1200
5	2017	95	25	30	45	100	0.7	40	60	100	0.5	1200
6	2016	80	20	25	55	100	0.6	35	65	100	0.5	1160
7	2015	95	15	25	60	100	0.6	30	70	100	0.4	1100
8	2014	75	10	25	65	100	0.5	25	75	100	0.4	1050
9	2013	85	5	20	75	100	0.4	20	80	100	0.3	1000
10	2012	80	0	20	80	100	0.3	10	90	100	0.2	950
11	2011	75	0	10	90	100	0.2	5	95	100	0.1	850
12	2010	75	0	0	100	100	0.0	0	100	100	0.0	800

3) After you delete the Diversity numbers, your table will now look like this:

	A	B	C	D	E
1	Year	Advertising expense	Ethnicity Diversity index	Gender Diversity index	EBIT
2	2019	95	0.7	0.5	1200
3	2018	90	0.7	0.5	1200
4	2017	95	0.7	0.5	1200
5	2016	80	0.6	0.5	1160
6	2015	95	0.6	0.4	1100
7	2014	75	0.5	0.4	1050
8	2013	85	0.4	0.3	1000
9	2012	80	0.3	0.2	950
10	2011	75	0.2	0.1	850
11	2010	75	0.0	0.0	800

4) Install "Analysis ToolPak", an Excel add-in

"Analysis ToolPak" is an add-in for Microsoft Excel that comes with Microsoft Excel. To be able to run regression using Excel, you need to first install "Analysis ToolPak", an Excel add-in program that provides data analysis tools. To load the Analysis ToolPak add-in, follow these steps:

On the File tab, click Options.

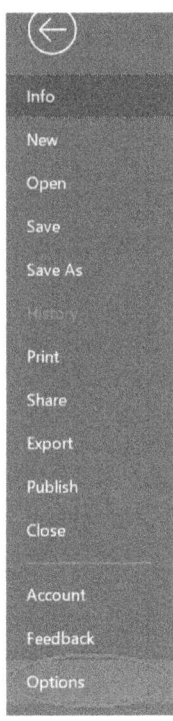

Info

0-Predictive HR Anaytics Excel

E: » Cedric articles » Amazon » HR analytics articles

Convert

Compatibility Mode

Some new features are disabled to prevent problems when working with previous versions of Office. Converting this file will enable these features, but may result in layout changes.

Protect Workbook ▼

Protect Workbook

Control what types of changes people can make to this workbook.

Check for Issues ▼

Inspect Workbook

Before publishing this file, be aware that it contains:
- Document properties, printer path, author's name, related dates and absolute path
- Hidden worksheets
- Macros, forms, or ActiveX controls
- Content that cannot be checked for accessibility issues because of the current file type

Under Add-ins, click Analysis ToolPak and click the "Go" button.

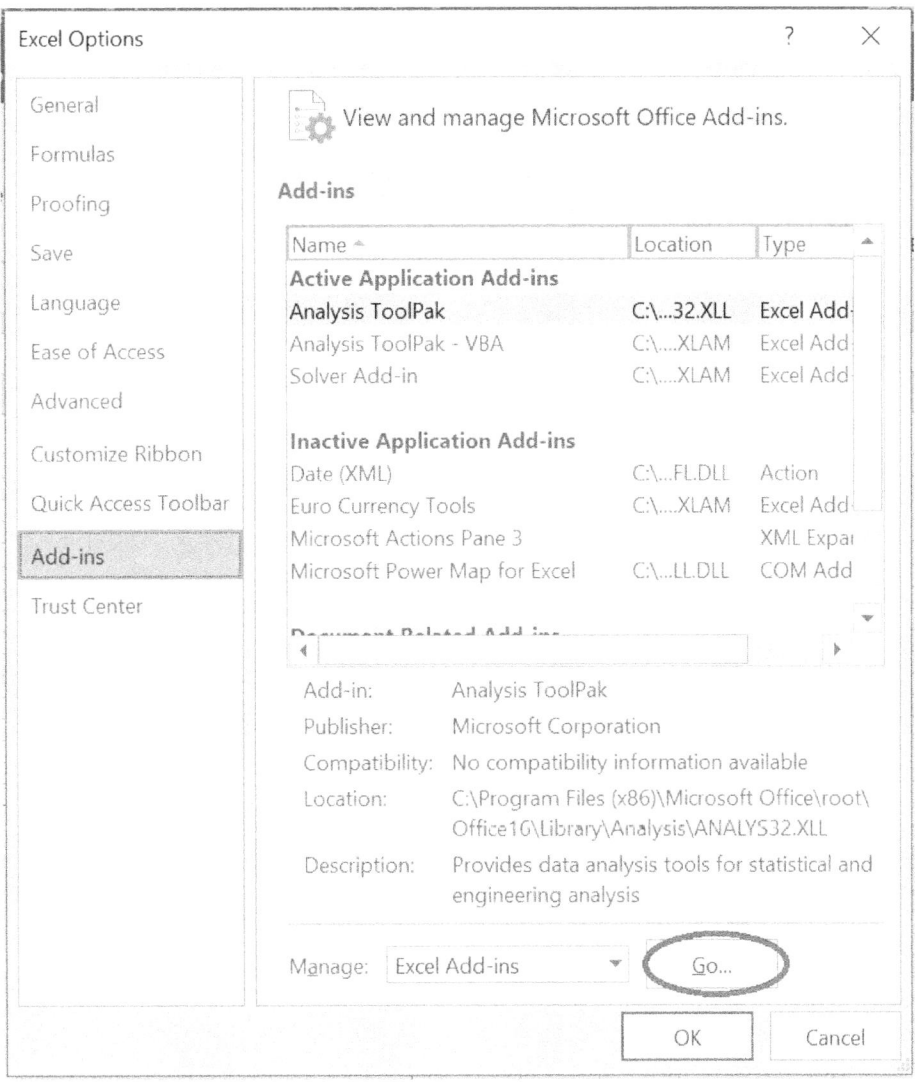

Click "Analysis ToolPak" and click on OK.

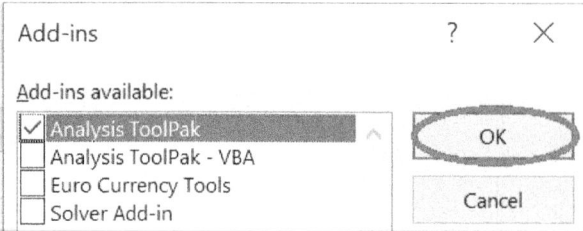

5) On the Data tab, in the Analysis group, you are now able to click on "Data Analysis".

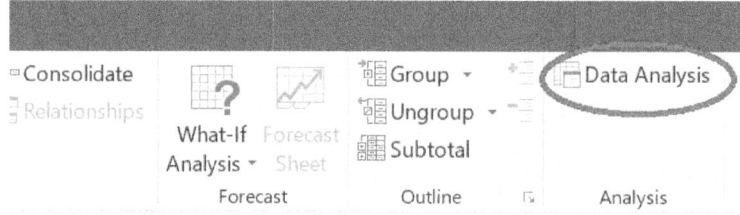

6) Select "Regression" and click "OK".

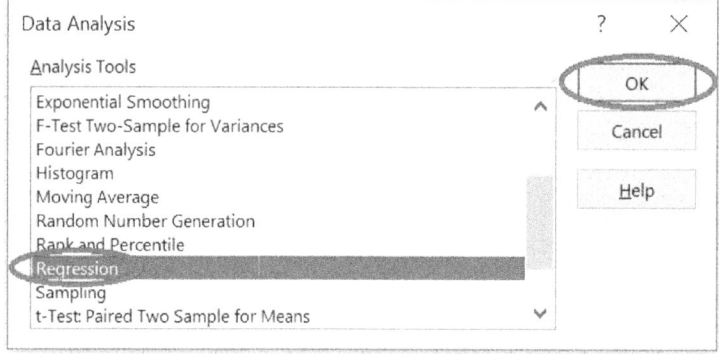

7) After you click OK in the "Data Analysis" dialog box, you will see a "Regression" dialog box.
8) For "Input Y Range", select cells (E1:E11). This is the predictor variable or dependent variable.
9) For "Input X Range", select cells (B1:D11). These are the explanatory variables or independent variables.
10) Check "Labels" box.
11) Click the "Output Range" box, and select cell A13.
12) Click "OK".

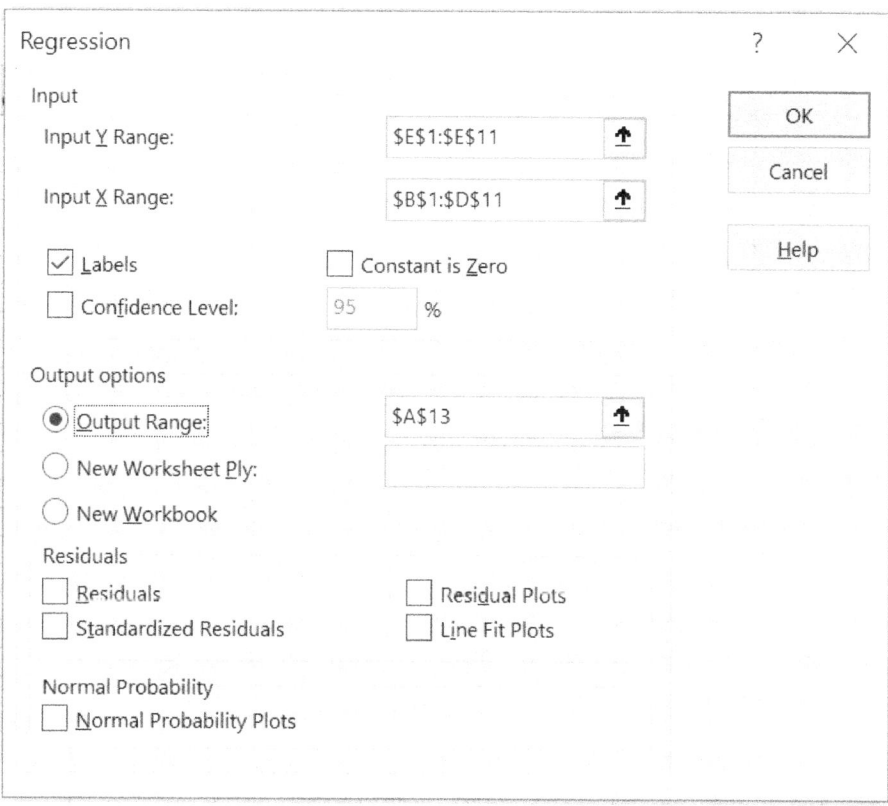

After you click "OK", Excel generates the following Summary Output. Round the numbers to 3 decimal places.

SUMMARY OUTPUT

Regression Statistics	
Multiple R	0.989
R Square	0.978
Adjusted R Square	0.967
Standard Error	27.055
Observations	10

ANOVA

	df	SS	MS	F	Significance F
Regression	3	192698.046	64232.682	87.750	0.000
Residual	6	4391.954	731.992		
Total	9	197090			

	Coefficients	Standard Error	t Stat	P-value	Lower 95%	Upper 95%	Lower 95.0%	Upper 95.0%
Intercept	679.151	113.802	5.968	0.001	400.688	957.613	400.688	957.613
Advertising expense	1.226	1.535	0.799	0.455	-2.529	4.982	-2.529	4.982
Ethnicity Diversity index	345.296	312.461	1.105	0.311	-419.270	1109.861	-419.270	1109.861
Gender Diversity index	329.534	395.650	0.833	0.437	-638.587	1297.656	-638.587	1297.656

R Square: In the output, R Square is 0.978, which means it is a good fit. 97% of the variation in EBIT (Output) is explained by the independent variables (Input), "Advertising expense", "Ethnicity Diversity Index" and "Gender Diversity Index". The closer R Square is to "1", the better the regression line fits the data.

Significance F and P-values: To determine if your results are statistically significant (i.e. reliable), check "Significance F" (0.001). If the value of "Significance F" is less than 0.05, it is statistically significant (i.e. reliable). If "Significance F" is bigger than 0.05, don't use this set of independent variables. Delete those variables with "P-value" that is bigger than 0.05 and run the regression again until "Significance F" drops below 0.05. Most or all of your P-values should be lower than 0.05. In our example below "P-value" is 679.151, 1.226, 345.296 and 329.534, for Intercept, "Advertising expense", "Ethnicity Diversity Index" and "Gender Diversity Index" respectively.

Coefficients

From the Summary Output, the regression line is:
EBIT (Output)
= 679.151 + (1.226 * Advertising expense) + (345.296 * Ethnicity Diversity Index) + (329.534 * Gender Diversity Index)

Based on the above regression formula,
- For each unit increase in Advertising expense, EBIT increase by 1.226.
- For each unit increase in Ethnicity Diversity Index, EBIT increase by 345.296.
- For each unit increase in Gender Diversity Index, EBIT increase by 329.534.

Coefficients can also be used for forecasting. For example, if "Advertising expense" is 80, "Ethnicity Diversity Index" is 2.8, "Gender Diversity Index" is 1.8, then **predicted "EBIT"**
= 679.151 + (1.226 * Advertising expense) + (345.296 * Ethnicity Diversity Index) + (329.534 * Gender Diversity Index)
= 679.151 + (1.226 * 80) + (345.296 * 2.8) + (329.534 * 1.8)
= $2,337

References:
(1) Vivian Hunt, Dennis Layton, and Sara Prince (2015), Why diversity matters, https://www.mckinsey.com/business-functions/organization/our-insights/why-diversity-matters (24 September 2018)

18.3) Multiple Regression Example: Predict an employee's performance rating based on their "Social Network Diversity Index", "Social Network Size" & "Skillsets"

According to Greg Newman, other than Human Capital (skillsets that enables employees to perform), Social Capital (relationships employees build to help them get work done) has a big influence on performance. [1]

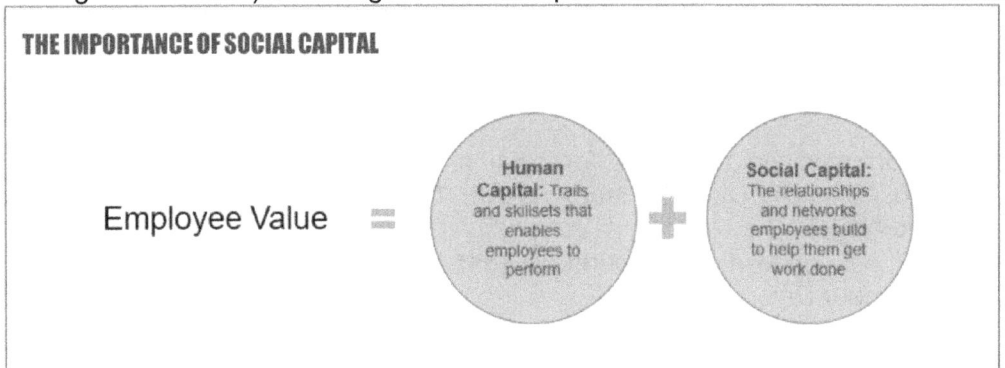

Source: Greg Newman (2017) Retain or let go? The data that you need to react correctly when an employee resigns.. https://www.linkedin.com/pulse/retain-release-data-you-need-react-correctly-when-employee-newman/ (12 June 2019)

In this example, **we want to predict the effect of an employee's "Social Network Diversity", "Social Network Size", and "Skillsets Score" on their "Performance Rating"**. The data below are events that already happened.

	A	B	C	D	E	F	G	H
1		Ethnicity Diversity						
2	Staff	Production friends	Sales friends	Finance friends	IT friends	Social Network Size	Skillsets Score	Performance Rating
3	Ann	8	6	6	7	27	9.0	9.0
4	Andy	6	6	6	6	24	9.0	9.0
5	Ben	10	5	4	3	22	8.0	8.0
6	Bond	12	3	3	2	20	7.5	7.0
7	Carl	12	3	2	1	18	7.0	6.0
8	Candy	13	2	1	1	17	6.0	5.0
9	Dave	14	2	1	0	17	5.0	4.0
10	Eddy	14	2	0	0	16	4.0	3.0
11	Faith	14	1	0	0	15	3.0	2.0
12	Gary	13	0	0	0	13	2.0	1.0

The Regression data analysis tool cannot analyze Diversity data in numbers of each Diversity (6 Production department friends: 5 Sales department friends: 4 Finance department friends), or in percentages of each Diversity (40% Production department friends : 33% Sales department friends: 27% Finance department friends). The "Diversity numbers" needs to be converted into a "Diversity Index", using the Simpson's Diversity Index formula.

1) Add a "Social Network Diversity Index" column in the table as shown below.

G3   fx =1-(((B3*(B3-1))+(C3*(C3-1))+(D3*(D3-1))+(E3*(E3-1)))/(F3*(F3-1)))

Staff	Ethnicity Diversity				Social Network Size	Social Network Diversity Index	Skillsets Score	Performance Rating
	Production friends	Sales friends	Finance friends	IT friends				
Ann	8	6	6	7	27	0.8	9.0	9.0
Andy	6	6	6	6	24	0.8	9.0	9.0
Ben	10	5	4	3	22	0.7	8.0	8.0
Bond	12	3	3	2	20	0.6	7.5	7.0
Carl	12	3	2	1	18	0.5	7.0	6.0
Candy	13	2	1	1	17	0.4	6.0	5.0
Dave	14	2	1	0	17	0.3	5.0	4.0
Eddy	14	2	0	0	16	0.2	4.0	3.0
Faith	14	1	0	0	15	0.1	3.0	2.0
Gary	13	0	0	0	13	0.0	2.0	1.0

The Diversity Index is derived using this formula:

$$D = 1 - \frac{\Sigma n(n-1)}{N(N-1)}$$

As an example, "Social Network Diversity Index" for cell G3
= 1-(((B3*(B3-1))+(C3*(C3-1))+(D3*(D3-1))+(E3*(E3-1)))/(F3*(F3-1)))
= 1-(((8*(8-1))+(6*(6-1))+(6*(6-1))+(7*(7-1)))/(27*(27-1)))
= 0.8

2) Now that you have the "Social Network Diversity Index", delete the columns with the Diversity numbers.

	A	B	C	D	E	F	G	H	I
1		Ethnicity Diversity							
2	Staff	Production friends	Sales friends	Finance friends	IT friends	Social Network Size	Social Network Diversity Index	Skillsets Score	Performance Rating
3	Ann	8	6	6	7	27	0.8	9.0	9.0
4	Andy	6	6	6	6	24	0.8	9.0	9.0
5	Ben	10	5	4	3	22	0.7	8.0	8.0
6	Bond	12	3	3	2	20	0.6	7.5	7.0
7	Carl	12	3	2	1	18	0.5	7.0	6.0
8	Candy	13	2	1	1	17	0.4	6.0	5.0
9	Dave	14	2	1	0	17	0.3	5.0	4.0
10	Eddy	14	2	0	0	16	0.2	4.0	3.0
11	Faith	14	1	0	0	15	0.1	3.0	2.0
12	Gary	13	0	0	0	13	0.0	2.0	1.0

3) After you delete the Diversity numbers, your table will now look like this:

	A	B	C	D	E
1	Staff	Social Network Size	Social Network Diversity Index	Skillsets Score	Performance Rating
2	Ann	27	0.8	9.0	9.0
3	Andy	24	0.8	9.0	9.0
4	Ben	22	0.7	8.0	8.0
5	Bond	20	0.6	7.5	7.0
6	Carl	18	0.5	7.0	6.0
7	Candy	17	0.4	6.0	5.0
8	Dave	17	0.3	5.0	4.0
9	Eddy	16	0.2	4.0	3.0
10	Faith	15	0.1	3.0	2.0
11	Gary	13	0.0	2.0	1.0

4) Install "Analysis ToolPak", an Excel add-in

"Analysis ToolPak" is an add-in for Microsoft Excel that comes with Microsoft Excel. To be able to run regression using Excel, you need to first install "Analysis ToolPak", an Excel add-in program that provides data analysis tools. To load the Analysis ToolPak add-in, follow these steps:

On the File tab, click Options.

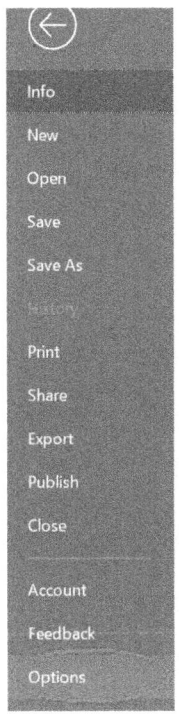

Under Add-ins, click Analysis ToolPak and click the "Go" button.

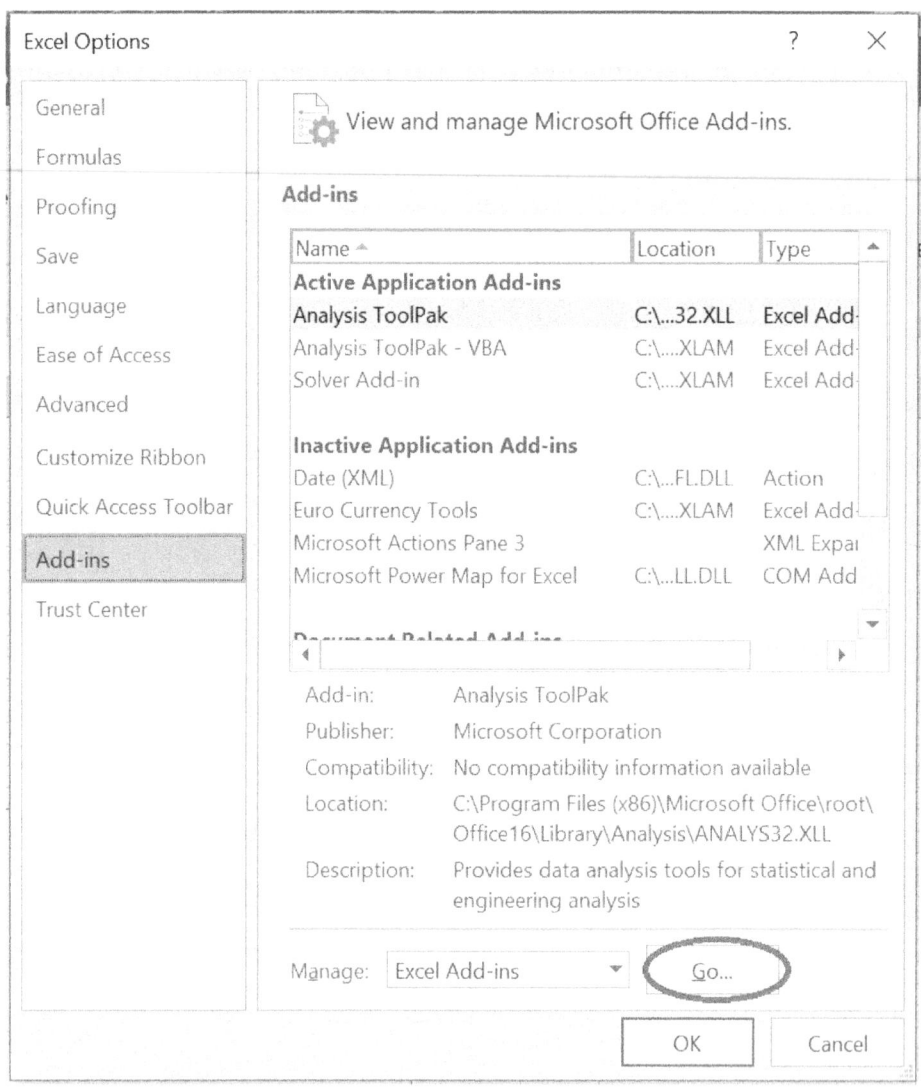

Click "Analysis ToolPak" and click on OK.

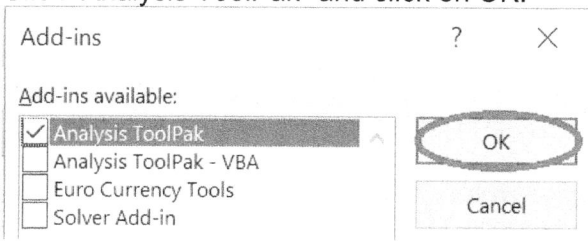

5) On the Data tab, in the Analysis group, you are now able to click on "Data Analysis".

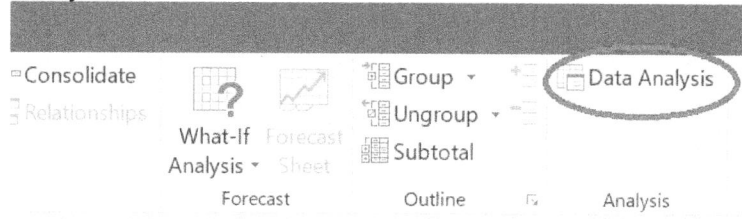

6) Select "Regression" and click "OK".

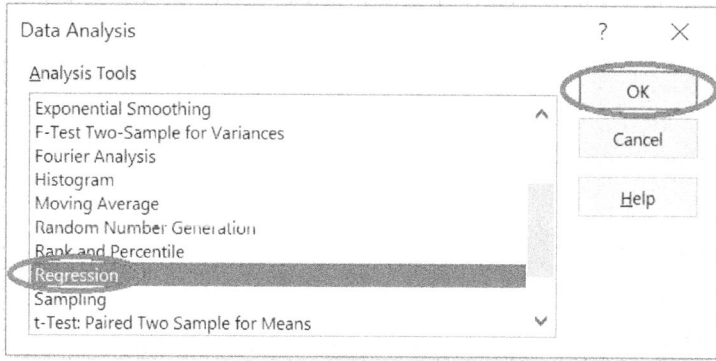

7) After you click OK in the "Data Analysis" dialog box, you will see a "Regression" dialog box.
8) For "Input Y Range", select cells (E1:E11). This is the predictor variable or dependent variable.
9) For "Input X Range", select cells (B1:D11). These are the explanatory variables or independent variables.
10) Check "Labels" box.
11) Click the "Output Range" box, and select cell A13.
12) Click "OK".

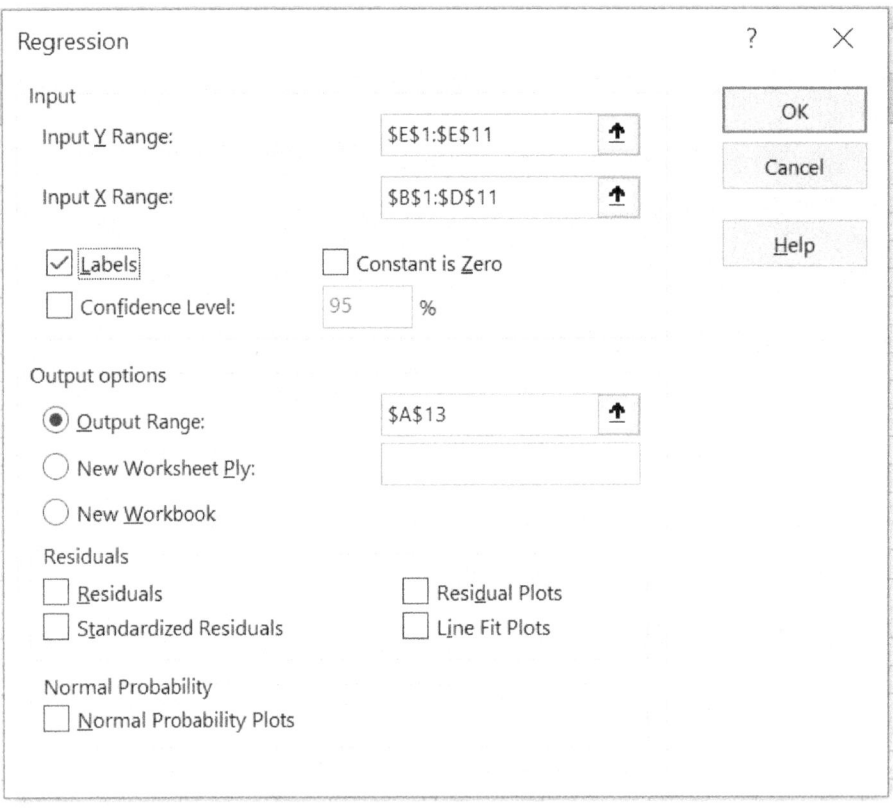

After you click "OK", Excel generates the following Summary Output. Round the numbers to 3 decimal places.

SUMMARY OUTPUT

Regression Statistics	
Multiple R	0.999
R Square	0.998
Adjusted R Square	0.997
Standard Error	0.154
Observations	10

ANOVA

	df	SS	MS	F	Significance F
Regression	3	74.258	24.753	1042.393	0.000
Residual	6	0.142	0.024		
Total	9	74.4			

	Coefficients	Standard Error	t Stat	P-value	Lower 95%	Upper 95%	Lower 95.0%	Upper 95.0%
Intercept	-1.014	0.689	-1.472	0.191	-2.699	0.672	-2.699	0.672
Social Network Size	0.099	0.033	2.998	0.024	0.018	0.179	0.018	0.179
Social Network Diversity Index	6.674	2.392	2.790	0.032	0.822	12.526	0.822	12.526
Skillsets Score	0.250	0.258	0.967	0.371	-0.382	0.882	-0.382	0.882

R Square: In the output, R Square is 0.998, which means it is a good fit. 99% of the variation in Performance Rating (Output) is explained by the independent variables (Input), "Social Network Diversity", "Social Network Size", and "Skillsets Score". The closer R Square is to "1", the better the regression line fits the data.

Significance F and P-values: To determine if your results are statistically significant (i.e. reliable), check "Significance F" (0.001). If the value of "Significance F" is less than 0.05, it is statistically significant (i.e. reliable). If "Significance F" is bigger than 0.05, don't use this set of independent variables. Delete those variables with "P-value" that is bigger than 0.05 and run the regression again until "Significance F" drops below 0.05. Most or all of your P-values should be lower than 0.05. In our example "P-value" is 0.191, 0.024, 0.032 and 0.371, for Intercept, "Social Network Size", "Social Network Diversity index", and "Skillsets Score" respectively.

Coefficients

SUMMARY OUTPUT

Regression Statistics	
Multiple R	0.999
R Square	0.998
Adjusted R Square	0.997
Standard Error	0.154
Observations	10

ANOVA

	df	SS	MS	F	Significance
Regression	3	74.258	24.753	1042.393	0.000
Residual	6	0.142	0.024		
Total	9	74.4			

	Coefficients	Standard Error	t Stat	P-value	Lower 95%	Upper 95%	Lower 95.0%	Upper 95.0%
Intercept	-1.014	0.689	-1.472	0.191	-2.699	0.672	-2.699	0.672
Social Network Size	0.099	0.033	2.998	0.024	0.018	0.179	0.018	0.179
Social Network Diversity Index	6.674	2.392	2.790	0.032	0.822	12.526	0.822	12.526
Skillsets Score	0.250	0.258	0.967	0.371	-0.382	0.882	-0.382	0.882

From the Summary Output, the regression line is:
Performance Rating (Output)
= -1.014 + (0.099*Social Network Size) + (6.674*Social Network Diversity Index) + (0.250*Skillsets Score)

Based on the above regression formula,
- For each unit increase in Social Network Size, Performance Rating increase by 0.099.
- For each unit increase in Social Network Diversity Index, Performance Rating increase by 6.674.
- For each unit increase in Skillsets Score, Performance Rating increase by 0.250.

Coefficients can also be used for forecasting. For example, if an employee's "Social Network Size" is 16, "Social Network Diversity Index" is 0.6, "Skillsets Score" is 8, then his **predicted "Performance Rating"**
= -1.014 + (0.099*Social Network Size) + (6.674*Social Network Diversity Index) + (0.250*Skillsets Score)
= -1.014 + (0.099*16) + (6.674*0.6) + (0.250*8)
= 6.57

References:
(1) Greg Newman (2017) Retain or let go? The data that you need to react correctly when an employee resigns.. https://www.linkedin.com/pulse/retain-release-data-you-need-react-correctly-when-employee-newman/ *(12 June 2019)*

19) Predict Employee Attrition & Absenteeism

Research have shown that employee attrition and absenteeism is influenced by factors such as Demographics (Age, Gender, Marital status, Tenure), Commute time, Performance rating, Salary market-ratio, Personality trait, Behavioral signs (avoids eye contact, smile less, increasingly tardy, participates less in meeting, vacation days not taken), Triggering events (e.g. like having kids, getting divorced or married, or having to care for a sick family member, close colleague left, reorganization), etc.

Attrition's impact on Revenue
- According to William Wolf, Credit Suisse's global head of talent acquisition and development, a one-point reduction in unwanted attrition rates saves the bank $75 million to $100 million a year. [1]

Career Growth's impact on Attrition
- Glassdoor found that workers who stay longer in the same job without a title change are significantly more likely to leave for another company for the next step in their career. Stagnating in a role for an additional 10 months raises the odds that employees will leave the company for their next role by about one percentage point, a statistically significant effect. [2]

Commute time's impact on Attrition
Many companies such as Xerox, KeyBank, Gate Gourmet, Kenexa, Workday, and Evolv have found a connection between commute time and new-hire retention and new hire success. [3]
- Kenexa, found that a lengthy commute raises the risk of attrition in call-center and fast-food jobs. It asks applicants for call-center and fast-food jobs to describe their commute by picking options ranging from "less than 10 minutes" to "more than 45 minutes." The longer the commute, the lower their recommendation score for these jobs. Applicants also can be asked how long they have been at their current address and how many times they have moved. People who move frequently have a higher chance of leaving. [4]
- Research by consultant Jeff Parks found that at one manufacturer, a 30 to 45-minute commute, the probability of quitting jumped to more than 92 percent. [5]
- Gate Gourmet had a new hire turnover rate of about 50 percent. It found that commute times that averaged 35 minutes to be the main cause of the new-hire performance and turnover problems. Thus, the firm changed its

recruitment criteria to emphasize on an applicant's public transportation accessibility and how far the employee lived from the workplace. As a result, the firm managed to lower unwanted turnover to 27 percent." [5]

Compensation and Benefits' impact on Attrition
- In 2011, two Hewlett-Packard (HP) scientists analyzed two years data and generated Flight Risk score. Flight Risk score predicts the probability of leaving of each of HP's 300,000 employees. Higher pay, promotions and better performance ratings, where negatively related to flight risk. However, there is a complicated relationship between these findings. For example, when an employee got a promotion but did not get a substantial pay raise, this employee would still be more likely to quit. In addition, the system informs the managers what the key risk factors of employee attrition are, and puts pressure on managers to develop strategies to retain their staff. There are privacy-related problems with this Flight Risk score, as employees may not be aware that their data was collected, and may not agree with how the data is being used. This is why access to these data should only granted to a senior management who receive training in interpreting Flight Risk scores so they would understand the potential ramifications and confidentiality issues that come with this data. [6]
- Google has always been on the forefront of predictive analytics. Through researched data and historical patterns Google found that new salespeople, who do not get a promotion within four years, are much more likely to leave the company. [7]
- A survey by TechnologyAdvice found that 56 percent of employees said perks were very or moderately important when evaluating a job, and 56 percent of employees say they would trade a salary increase for certain on-the-job perks. [8]
- According to a study by Culture Amp on retention intention at Box, a worker's pay or relationship with his boss matters far less than how connected the worker feels to his team.[9]

Corporate Culture's impact on Attrition
Research have shown that there is a relationship between Corporate Culture and employee turnover:
- Glassdoor found that workplace culture matters for employee retention. When employees switch employers, Glassdoor found they usually move to companies with higher Glassdoor ratings. In particular, Glassdoor found that raising a company's overall rating on Glassdoor by one star (on a one-to-five

scale) was associated with a four-percentage-point higher chance that employees would stay for their next role. [10]

Demographics' impact on Attrition
- LinkedIn's data shows that women job hopped at a higher rate than men. In five years after graduation for those who completed college between 2006 and 2010, women held 3 jobs compared to men's 2.71 jobs. For those who graduated between 1986 and 1990, women held 1.64 jobs in the five years after college, compared to men's 1.57 jobs. A common premise is that working women change jobs to something with more flexible hours when they have a family, even if that means changing multiple times to find the flexible job. [11]
- The median number of years U.S. workers stay with their employers is 4.6, according to the BLS. For ages 25 to 34 that median number is only three years, while those age 55 to 64 stay at the same job for about 10 years. [12]
- The tenure of general managers impacts the retention of employees. As the length of employment for general manager increases, employee turnover drops. [13]

Demographics' impact on Attrition

- Median employee tenure in January 2018 was higher for older workers than younger ones. For example, the median tenure of workers ages 55 to 64 (10.1 years) was more than 3 times that of workers ages 25 to 34 (2.8 years). (14)

Median years of tenure with current employer for employed wage and salary workers by age and sex, January 2018

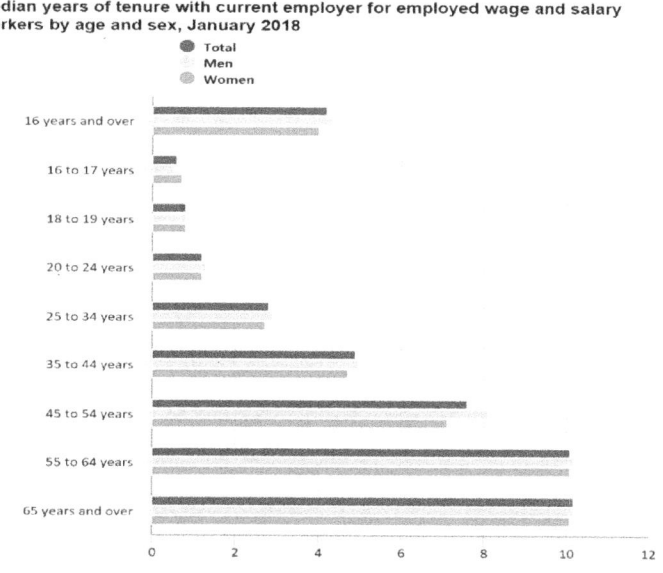

Age and sex	Total	Men	Women
20 to 24 years	1.2	1.3	1.2
25 to 34 years	2.8	2.9	2.7
35 to 44 years	4.9	5.0	4.7
45 to 54 years	7.6	8.1	7.1
55 to 64 years	10.1	10.2	10.1
65 years and over	10.2	10.2	10.1

Source: Bureau of Labor Statistics, U.S. Department of Labor, The Economics Daily, Median tenure with current employer was 4.2 years in January 2018 on the Internet at https://www.bls.gov/opub/ted/2018/median-tenure-with-current-employer-was-4-point-2-years-in-january-2018.htm (visited November 21, 2018).

Diversity and Inclusion's impact on Absenteeism
- Deloitte found that diversity and inclusion correlates with actual absenteeism. Data from one organization showed that if 10 percent more employees feel included, the company will increase work attendance by about one day per year per employee. (15)

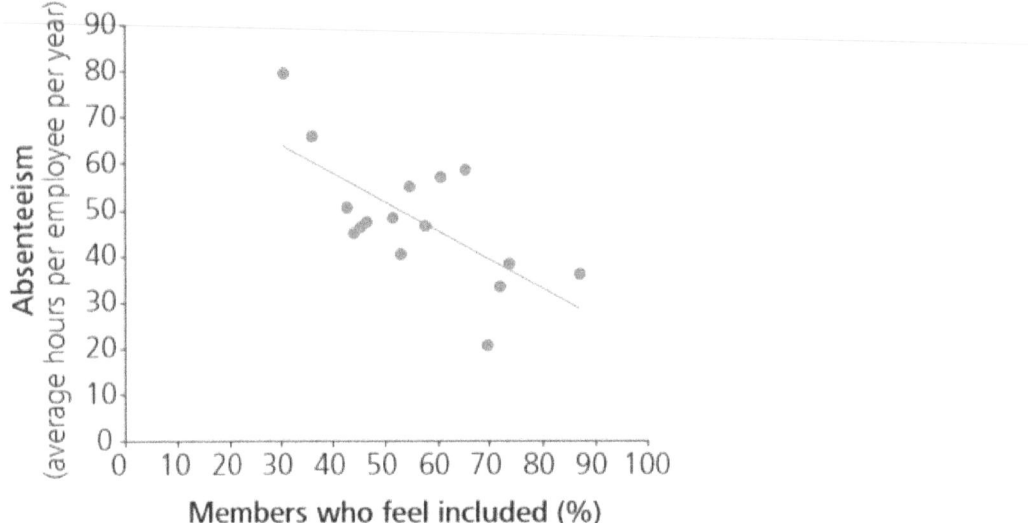

Source: Deloitte Australia (Deloitte) and the Victorian Equal Opportunity and Human Rights Commission (2013) A new recipe to improve business performance, https://www2.deloitte.com/content/dam/Deloitte/au/Documents/human-capital/deloitte-au-hc-diversity-inclusion-soup-0513.pdf (9 November 2018)

Engagement's impact on Attrition & Absenteeism
- A study by Praful Tickoo, the head of people analytics at Genpact and his team, discovered that they can predict attrition loss six months in advance! This is because people usually leave a company because of months of frustration, and they will show signs of unhappiness early. Through statistical analysis, Praful's team found that people who left the company became significantly less engaged in their communications up to six months in advance. Thus, if companies can identify these issues early, they can try to address the issue before the employee resigns. (16)
- Marks & Spencer found that absenteeism in stores at the top quartile of engagement scores were twenty-five percent lower compared to those in the bottom quartile. (17)

Performance Management's impact on Attrition
- According to a study by Culture Amp on retention at Credit Suisse, managers' performance and team size are powerful influences, with a spike in attrition among employees working on large teams with low-rated managers. [18]
- According to Gallup, 'clear expectations are the most basic and fundamental employee need'. 6 in 10 employees indicate that they know what is expected of them at work, but if that ratio can be increased to 8 in 10, your business can achieve a 14% reduction in staff turnover and a 7% increase in productivity. [19]

Personality Trait's impact on Attrition
- Xerox used to hire applicants who had relevant call center experience to staff its call centers. But an analytics algorithm showed that experience does not matter for top performers in call centers. It showed that what matters for a call center worker who won't quit before the company recovers its investment in training – is personality! It further showed that creative people usually stay for at least six months, whereas Inquisitive people usually don't! The analytics data showed that the ideal call-center worker lives near the job, has reliable transportation and uses one or more social networks, but not more than four. They tend not to be overly inquisitive or empathetic, but is creative. Applicants for the job takes a 30-minute test that screens them for personality traits and puts them through job scenarios. Then the program produces a score: red for low potential, yellow for medium potential or green for high potential. Xerox accepts some yellows if it thinks it can train them, but mostly hires greens. After a half-year trial that cut attrition by 20 percent, Xerox now leaves all hiring for its call-center jobs to the analytics software that asks applicants to choose between statements like: "I ask more questions than most people do" and "People tend to trust what I say." Xerox is an example of the insights that can be gained through leveraging predictive analytics. [20]

Training & Development's impact on Absenteeism

- While only 4 in 10 employees say that they have had opportunities to learn and grow, Gallup suggests that improving that ratio to 8 in 10 can allow your business to achieve a 44% drop in absenteeism and a 16% jump in productivity. [21]
- Bridge survey data finds that offering career training and development would keep 86 percent of millennials from leaving their current position. [22]

Triggering events for Attrition

Triggering events such as having kids, getting divorced or married, or having to care for a sick family member are important predictors of turnover:

- A major lifestyle change occurs. E.g. a new baby, getting divorced or married, the last child graduated from university, shifted house, care for a sick family member.
- When an employee's close colleague or manager leaves.
- Competitors firms recently got a large project or contract.
- When your employee completed their degree.
- Employees who work in a high turnover team or job.
- Employees who doesn't want to refer others to join the company.

References:
(1) Rachel Emma Silverman and Nikki Waller (2015) The Algorithm That Tells the Boss Who Might Quit. https://www.visier.com/the-algorithm-that-tells-the-boss-who-might-quit/
(26 November 2018)
(2) Dr. Andrew Chamberlain (2017) Why Do Employees Stay? A Clear Career Path and Good Pay, for Starters. https://www.glassdoor.com/research/why-do-employees-stay-a-clear-career-path-and-good-pay-for-starters/ (26 November 2018)
(3) Dr. John Sullivan (2015) How Commute Issues Can Dramatically Impact Employee Retention https://www.tlnt.com/how-commute-issues-can-dramatically-impact-employee-retention/ (21 November 2018)
(4) Joseph Walker, (2012) Meet the New Boss: Big Data. The Wall Street Journal. https://winintelligence.org/meet-the-new-boss-big-data/
(24 September 2018)
(5) Dr. John Sullivan (2015) How Commute Issues Can Dramatically Impact Employee Retention https://www.tlnt.com/how-commute-issues-can-dramatically-impact-employee-retention/ (21 November 2018)
(6) Erik van Vulpen, Predictive Analytics in Human Resources - Tutorial and 7 case studies, https://www.analyticsinhr.com/blog/predictive-analytics-human-resources/
(24 September 2018)
(7) Tom McKeown, Snapshots in Time (2017), https://www.trendata.com/snapshots-in-time/
(24 September 2018)

(8) Savita V Jayaram (2016), The Correlation between Benefits and Employee Retention http://www.hrinasia.com/employee-retention/the-correlation-between-benefits-and-employee-retention/ (27 September 2018)

(9) Rachel Emma Silverman and Nikki Waller (2015) The Algorithm That Tells the Boss Who Might Quit. https://www.visier.com/the-algorithm-that-tells-the-boss-who-might-quit/ (26 November 2018)

(10) Dr. Andrew Chamberlain (2017) Why Do Employees Stay? A Clear Career Path and Good Pay, for Starters. https://www.glassdoor.com/research/why-do-employees-stay-a-clear-career-path-and-good-pay-for-starters/ (26 November 2018)

(11) Vivian Giang (2016) Why Women Job Hop More Than Men. https://www.fastcompany.com/3058996/why-women-job-hop-more-than-men (22 November 2018)

(12) Caroline Zaayer Kaufman (2018) Have you had a 'normal' number of jobs? https://www.monster.com/career-advice/article/have-had-normal-number-jobs (22 November 2018)

(13) Jac FITZ-ENZ (2010), The New HR Analytics: Predicting the Economic Value of Your Company's Human Capital Investments, AMACOM

(14) Bureau of Labor Statistics, U.S. Department of Labor, The Economics Daily, Median tenure with current employer was 4.2 years in January 2018 on the Internet at https://www.bls.gov/opub/ted/2018/median-tenure-with-current-employer-was-4-point-2-years-in-january-2018.htm (visited November 21, 2018).

(15) Deloitte Australia (Deloitte) and the Victorian Equal Opportunity and Human Rights Commission (2013) A new recipe to improve business performance, https://www2.deloitte.com/content/dam/Deloitte/au/Documents/human-capital/deloitte-au-hc-diversity-inclusion-soup-0513.pdf (9 November 2018)

(16) Josh Bersin (2018). What Emails Reveal About Your Performance At Work. https://joshbersin.com/2018/10/what-emails-reveal-about-your-performance-at-work/# (12 November 2018)

(17) Culture Amp (2018) Why Employee Engagement Matters. https://www.cultureamp.com/resources/guides/foundation-guides/why-employee-engagement-matters.html#ref-3 (14 November 2018)

(18) Rachel Emma Silverman and Nikki Waller (2015) The Algorithm That Tells the Boss Who Might Quit. https://www.visier.com/the-algorithm-that-tells-the-boss-who-might-quit/ (26 November 2018)

(19) Tanvir Haque (2018) Four Strategies To Boost Employee Engagement That Can Improve Your Bottom Line. https://www.entrepreneur.com/article/312697 (20 November 2018)

(20) Joseph Walker, (2012) Meet the New Boss: Big Data. The Wall Street Journal. *https://winintelligence.org/meet-the-new-boss-big-data/* (24 September 2018)

(21) Tanvir Haque (2018) Four Strategies To Boost Employee Engagement That Can Improve Your Bottom Line. https://www.entrepreneur.com/article/312697 (20 November 2018)

(22) Bride (2018) Millennials Are Most Likely to Stay Loyal to Jobs With Development Opportunities . https://www.instructure.com/bridge/news/press-releases/millennials-are-most-likely-stay-loyal-jobs-development-opportunities?newhome=bridge (19 June 2019)

19.1) Case 1: AIHR – Turnover Predictors

The below infographic from AIHR, categorizes why people quit into ten Turnover Predictors [1]:

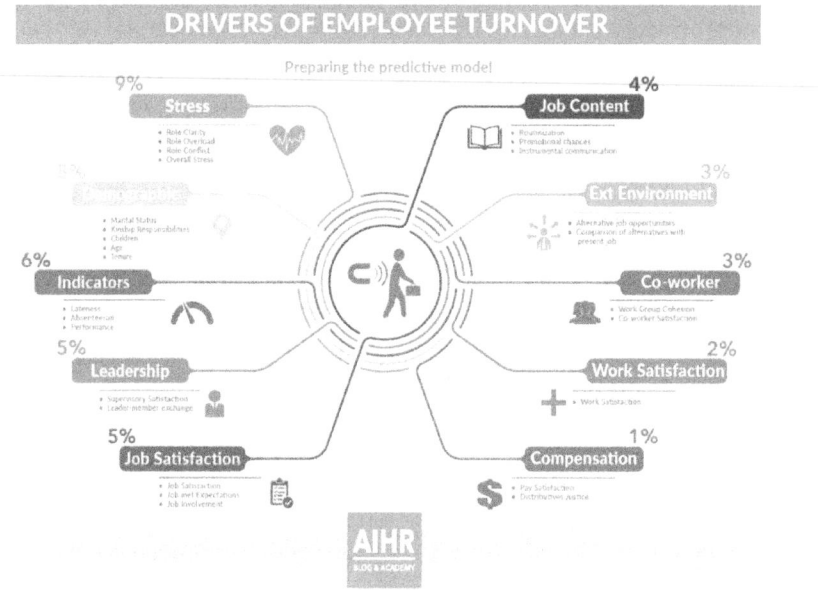

Source: AIHR

- **Stress:**

Stress makes people leave their jobs, are usually caused by role ambiguity, role overload, and role conflict (occurs when a person has two contradictory positions).

- **Demographic:**

Age is negatively correlated to turnover - younger employees tend job hop more frequently than older employees. Tenure is another indicator of turnover - employees are more likely to leave in their fourth or fifth year, compared to their first year. When employees work for a company a very long time, they are less likely to change jobs. Married people, especially those with children are less likely to change jobs compared to singles, due to extra responsibilities of marriage and children. The Turnover Predictors Variables interact with each other, and some effects will be enhanced while others will be reduced. For example, marriage is a significant predictor when people are younger. When a man and a woman, gets married in their early 30s and plan to have children, they make different choices. The woman is more likely to quit when her work load is high compared to the man. The woman usually wants a more relaxed and flexible work environment so she can take extra time off to raise the kids. She may even want to stop working for a few years to devote extra attention to the kids. The man, however, will be less likely to quit as he has the responsibility of providing the financial stability to raise a family.

- **Indicators:**

Indicators such as Lateness, Absenteeism, and Performance are predictors of turnover. Lateness - when people consistently arrive late at work, it could be due to demotivation. Absenteeism - employees who are absent frequently are more likely to leave, because they take sick leave for interview, or because of lack of motivation. Absenteeism is the strongest indicator for turnover intentions, together with tenure. Performance - employees with a low performance are likely to leave. But, when an employee performs exceptionally well over a long period of time, they are more likely to leave because they lack challenge and change.

- **Leadership:**

"People leave their bosses, not their jobs" holds true in research. When an employee is happy with their supervisor, they tend to stay with the company longer.

- **Job satisfaction:**

Job meeting expectations is about whether the job meets the expectations people had, before they started. When expectations and reality are really misaligned, people might even leave within a few months.

- **Job content:**

Job content is about how people experience their job. Routinization leads to turnover because nobody likes to do the same thing every day.

- **External environment:**

People constantly compare their situation with others, but are unlikely to leave when there are few alternative job opportunities.

- **Co-workers:**

Co-worker satisfaction is a factor that predicts turnover. People love or hate their jobs because of their colleagues.

- **Work satisfaction:**

Where job satisfaction focuses more narrowly on one's job, work satisfaction looks at a person's work from a broader perspective.

- **Compensation**

Pay satisfaction is a significant predictor of turnover because people love to compare. People are more likely to change jobs when a friend or colleague with similar job earns much more. It is not the de facto pay that matters, but the person's satisfaction with this pay.

References:
(1) Erik van Vulpen, What Drives Employee Turnover? Part 2, https://www.analyticsinhr.com/blog/what-drives-employee-turnover/ (26 September 2018)

19.2) Case 2: FlightNetwork "At-Risk" Employees Criteria

Omer Aziz, Chief HR Officer at FlightNetwork, created his own method to predict employee turnover. Aziz simplified the "at-risk" employees into five criteria [1]:
- No title changes in more than 1 year.
- No salary increases in more than 1 year.
- Rated highly in the last 2 performance reviews.
- Have a commute more than 1 hour and 15 minutes to work (according to Aziz, this is the "tipping point" in Toronto, Canada. You can use home address postal codes to determine the commute times that may be the "tipping point" in your city).
- Had a life change in the past year (e.g. like having kids, getting divorced or married, or having to care for a sick family member)

Thus, you might have an "at-risk" employee, if you have an employee who needs to commute for 1 hour every day, and she recently had a new born. Her long commute may now be an issue, as it now affects her need to spend more time with her family. Out of the 270 employees, 12 names met the first four criteria above. Aziz contacted their managers to explained his methodology, and to warn them who might be flight risks. A few days later, one of the at-risk employee scheduled a meeting with his manager to discuss a pay raise. The manager who was pre-warned by Aziz, was ready and negotiated a compensation increase for the employee. Satisfied with his new salary, the employee stayed [1].

References:
(1) Lyssa Test, How to Predict Employee Turnover (2018) How to Predict Employee Turnover, https://blog.namely.com/blog/how-to-predict-employee-turnover?fbclid=IwAR1Nn7ZiEUc8DMMtSZs5sq-LwHpwE17asdvsG3OnBPUZGbkEBdiLW55_VJU

19.3) Case 3: Retention Analytics at Neilson Holdings

In 2015, one of Nielson Holdings PLC's biggest business leader asked Piyush Mathur, SVP of People Analytics: do you know why people are leaving my team? Mathur and his team set out to build a basic attrition model to investigate what was causing the rising company-wide attrition. Within months, Mathur and his team identified the primary drivers of voluntary attrition, and slashed Nielsen's voluntary attrition by nearly fifty percent, saving Nielson Holdings PLC millions of dollars [1].

	Make sure you are answering a critical question for your business	
	Belief	Attrition seems high
	Business Case	Every point of attrition costs us $5 million
	Proof	One US Bus. voluntary turnover is at XX%, which is higher than the company average

nielsen

Source: Piyush Mathur, Douglas Shagam, The Nielsen Company (2016) People analytics: Breaking myths with agility and passion | Talent Connect 2016. https://www.slideshare.net/linkedin-talent-solutions/people-analytics-breaking-myths-with-agility-and-passion-talent-connect-2016 (2 November 2018)

Preventing attrition became Mathur and Shagam's main objective. First, Mathur need to derive a simple model to measure and predict attrition, put in place programs to fix it, and finally measure its impact and share the findings. Instead of spending time to get perfect data, Mathur built an attrition model with 20 points of employee data such as gender, tenure, age, and manager rating. Over time, they refined it to include data such as time and participation in CSR programs [1].

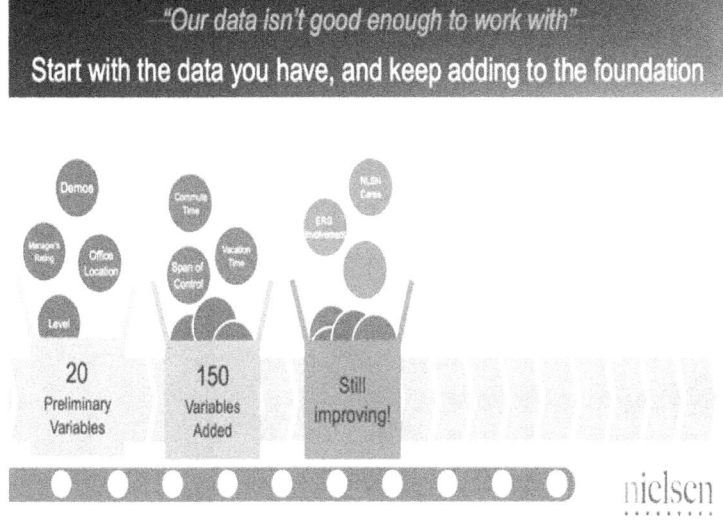

Source: Piyush Mathur, Douglas Shagam, The Nielsen Company (2016) People analytics: Breaking myths with agility and passion | Talent Connect 2016. https://www.slideshare.net/linkedin-talent-solutions/people-analytics-breaking-myths-with-agility-and-passion-talent-connect-2016 (2 November 2018)

The attrition model provided insights on how to retain Nielsen's employees [1]:
- Employee's first-year matters most. The possibility of employees leaving is high if they don't reach their first performance review.
- Ethnicity and Gender has no impact on tenure.
- Promotion and lateral move retain employees.

As insights are meaningless without action Mathur's team set up several programmes to cut attrition [1]:
- After identifying employees' attributes with the highest possibility of leaving, the company leaders set up chats with them. As a result, 40 per cent of the flight-risk employees was transferred to new roles.
- A "Golden Year" program was created to track an employee's first year as attrition is the highest during this period
- A "Ready to Rotate" group as created to empower employees to find new roles within Nielsen, as Mathur's team discovered that lateral moves increases an employee's probability of staying with the company by 48%.

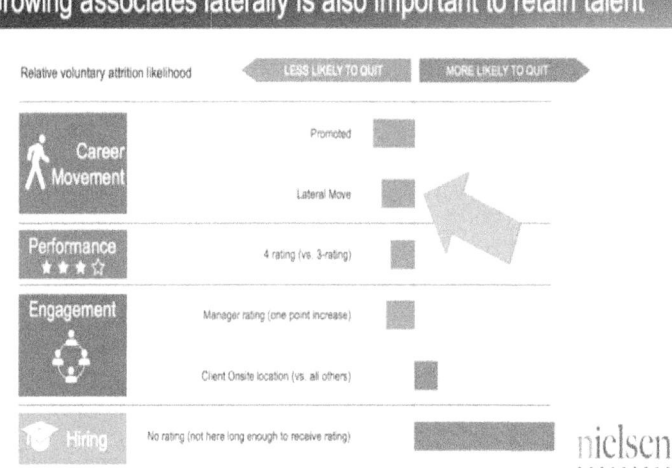

Source: Piyush Mathur, Douglas Shagam, The Nielsen Company (2016) People analytics: Breaking myths with agility and passion | Talent Connect 2016. https://www.slideshare.net/linkedin-talent-solutions/people-analytics-breaking-myths-with-agility-and-passion-talent-connect-2016 (2 November 2018)

References:
(1) Keenan Steiner (2017) People Analytics Isn't as Hard as You Think—Nielsen Proves Why. https://business.linkedin.com/talent-solutions/blog/employee-retention/2017/how-nielsen-used-people-analytics-to-increase-retention-and-saved-millions-of-dollars (2 November 2018)

19.4) Case 4: iNostix – Flight Risk Scoring

Traditional HR metrics such as employee turnover rate doesn't tell you where you stand with regards to future turnover risks. One of the hottest topics is the ability to predict critical organizational risks such as key employee turnover, work accidents or absenteeism. Instead of the usual reporting groups such as department or business unit, Predictive Employee Turnover Analytics (PETA) constructs groups based on the risk profiles. In the table below, PETA computes risk scores for each employee and sums them into high risk groups, such as age groups, potential groups, salary groups, gender groups, tenure groups, performance groups. [1]

Flight Risk Scoring

Characteristic Name	Attribute	'Flight Risk' Scoring (=Prediction)
Age 1	26 - 29	240
Potential 1	"Rough Diamond" (HiPo/LoPe)	225
Salary 1	45.000 – 55.000	185
Gender 1	Male	180
Tenure 1	1 – 2 year	160
Age 2	Up to 26	140
Salary 2	65.000 – 75.000	135
Potential 2	"Talent Risk" (LoPo/LoPe)	130
Tenure 2	6 – 10 year	125
Performance 1	2/5	110
Performance 2	5/5	90
Salary 3	55.000 – 65.000	90

Source: Luk Smeyers, iNostix (2016) 5 Reasons to Start with Predictive Employee Turnover Analysis – HR Analytics. https://www.inostix.com/blog/en/5-reasons-to-start-with-predictive-employee-turnover-analysis-hr-analytics/ (1 February 2016)

References:
(1) Luk Smeyers, iNostix (2016) 5 Reasons to Start with Predictive Employee Turnover Analysis – HR Analytics. https://www.inostix.com/blog/en/5-reasons-to-start-with-predictive-employee-turnover-analysis-hr-analytics/ (1 February 2016)

19.5) Case 5: IBM Kenexa Talent Insights

IBM Kenexa Talent Insights, a data discovery application allows HR and managers to explore, analyze, and navigate talent data so they can observe patterns or gain actionable insights. [1] In the diagram below, companies can analyze employee tenure and engagement by hire-source, educational institute, and former employer to see if there is a relationship.

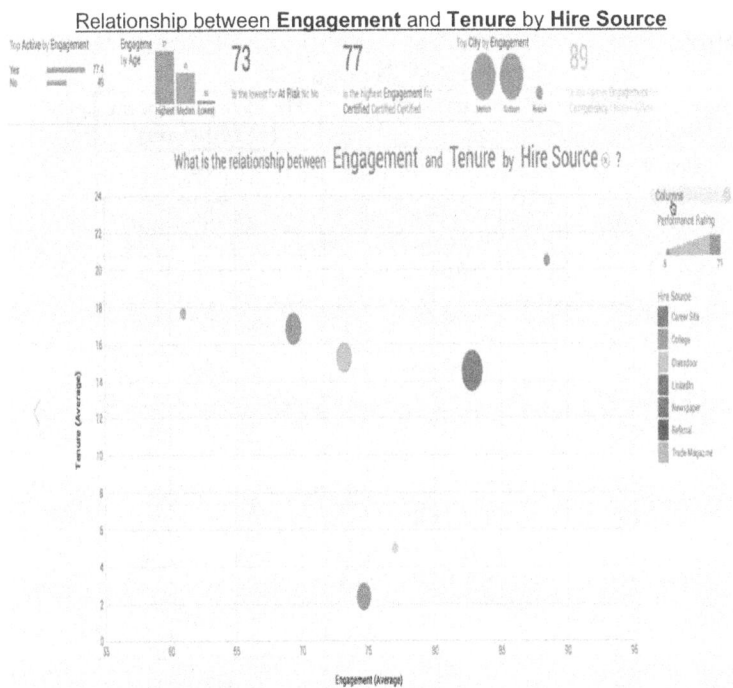

Source: IBM (2017) IBM Kenexa Talent Insights Powered by Watson Analytics
https://www.youtube.com/watch?v=AXoiyi2XlEs (1 March 2017)

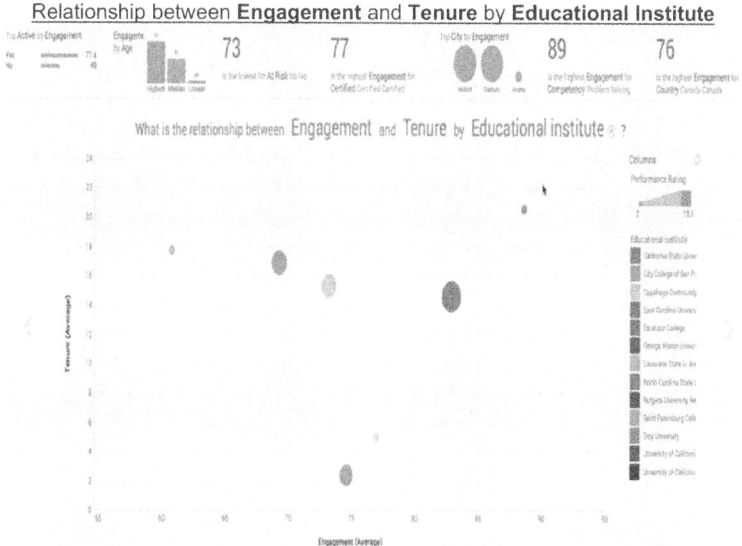

Source: IBM (2017) IBM Kenexa Talent Insights Powered by Watson Analytics
https://www.youtube.com/watch?v=AXoiyi2XIEs (1 March 2017)

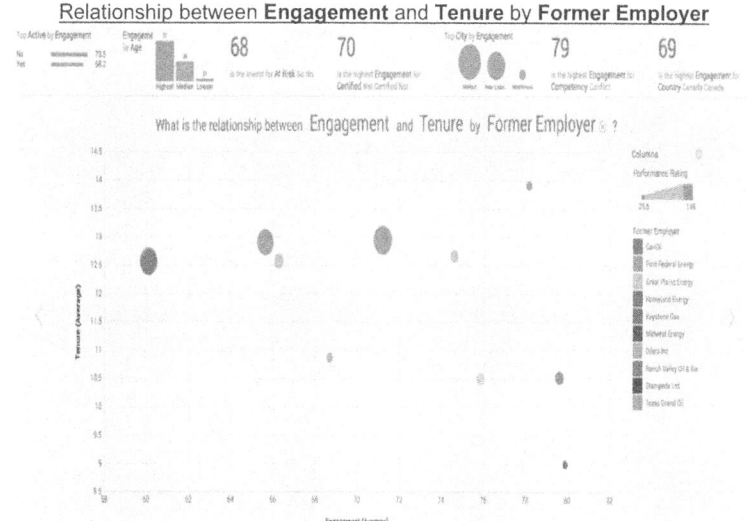

Source: IBM (2017) IBM Kenexa Talent Insights Powered by Watson Analytics
https://www.youtube.com/watch?v=AXoiyi2XIEs (1 March 2017)

References:
(1) IBM (2017) IBM Kenexa Talent Insights Powered by Watson Analytics
https://www.youtube.com/watch?v=AXoiyi2XIEs (1 March 2017)

19.6) Probability Tree Example: Predict Employee Resignation

It is important to understand the drivers of employee dissatisfaction as great talent is scarce, hard to keep and highly solicited by competitors. An analytical model can then be built to predict employee resignation. Decision trees can be constructed to provide insights into which employees might churn.

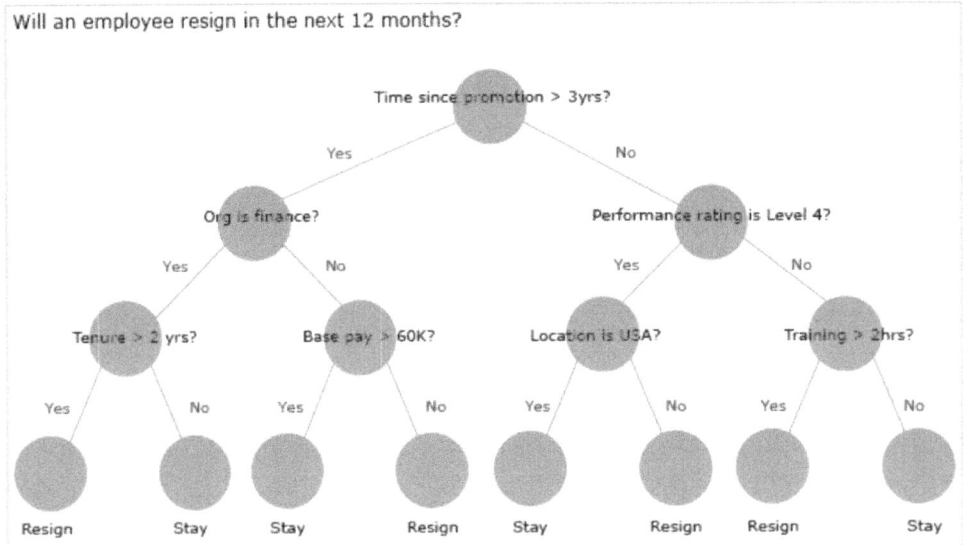

Source: Ian Cook (2019) Fact or Hype: Validating Predictive People Analytics and Machine Learning https://www.visier.com/clarity/predictive-people-analytics-machine-learning/ (6 June 2019)

DECISION TREE MODEL FOR EMPLOYEE CHURN

A Decision tree model is a predictive and analytical model of computation for sequential decision problems under uncertainty

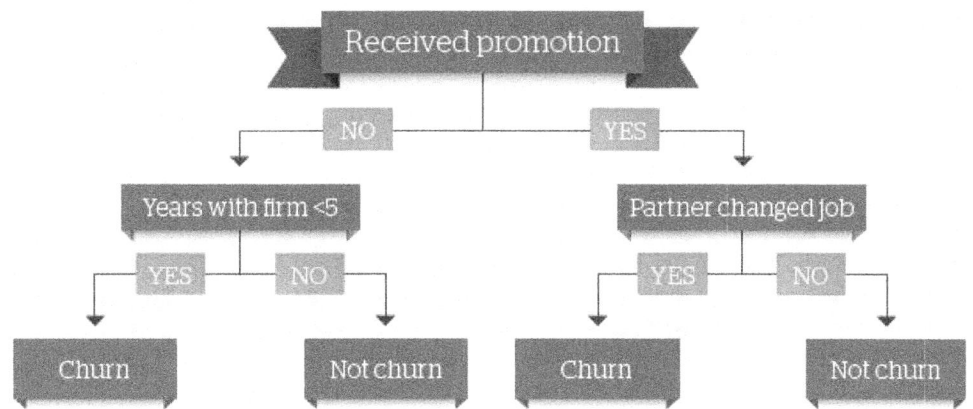

Source: Prof. Dr. Bart Baesens (2019) Strategic importance of HR can shoot up with Analytics https://www.peoplematters.in/article/predictive-hr-analytics/strategic-importance-hr-can-shoot-analytics-12599 (6 June 2019)

19.7) Correlation Example: Predict Employee Flight Risk

Correlation can be used to understand the factors that drives employee resignation, which you can then use to predict your employee flight risk.

In this example, we use correlation to find out what variables are good predictors of employee turnover. The data below represents events that have already happened.

	A	B	C	D	E	F
1	Staff name	Travel time (Minutes)	Marital status (1 = S; 2 = M)	Salary market-ratio	Gender (1 = M; 2 = F)	Status (0 = Left; 1 = Stay)
2	Anne	60	1	0.60	1	0
3	Bond	60	1	0.60	2	0
4	Charle	60	1	0.60	1	0
5	Dan	60	1	0.60	2	0
6	Eric	30	1	1.00	1	0
7	Fanny	30	1	1.00	2	0
8	Gaia	30	1	1.00	1	0
9	Han	30	1	1.00	2	0
10	Ian	5	2	1.30	1	1
11	John	5	2	1.30	2	1
12	Kate	5	2	1.30	1	1

1) Install "Analysis ToolPak", an Excel add-in

"Analysis ToolPak" is an add-in for Microsoft Excel that comes with Microsoft Excel. To be able to run regression using Excel, you need to first install "Analysis ToolPak", an Excel add-in program that provides data analysis tools. To load the Analysis ToolPak add-in, follow these steps:

- On the File tab, click Options.

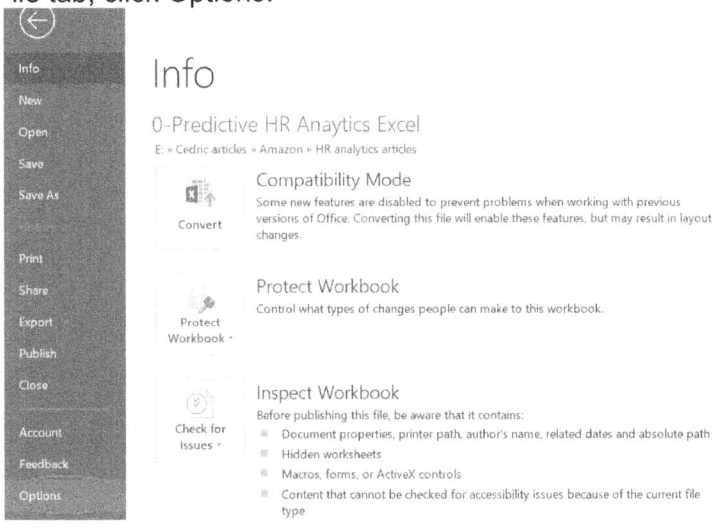

- Under Add-ins, click Analysis ToolPak and click the "Go" button.

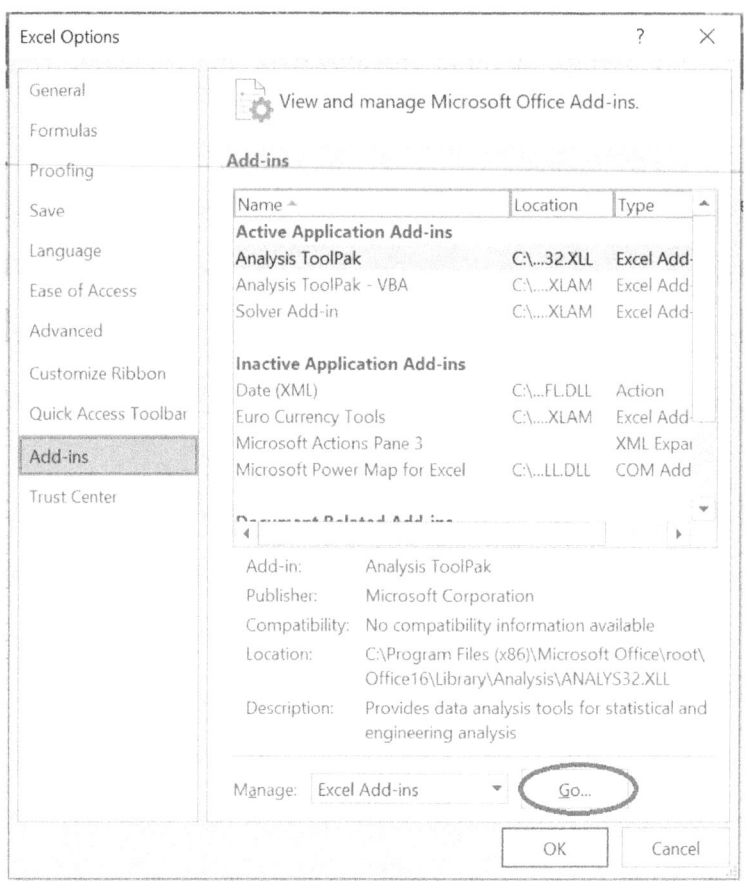

- Click "Analysis ToolPak" and click on OK.

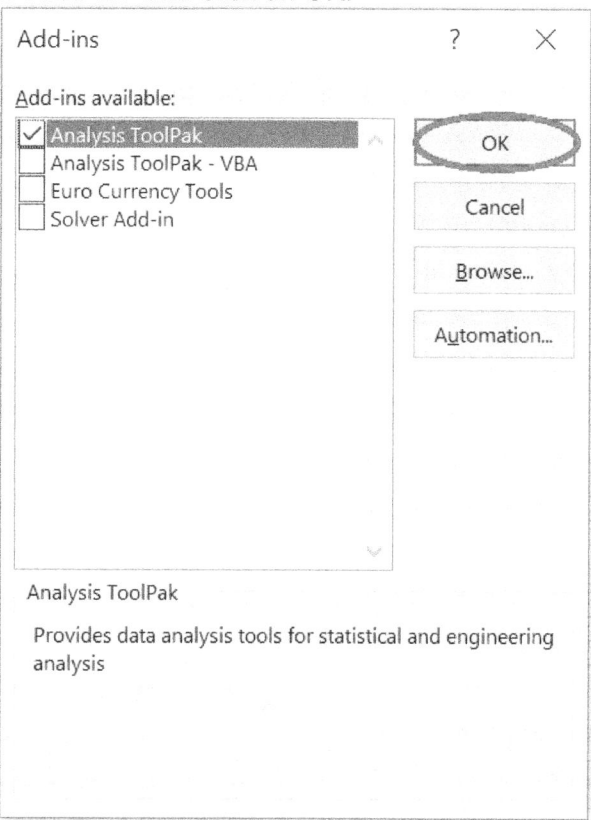

- On the Data tab, in the Analysis group, you are now able to click on "Data Analysis".

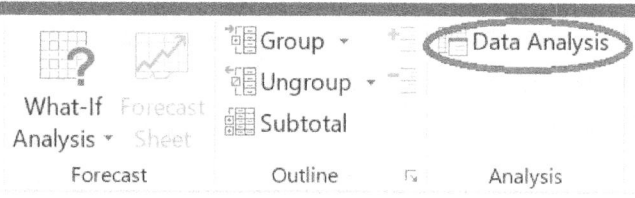

2) Copy the example data in the following table, and paste it in cell A1 of a new Excel worksheet.

	A	B	C	D	E	F
1	Staff name	Travel time (Minutes)	Marital status (1 = S; 2 = M)	Salary market -ratio	Gender (1 = M; 2 = F)	Status (0 = Left; 1 = Stay)
2	Anne	60	1	0.60	1	0
3	Bond	60	1	0.60	2	0
4	Charle	60	1	0.60	1	0
5	Dan	60	1	0.60	2	0
6	Eric	30	1	1.00	1	0
7	Fanny	30	1	1.00	2	0
8	Gaia	30	1	1.00	1	0
9	Han	30	1	1.00	2	0
10	Ian	5	2	1.30	1	1
11	John	5	2	1.30	2	1
12	Kate	5	2	1.30	1	1

3) Select "Correlation" and click "OK".

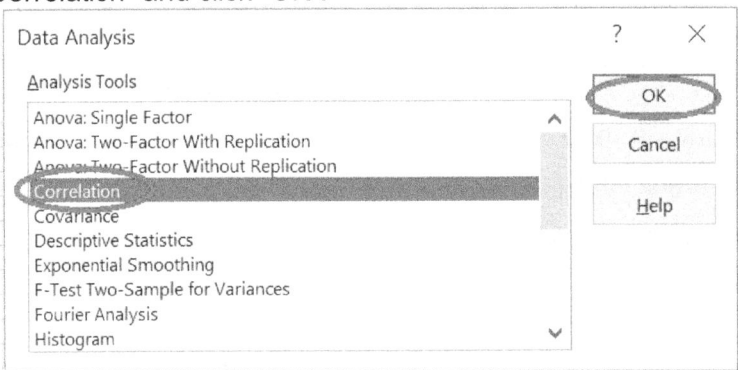

4) After you click OK in the "Data Analysis" dialog box, you will see a "Correlation" dialog box.
5) For "Input Range", select cells (B1:F12).
6) Check "Labels in first row"
7) For "Output Range", select cells (A14).
8) Click "OK"

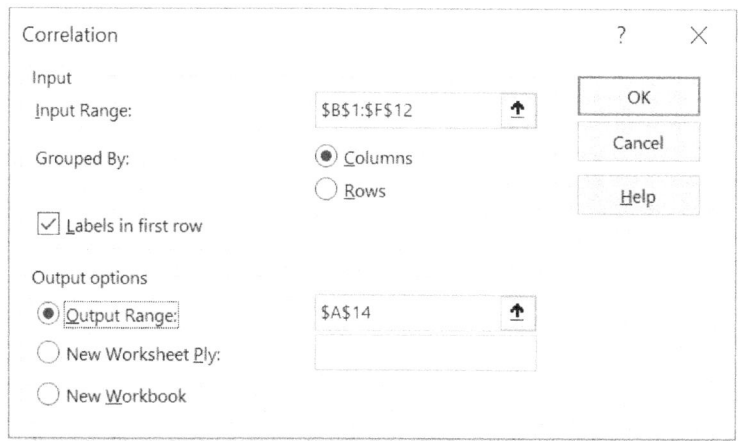

After you click "OK", Excel generates the following Correlation analysis.

	Travel time (Minutes)	Marital status (1 = S; 2 = M)	Salary market-ratio	Gender (1 = M; 2 = F)	Status (0 = Left; 1 = Stay)
Travel time (Minutes)	1				
Marital status (1 = S; 2 = M)	-0.81	1			
Salary market-ratio	-1.00	0.79	1		
Gender (1 = M; 2 = F)	0.12	-0.15	-0.12	1	
Status (0 = Left; 1 = Stay)	-0.81	1	0.79	-0.15	1

A negative correlation coefficient means that an increase in X is associated with a decrease in Y. Similar to a positive correlation, a negative correlation shows a connection between two variables, and the relative strengths are the same. In other words, a correlation coefficient of 0.85 has the same strength as a correlation coefficient of -0.85. Correlation coefficients are always values between -1 and 1, where "-1" means that there is a perfect linear negative correlation, while "1" shows a perfect linear positive correlation. A correlation coefficient of zero, or near to zero, means that there is no meaningful relationship between variables. Correlation coefficient of 0.91 or -0.92 shows a very strong positive and negative correlation respectively. However, correlation does not mean causation. An example of negative correlation is the amount of snowfall and the temperature. As the temperature increases, the amount of snowfall decreases. An example of positive correlation is the relationship between temperature and ice cream sales. As temperature increases, so do ice cream sales.

9) Observations from the above Excel Correlation analysis:

	Travel time (Minutes)	Marital status (1 = S; 2 = M)	Salary market -ratio	Gender (1 = M; 2 = F)	Status (0 = Left; 1 = Stay)
Travel time (Minutes)	1				
Marital status (1 = S; 2 = M)	-0.81	1			
Salary market -ratio	-1.00	0.79	1		
Gender (1 = M; 2 = F)	0.12	-0.15	-0.12	1	
Status (0 = Left; 1 = Stay)	-0.81	1	0.79	-0.15	1

From the Excel Correlation analysis, these variables are good predictors of employee turnover as they have strong correlation of below -0.75 and above 0.75:
- Travel time
- Marital status
- Salary market-ratio

From the Excel Correlation analysis, Gender has very little impact on employee turnover as they have very weak correlation of between -0.20 to 0.20.

19.8) Logistic Regression Example: Predict Staff Resignation

In the data below, we want to predict the non-numeric dependent variable, Staff Resignation (i.e. whether the Staff "Resign" or "Stay") using Excel Logistic regression. The data represents events that have already happened. As the variables can only take 1 or 0,
- For the "Resign" column: Stay = 0, Resign = 1.
- For the "Travel time" column: Far = 1, Near = 0.

Resign	Compa-ratio	Age	Travel time
0	1.3	55	1
0	1.2	34	0
0	1	50	0
1	1.2	39	0
1	0.5	34	1
1	0.4	45	1

1) Our objective is to create an equation with coefficients b_0 to b_3 and then enter values for Compa-ratio, Age, and Travel time to predict Staff Resignation (Resign or Stay). We have 4 coefficients (Constant, Compa-ratio, Age, Travel time). The logistic equation is:

$$\text{Logit(Resign)} = b_0 + b_1 * \text{Compa-ratio} + b_2 * \text{Age} + b_3 * \text{Travel time}$$

2) Logit is a function that takes a probability of an event as input and returns the logarithm of the odds of that event as output. We need to first assign an arbitrary value (e.g. 0.000) for these coefficients (b0, b1, b2, b3) - Later you will be shown how to use the Excel Solver to replace these starting arbitrary coefficients (e.g. 0.000) with optimized coefficients to create an equation to predict probability.

	A	B	C	D	E	F	G	H
1						b_0 (Intercept)=		0.000
2						$b_1=$		0.000
3						$b_2=$		0.000
4						$b_3=$		0.000
5								
6								
7		Resign	Compa-ratio	Age	Travel time			
8		0	1.3	55	1			
9		0	1.2	34	0			
10		0	1	50	0			
11		1	1.2	39	0			
12		1	0.5	34	1			
13		1	0.4	45	1			

3) Here, you need to calculate a Logit for each record. Enter the Logit formula below for all the data records, in your Excel spreadsheet:

$$\text{Logit(Resign)} = b_0 + b_1*\text{Compa-ratio} + b_2*\text{Age} + b_3*\text{Travel time}$$

N   fx   =$H$1+$H$2*B8+$H$3*C8+$H$4*D8

	A	B	C	D	E	F	G	H	I	J
1							b_0 (Intercept)=	0.000		
2							b_1=	0.000		
3							b_2=	0.000		
4							b_3=	0.000		
5										
6										
7		Resign	Compa-ratio	Age	Travel time	Logit (L)				
8		0	1.3	55	1	=$H$1+				
9		0	1.2	34	0	0.000				
10		0	1	50	0	0.000				
11		1	1.2	39	0	0.000				
12		1	0.5	34	1	0.000				
13		1	0.4	45	1	0.000				

4) Here, you need to calculate e^L for each record. The number e is the base of the natural logarithm. It is approximately equal to 2.71828163. e^L must be calculated for each record. Enter the Exponential formula below, in your Excel spreadsheet:

	A	B	C	D	E	F	G	H
1						b_0 (Intercept)=		0.000
2						b_1=		0.000
3						b_2=		0.000
4						b_3=		0.000
5								
6								
7		Resign	Compa-ratio	Age	Travel time	Logit (L)	e^L	
8		0	1.3	55	1	0.000	=EXP(E8)	
9		0	1.2	34	0	0.000	1.000	
10		0	1	50	0	0.000	1.000	
11		1	1.2	39	0	0.000	1.000	
12		1	0.5	34	1	0.000	1.000	
13		1	0.4	45	1	0.000	1.000	

Formula bar: =EXP(E8)

5) Here, you need to calculate P(X) for each record. P(X) is the probability of event X occurring. Enter the formula for probability of the event (i.e. Resign) in your Excel spreadsheet, using the formula below:

$$P(X) = e^L / (1 + e^L)$$

	A	B	C	D	E	F	G	H
1							b_0 (Intercept)=	0.000
2							b_1=	0.000
3							b_2=	0.000
4							b_3=	0.000
5								
6								
7		Resign	Compa-ratio	Age	Travel time	Logit (L)	e^L	P (X)
8	0	1.3	55	1	0.000	1.000	=F8/(1+F8)	
9	0	1.2	34	0	0.000	1.000	0.500	
10	0	1	50	0	0.000	1.000	0.500	
11	1	1.2	39	0	0.000	1.000	0.500	
12	1	0.5	34	1	0.000	1.000	0.500	
13	1	0.4	45	1	0.000	1.000	0.500	

Formula bar: =F8/(1+F8)

6) Here, you need to calculate LL, the Log-Likelihood Function. The log-likelihood function computes a probability based on the input variables values. Enter the log-likelihood formula below, in your Excel spreadsheet:

N   fx   =A8*LN(G8)+(1-A8)*(LN(1-G8))

	A	B	C	D	E	F	G	H
1						b_0 (Intercept)=		0.000
2						b_1=		0.000
3						b_2=		0.000
4						b_3=		0.000
5								
6								
7	Resign	Compa-ratio	Age	Travel time	Logit (L)	e^L	P (X)	Log-Likelihood (LL)
8	0	1.3	55	1	0.000	1.000	0.500	=A8*LN(G8
9	0	1.2	34	0	0.000	1.000	0.500	-0.693
10	0	1	50	0	0.000	1.000	0.500	-0.693
11	1	1.2	39	0	0.000	1.000	0.500	-0.693
12	1	0.5	34	1	0.000	1.000	0.500	-0.693
13	1	0.4	45	1	0.000	1.000	0.500	-0.693
14							sum of LL:	-4.159

7) The sum that we wish to maximize is the total of log-likelihood (LL):

							fx	=SUM(H8:H13)

	A	B	C	D	E	F	G	H
1							b_0 (Intercept)=	0.000
2							b_1=	0.000
3							b_2=	0.000
4							b_3=	0.000
5								
6								
7		Resign	Compa-ratio	Age	Travel time	Logit (L)	e^L	P (X)	Log-Likelihood (LL)
8	0	1.3	55	1	0.000	1.000	0.500	-0.693
9	0	1.2	34	0	0.000	1.000	0.500	-0.693
10	0	1	50	0	0.000	1.000	0.500	-0.693
11	1	1.2	39	0	0.000	1.000	0.500	-0.693
12	1	0.5	34	1	0.000	1.000	0.500	-0.693
13	1	0.4	45	1	0.000	1.000	0.500	-0.693
14							sum of LL:	-4.159

The objective of Logistic Regression is find the coefficients of the Logit (b_0, b_1, b_2 + ...+ b_k) that maximize LL, the Log-Likelihood Function in cell H14, to produce the Maximum Log-Likelihood (MLL) Function. The only values we can change are the guesses for the coefficient b_0 through b_3, which we have assigned an arbitrary value of 0.000. We don't have to optimize them ourselves, as we can use Solver, an Excel add-in, that adjusts the coefficient to maximize or minimize the value in the cell.

8) The Excel Solver is an add-in that is included in Excel. But it must be manually activated by you before it can be utilized for the first time. To use Solver, on the File tab, click Options.

9) Under Add-ins, click Excel Add-ins and click the "Go" button.

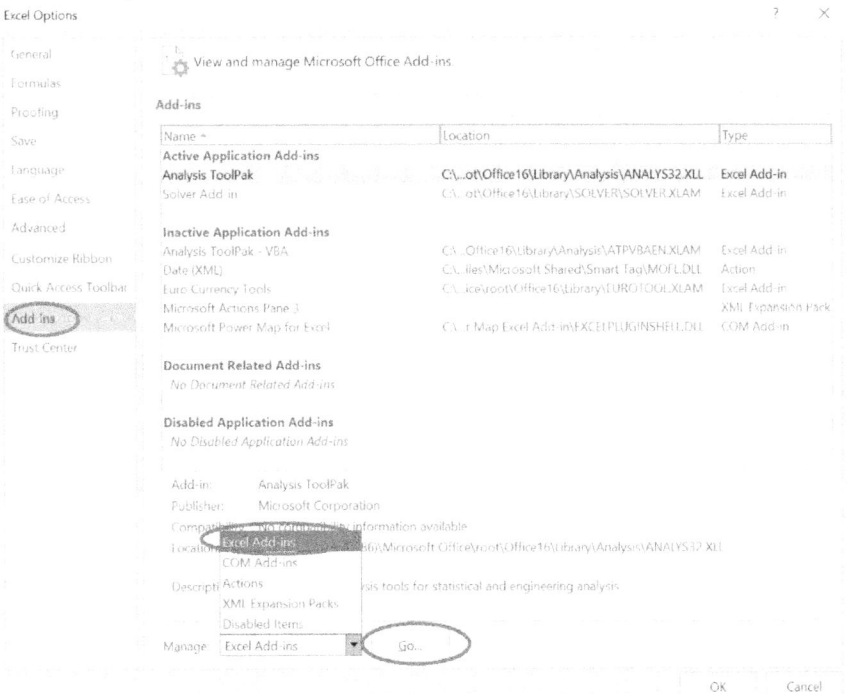

10) Click "Solver Add-In" and click on OK.

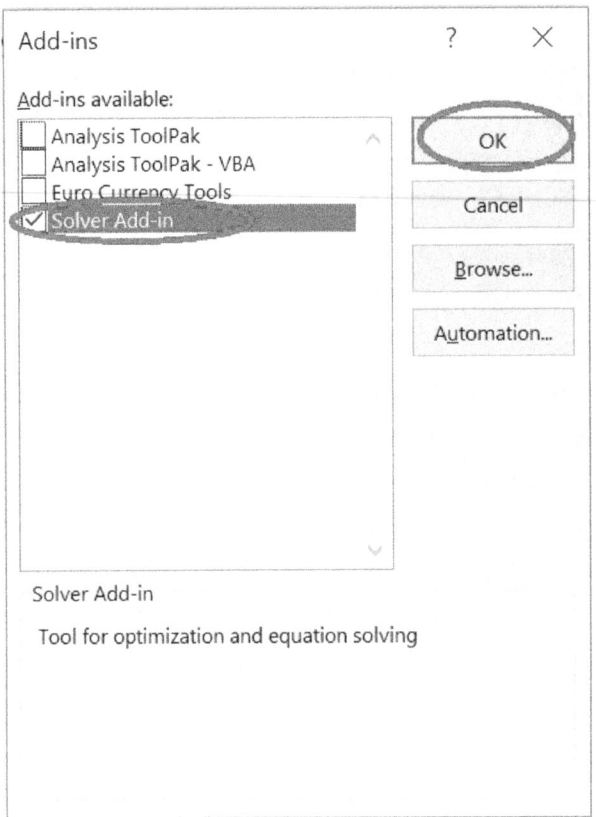

11) On the Data tab, in the Analysis group, you are now able to click on "Solver". Click the Solver button.

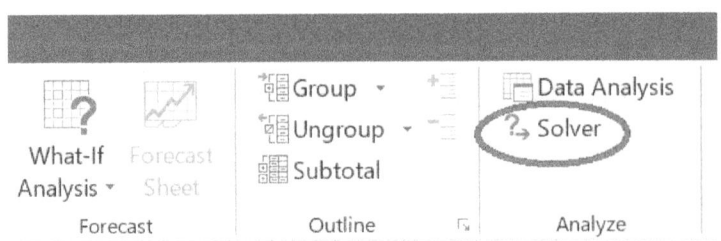

12) The objective is to maximize the sum of the log-likelihood column (LL), by changing the values in H1:H4, representing coefficients b_0-b_3.
- For "Set Objective", select cell (H14), the sum of the log-likelihood column (LL).
- Uncheck the box labeled "Make Unconstrained Variables Non-Negative".
- For "Select a Solving Method", select "GRG Nonlinear" because we are not performing a linear optimization.
- Click "Solve"

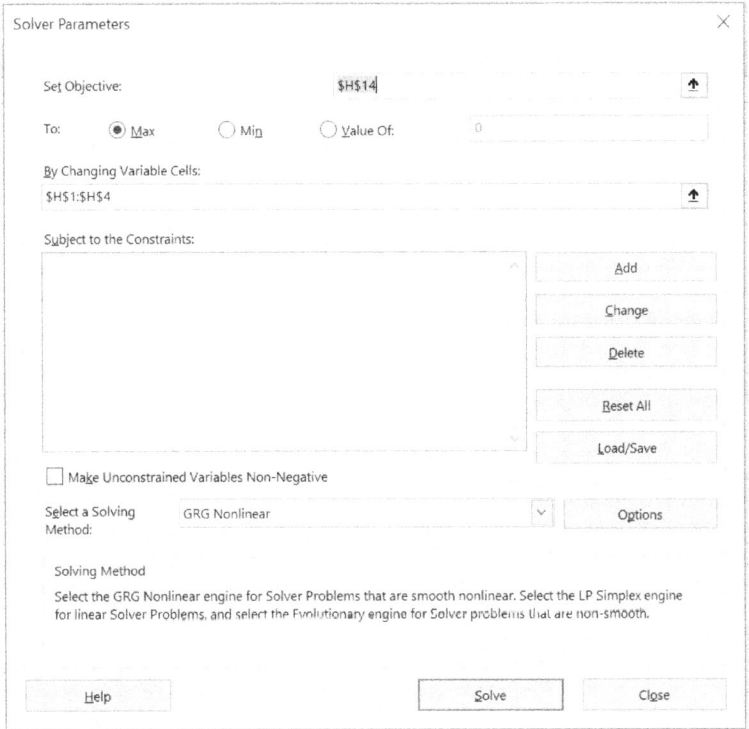

13) After clicking "Solve", you will see the screen "Solver Results". Check "Keep Solver Solution".

14) After clicking "Solve" you get new values in H1:H4, representing coefficients b_0-b_3.

	A	B	C	D	E	F	G	H
1						b_0 (Intercept)=		12.117
2						b_1=		-5.452
3						b_2=		-0.163
4						b_3=		0.814
5								
6								
7		Resign	Compa-ratio	Age	Travel time	Logit (L)	e^L	P (X)	Log-Likelihood (LL)
8	0	1.3	55	1	-3.136	0.043	0.042	-0.043
9	0	1.2	34	0	0.023	1.024	0.506	-0.705
10	0	1	50	0	-1.499	0.223	0.183	-0.202
11	1	1.2	39	0	-0.793	0.452	0.312	-1.166
12	1	0.5	34	1	4.654	104.986	0.991	-0.009
13	1	0.4	45	1	3.403	30.056	0.968	-0.033
14							sum of LL:	-2.158

15) To predict whether a Staff whose compa-ratio is 0.7, age is 30, and Travel time is 1 (far) will resign, copy those figures that are circled in your spreadsheet.

E18   fx   =$H$1+$H$2*B18+$H$3*C18+$H$4*D18

	A	B	C	D	E	F	G	H	I
1							b_0 (Intercept)=	12.117	
2							b_1=	-5.452	
3							b_2=	-0.163	
4							b_3=	0.814	
5									
6									
7	Resign	Compa-ratio	Age	Travel time	Logit (L)	e^L	P (X)	Log-Likelihood (LL)	
8	0	1.3	55	1	-3.136	0.043	0.042	-0.043	
9	0	1.2	34	0	0.023	1.024	0.506	-0.705	
10	0	1	50	0	-1.499	0.223	0.183	-0.202	
11	1	1.2	39	0	-0.793	0.452	0.312	-1.166	
12	1	0.5	34	1	4.654	104.985	0.991	-0.009	
13	1	0.4	45	1	3.403	30.054	0.968	-0.033	
14							sum of LL:	-2.158	
15									
16									
17		Compa-ratio	Age	Travel time	Logit (L)	e^L	P (X)		
18		0.7	30	1	4.217				
19									

You will notice that the formula is cell E18 is:

Logit(L)
=$H$1+$H$2*B18+$H$3*C18+$H$4*D18
= b_0 + b_1*Compa-ratio + b_2*Age + b_3*Travel time
= 12.117 − 5.452*Compa-ratio − 0.163*Age + 0.814*Travel time

Just like a multiple linear regression, you need to enter the new b_0, b_1, b_2 and b_3 coefficient values into your logistic regression equation to predict a value. But unlike a linear regression that predicts values like sales amount, the logistic regression equation predicts probabilities.

16) Next, you need to calculate e^L. The number e is the base of the natural logarithm. It is approximately equal to 2.71828163. Enter the Exponential formula below, in your Excel spreadsheet:

F18   fx   =EXP(E18)

	A	B	C	D	E	F	G	H
1							b_0 (Intercept)=	12.117
2							b_1=	-5.452
3							b_2=	-0.163
4							b_3=	0.814
5								
6								
7	Resign	Compa-ratio	Age	Travel time	Logit (L)	e^L	P (X)	Log-Likelihood (LL)
8	0	1.3	55	1	-3.136	0.043	0.042	-0.043
9	0	1.2	34	0	0.023	1.024	0.506	-0.705
10	0	1	50	0	-1.499	0.223	0.183	-0.202
11	1	1.2	39	0	-0.793	0.452	0.312	-1.166
12	1	0.5	34	1	4.654	104.985	0.991	-0.009
13	1	0.4	45	1	3.403	30.054	0.968	-0.033
14							sum of LL:	-2.158
15								
16								
17		Compa-ratio	Age	Travel time	Logit (L)	e^L	P (X)	
18		0.7	30	1	4.217	67.800		

17) Next, you need to calculate P(X). P(X) is the probability of event X occurring. Enter the formula for probability of the event (i.e. Buy) in your Excel spreadsheet, using the formula below:

$$P(X) = e^L / (1 + e^L)$$

G18   fx   =F18/(1+F18)

	A	B	C	D	E	F	G	H
1							b_0 (Intercept)=	12.117
2							b_1=	-5.452
3							b_2=	-0.163
4							b_3=	0.814
5								
6								
7	Resign	Compa-ratio	Age	Travel time	Logit (L)	e^L	P (X)	Log-Likelihood (LL)
8	0	1.3	55	1	-3.136	0.043	0.042	-0.043
9	0	1.2	34	0	0.023	1.024	0.506	-0.705
10	0	1	50	0	-1.499	0.223	0.183	-0.202
11	1	1.2	39	0	-0.793	0.452	0.312	-1.166
12	1	0.5	34	1	4.654	104.985	0.991	-0.009
13	1	0.4	45	1	3.403	30.054	0.968	-0.033
14							sum of LL:	-2.158
15								
16								
17		Compa-ratio	Age	Travel time	Logit (L)	e^L	P (X)	
18		0.7	30	1	4.217	67.800	0.985	

From the spreadsheet, if Staff compa-ratio is 0.7, Age is 30, and Travel time is 1 (far), the probably that he will resign is 0.985, which is closer to 1 than to 0. Closer to 1 means it is probably "Resign", while closer to 0 means it is probably "Stay".

20) Predict Performance

Research have shown that Performance (e.g. Profitability, Income Income, Service, Sales, Total Share Holder Returns, Productivity) is affected by factors such as: Communications, Compensation, Culture, Demographics, Diversity & Inclusion, Engagement, Performance Management, Personality traits, Quality, Health and Safety, Social Network, Staff Attrition, Training & Development.

Communications' impact on Performance & Profitability:
- **Profitability:** Regular communication between leaders and employee is important. While only 3 in 10 employees say that they receive regular feedback, research suggests that by improving that ratio to 6 in 10, your company could see a 26% drop in absenteeism and an 11% jump in profits. [1]
- **Performance:** We all spend a lot of time communicating with colleagues at work. Some respond fast, others do it when they have time, while others need nudges to respond. Do these behaviors correlate with our performance at work? After analyzing months of communication patterns using messaging metadata (data about the messages, not the messages themselves), Genpact is able to statistically prove that certain types of communication behavior directly correlates to business performance. [2]
- **Performance:** A company found that the best performing engineers have more meetings and relationships than others. To make the engineers to walk around more to build relationship with others, HR rearranged the engineering building and move the cafeteria across the campus. [2]
- **Performance:** Praful Tickoo, Genpact's head of people analytics found that there is a 74% statistical correlation between communication patterns and the best performance rating for individual (using a 9-box performance process). The best performing leaders use simpler words to communicate, respond faster, and communicate more frequently. The more relationships they have, the higher performer they were. [2]

Compensation's impact on Sales:
- **Sales:** The bigger the pay spread, the better a company's performance. There is a correlation of 0.4 between net income growth and standard deviation of merit increase. [3]

Culture's impact on Service:
- **Service:** Research by Ployhart et al. (2011) demonstrated that stores with higher "service climate" and "perceived autonomy" had higher customer service performance. "Service climate" was measured by getting employees to rate their own restaurant's service climate using a short scale. A climate for service is an organizational climate where good service is valued, facilitated, and rewarded. When an organization has a climate for service, employees know that good customer service is desired, rewarded and encouraged. This type of organizational climate motivates employees to provide better service. "Perceived autonomy" is the degree which an employee can influence decisions at work. Perceived autonomy leads to higher customer service because employees with more autonomy feel responsible and thus are more assertive. Furthermore, autonomy means more decision latitude, which enables these employees to solve problems faster. [4]

Demographics' impact on Sales:
- **Sales:** Retention drives sales. Companies with the lowest sales has General Managers with the shortest tenure. [5]

Diversity and Inclusion's impact EBIT, Market Share & Performance:
Many researches have shown that companies with diverse gender or racial and ethnic employee profile are more likely to have financial returns above their national industry medians. Diverse sales force has a wider range of perspectives, and can relate better to the needs of the diverse customers they are selling to, as some customers may prefer to see themselves represented at companies they buy from.

- **Performance Ratings:** Deloitte discovered that the more included an employee feels, the more likely they will get a higher performance rating. In contrast, the more an employee feels excluded (e.g. that they are not treated respectfully or they do not feel belong to the team), the less likely they do their best. [6]

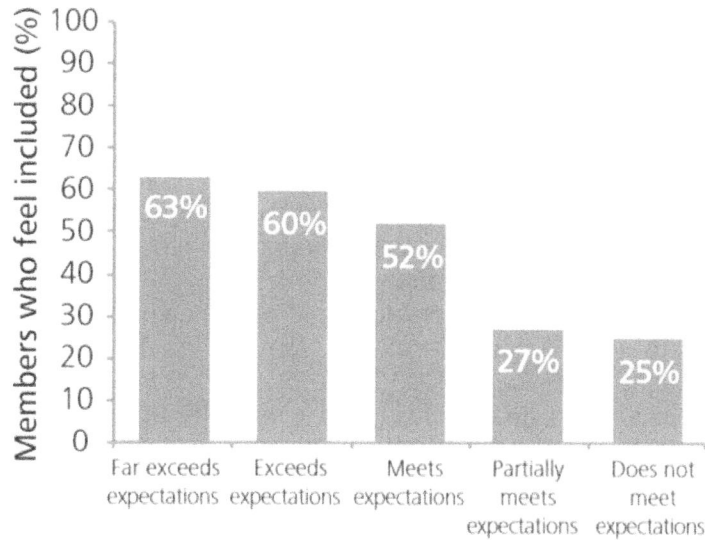

Source: Deloitte Australia (Deloitte) and the Victorian Equal Opportunity and Human Rights Commission (2013) A new recipe to improve business performance, https://www2.deloitte.com/content/dam/Deloitte/au/Documents/human-capital/deloitte-au-hc-diversity-inclusion-soup-0513.pdf (9 November 2018)

- **Sales:** For elMejorTrato.com, hiring for diversity has led to creating new products. They hired team members aged 55 and older and realized that their service, with some modifications, could also be useful for a market segment older than 50 years. Sales jumped 24.4 percent the following year. [7]
- **Market Share:** A study by the Harvard Business Review (HBR) found that a team with a member who shares a client's ethnicity is 152% more likely to understand that client than another team. HBR found that diverse companies are 45% likelier to report that their firm's market share grew over the previous year and 70% likelier to report that the firm captured a new market. [8]
- **Sales:** According to report in the American Sociological Review, workplace diversity is one of the most important predictors of a business' sales revenue, customer numbers and profitability. Cedric Herring found that companies with the highest racial diversity has 15 times more sales revenue on average than those with the lowest racial diversity. Every percentage increase in the rate of racial or gender diversity, results in an increase in sales revenues of 9 and 3 percent, respectively. Companies with a more diverse workforce reported higher customer numbers than those organizations with less diversity. The difference is even bigger for gender diversity rates. [9]
- **EBIT:** McKinsey's report, Diversity Matters, studied 366 public companies across various industries in Canada, Latin America, United Kingdom, and United States. The research found that [10]:
  - Companies in the top quartile for racial and ethnic diversity are 35 percent more likely to have financial returns above their respective national industry medians.
  - Companies in the top quartile for gender diversity are 15 percent more likely to have financial returns above their respective national industry medians.
  - In the United States, for every 10 percent increase in racial and ethnic diversity on the senior-executive team, earnings before interest and taxes (EBIT) rise 0.8 percent.
  - In the United Kingdom, for every 10 percent increase in gender diversity on the senior-executive team, EBIT rose by 3.5 percent.

Engagement's impact on Service, Profitability, Total Shareholder Returns:

- **Operating & Net Profit Margin:** Operating margin was 2.1% in low engagement companies versus 3.75% in high engagement companies; net profit margin was -1.38% in low engagement companies versus 2.06% in high engagement companies. [11]
- **Profit Margin:** Standard Chartered Bank reported that in 2007 they found that branches with a statistically significant increase in employee engagement (0.2 or more on a scale of five) had a 16 per cent higher profit margin growth compared to branches with decreased levels of employee engagement. [11]
- **Total Shareholder Returns:** A study conducted across 39 organizations showed that organizations with highly engaged employees achieve seven times greater 5-year total shareholder return (TSR) than organizations whose employees are less engaged. In companies where 60 to 70 percent of employees were engaged, average total shareholder's return (TSR) is 24.2 percent; in companies with only 49 to 60 percent of their employees engaged, TSR fell to 9.1 percent; companies with engagement below 25 percent suffered negative TSR. [11]
- **Service:** Sears found that a 5-point improvement in employee attitudes leads to a 1.3-point improvement in customer satisfaction, which in turn leads to a 0.5% improvement in revenue. [11]
- **Net Income:** Best Buy can accurately predict how employee engagement impacts the performance of their stores. A 0.1% increase in employee engagement results in an increase of over $100,000 in the store's annual income. The enormous impact of engagement prompted Best Buy to make its engagement surveys quarterly instead of annually. [12]

- **Service:** ISS found that Employee Engagement correlates strongly with Customer Experience. The strength of this correlation was 0.55. Diagram below shows the customer net promoter score (cNPS) and the employee net promoter score (eNPS). Net promoter score (NPS) is a tool to measure the loyalty of a firm's customers and employees. The NPS was based on the question: "How likely is it that you would recommend a company, product or service to a friend or colleague?" This answer was scored based on a 0-10 scale. Promoters (loyal enthusiasts) are those who respond with a score of 9 or 10. Passives are those who respond with a score of 7 or 8. Detractors (unhappy customers) are those who respond with a score of 0 to 6. Calculate the NPS by subtracting the percentage who are detractors from the percentage who are promoters. The drivers behind customer experience are: Motivation and engagement of service staff, Amount of training and quality of service staff, and the ability to act on customer expectations by customer service staff. With the findings, ISS recommendations was to monitor metrics on service employee engagement in areas such as the eNPS (employee net promoter score), employee churn, absenteeism and training hours.[13]

Correlation between eNPS and cNPS

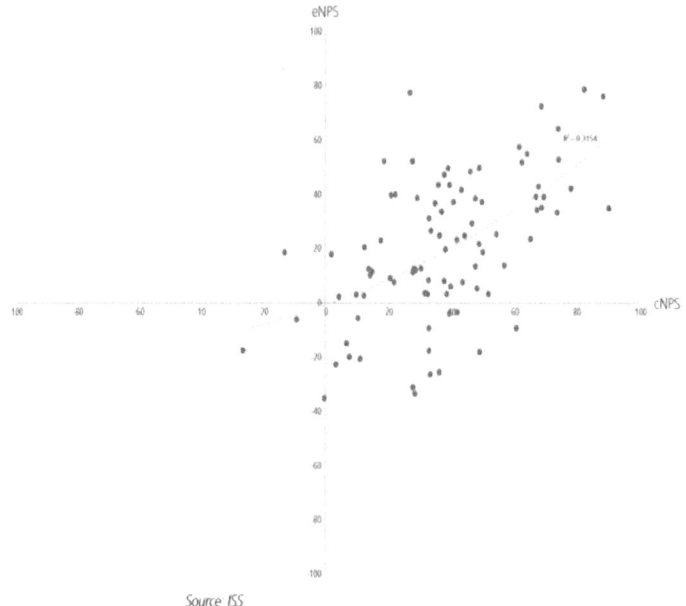

Source: Morten Kamp Andersen (proacteur), Simon Svegaard (ISS) & Peter Ankerstjerne (ISS). (2015), Linking Customer Experience with Service Employee Engagement, ISS White Paper

- **Profitability:** A study by ISS found that profitability is high, when employee engagement and customer advocacy are high. The average profitability in countries scoring highest on both employee engagement and customer advocacy was 7.75 percent, versus 4.52 percent for the lowest scoring groups. ISS tested this link by combining selected questions from their Employee Engagement Survey and their Customer Experience Survey with their profitability. In the diagram below, the eNPS (employee net promoter score) and cNPS (customer net promoter score) scores are depicted on the X and Y axis respectively, while the average profitability is shown in the boxes. The figure shows that if the employees and the customers report a high level of satisfaction (as measured by the NPS), then the country will have significantly higher margins than if either of the two scores are low. Implication of this study is that companies should reinvest more in customer service improvements, employee training, on-boarding processes, and service innovation. [14]

Link between margins and employee & customer net promoter scores at a country level

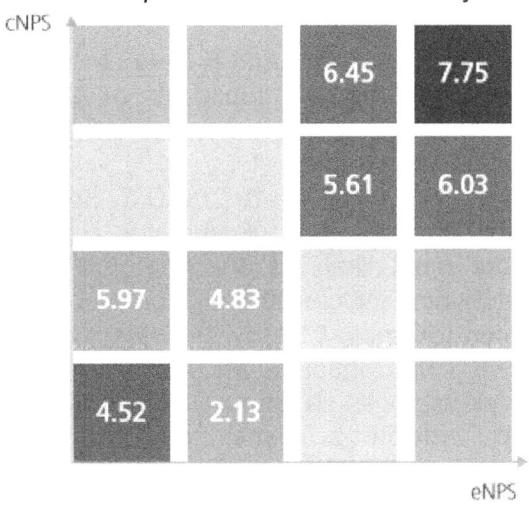

Source: ISS Prospectus, 2014

- Source: Morten Kamp Andersen (proacteur), Simon Svegaard (ISS) & Peter Ankerstjerne (ISS). (2015), Linking Customer Experience with Service Employee Engagement, ISS White Paper, pp 10-11.

Ethics' impact on Sales:
- **Sales margin:** In a Wall Street Journal article, they conducted an experiment that measured what people would pay for an item based on ethics. They divided each testing group into three categories: highly ethical, control, and unethical. Those presented an brand of coffee beans regarded as highly ethical were willing to spend $9.71 a pound. Conversely, those presented with a brand of coffee beans considered much less ethical were only willing to spend $5.89. That's a staggering $3.82! If you sell 1000 pounds of coffee beans, the difference in revenue is $3,820. Your cost of goods sold are $4.23. By being ethical, you have $5.83 margin whereas the less ethical brand has only $1.66. Which situation would you rather be in? [15]

Performance Management's impact on Productivity:
- **Productivity:** 6 in 10 employees indicate that they know what is expected of them at work, but if that ratio can be increased to 8 in 10, your business can achieve a 14% reduction in staff turnover and a 7% increase in productivity. [16]

Personality Traits' impact on Performance & Service:
Although most companies evaluate job applicants based on their knowledge, skills and experience, increasingly, many companies are using personality test to determine whether an applicant is a good fit.
- **Performance – "Grit" personality trait:** The U.S. Special Forces found that 'grit' and the ability to do more than 80 pushups are two key predictors of which candidates are most likely to succeed. Grit was actually a more accurate predictor of training success than IQ for the U.S. Special Forces. [17]
- **Service – "Conscientiousness" and "Extraversion" personality trait:** Personality traits like conscientiousness and extraversion provide better customer service experience. A more conscientious and extroverted employee tend to provide a better customer service experience. Conscientiousness people are hardworking, responsible, organized, and dependable. Extraverted people are sociable, active and talkative. They usually full of high-energy when they interact with people (customers). [18]
- **Performance – "Conscientiousness" personality trait:** Most research have found that that conscientiousness is the personality trait most often correlated with job performance for almost all functions. Conscientiousness refers to self-motivated individuals who sets and achieves their own ambitious work goals and always completes the assigned tasks on time.:
  - An Italian study discovered that conscientiousness predicts academic achievement. [19]

- ➤ The National Institute on Aging discovered that conscientiousness is linked to income and job satisfaction. [20]
- **Performance – "Conscientiousness" & "Agreeableness" personality trait:** Sackett and Walmsley studied the model for measuring personality known as the Big Five. In the Big Five model, a person's personality can be described using five personality traits (Openness to experience, Conscientiousness, Extroversion, Agreeableness, Neuroticism). After analysis, Sackett and Walmsley discovered that conscientiousness (i.e. being dependable, persevering, and orderly) was the most highly sought-after personality attribute for job applicants. But conscientiousness does not refer to employees who works long hours in the office. Conscientiousness refers to self-motivated individuals who sets and achieves their own ambitious work goals and always completes the assigned tasks on time. Agreeableness (i.e. being cooperative, flexible and tolerant) was the second most highly sought-after personality trait. The researchers also found that the conscientiousness trait was most closely associated with overall job performance for a wide variety of jobs from construction to health care, with agreeableness coming in second. Although conscientiousness is the most valued personality trait, the researchers cautioned that different jobs may have different rankings for the personality traits. For instance, a company that is recruiting a salesperson would prefer extraversion and friendliness traits to ensure that the salesperson is likely to work well with customers. [21]
- **Performance – "Creativity" personality trait:** Xerox used to hire applicants who had relevant call center experience to staff its call centers. But an analytics algorithm showed that experience does not matter for top performers in call centers. It showed that what matters for a call center worker who won't quit before the company recovers its investment in training – is personality! It further showed that creative people usually stay for at least six months, whereas Inquisitive people usually don't! [22]

Quality, Health and Safety's impact on Profitability:
- **Profitability**: Quality/safety outcomes have a positive correlation (.21) with profitability, so the effect of turnover on profitability through quality/safety is approximately -.025 (the product of .21 times -.12), which is a large portion of the total effect (-.03) of turnover on profits. [23]

Social Network's impact on Performance & Sales:
- **Performance:** Genpact, a global professional services firm, found that the best performing engineers have more meetings and relationships than others. [24]
- **Sales:** Most people also think that the time salespeople spend with customers is the most important factor affecting their sales. Interestingly, VoloMetrix found that the size and amount of a salesperson's network inside their company is a more important leading indicator of sales, than the time salespeople spend with customers. Their research discovered that regardless of what you are sell, who you are sell it to, or your work location anywhere in the world, success in sales correlates highly with three actionable metrics. Merely increasing the time your underperforming salesperson spend with customers will not increase sales much. The second and third factor are more important. Corporate buyers want a salesperson who is credible, who understands their needs, and who is able to address their concerns quickly and competently. Doing this well requires a salesperson is able to get access to management when needed, and have a comprehensive understanding of what their company can offer to the buyer above and beyond the current sales. [25]
  - Spending sufficient time with customers.
  - Having a big network in your own organization.
  - Spending time with your manager and other senior people in your organization.

Staff Attrition's impact on Profitability & Revenue:
- **Profitability**: A research published in the Journal of Management found that the correlation between organizational turnover and profits is -.03, so turnover does relate to lower profits, but the direct relationship is rather modest (explaining about 0.1% of profit variation). An interesting finding was that the effects of turnover on profitability were not much different whether turnover was voluntary or involuntary. Dismissing poor employees may remove bad apples, but the research suggests that the level of such turnover has about the same negative association with profits as when employees decide to leave. If you're constantly fixing selection mistakes by dismissing poor employees, that's a drag on your organization's performance, just as when good employees leave. [26]
- **Revenue**: According to William Wolf, Credit Suisse's global head of talent acquisition and development, a one-point reduction in unwanted attrition rates saves the bank $75 million to $100 million a year. [27]

Training and Development's impact on Productivity & Service:

- **Productivity:** While only 4 in 10 employees say that they have had opportunities to learn and grow in the last year, Gallup suggests that improving that ratio to 8 in 10 could allow your business to achieve a 44% drop in absenteeism and a 16% jump in productivity. [28]
- **Service:** Training improves employee's skills, and more skilled employees tend to provide better service (Ployhart et al., 2011). However, this does not apply for all jobs. Interestingly, a research by Ployhart et al. (2011) that focused on restaurants, showed that service training does not lead to higher service levels. Employees with less complex service jobs (e.g. waiting tables) seem to benefit less from training compared to employees with more complicated service jobs. This is probably because an employee who attended training does not necessarily mean that they will learn or apply their new skills to the job. It might also be that relatively easy service training such as serving food in a restaurant don't impact employees' customer service levels, as compared to more complicated and time-intensive service offering like software sales. [29]

References:
(1) Tanvir Haque (2018) Four Strategies To Boost Employee Engagement That Can Improve Your Bottom Line. https://www.entrepreneur.com/article/312697 (20 November 2018)
(2) Josh Bersin (2018). What Emails Reveal About Your Performance At Work. https://joshbersin.com/2018/10/what-emails-reveal-about-your-performance-at-work/# (12 November 2018)
(3) Jac FITZ-ENZ (2010), The New HR Analytics: Predicting the Economic Value of Your Company's Human Capital Investments, AMACOM
(4) Erik van Vulpen (2018), How 11 Factors Influence Customer Service Performance. AIHR. https://www.analyticsinhr.com/blog/factors-influencing-customer-service-performance/ (15 November 2018)
(5) Jac FITZ-ENZ (2010), The New HR Analytics: Predicting the Economic Value of Your Company's Human Capital Investments, AMACOM. pp. 63 – 65.
(6) Source: Deloitte Australia (Deloitte) and the Victorian Equal Opportunity and Human Rights Commission (2013) A new recipe to improve business performance, https://www2.deloitte.com/content/dam/Deloitte/au/Documents/human-capital/deloitte-au-hc-diversity-inclusion-soup-0513.pdf (9 November 2018)
(7) Tim Beyers (2018), Diversity at Work: 3 Best Practices For Small Business Growth, https://www.capitalone.com/small-business/sparkiq/article/diversity-at-work-3-best-practices-for-small-business-growth/ (24 September 2018)
(8) Sylvia Ann Hewlett, Melinda Marshall, Laura Sherbin, How Diversity Can Drive Innovation, (2013) https://hbr.org/2013/12/how-diversity-can-drive-innovation
(24 September 2018)
(9) American Sociological Association (2009), Research links diversity with increased sales revenue and profits, more customers,
https://www.eurekalert.org/pub_releases/2009-03/asa-rld033009.php
(24 September 2018)
(10) Vivian Hunt, Dennis Layton, and Sara Prince (2015), Why diversity matters, https://www.mckinsey.com/business-functions/organization/our-insights/why-diversity-matters (24 September 2018)
(11) Kevin Kruse (2012) Why Employee Engagement? (These 28 Research Studies Prove the Benefits) https://www.forbes.com/sites/kevinkruse/2012/09/04/why-employee-engagement/#36bbedf63aab (20 November 2018)
(12) Mohit Sharma, Talent Analytics: From Buzzword to Reality (2018), https://sightsinplus.com/2018/05/25/talent-analytics-from-buzzword-to-reality/
(24 September 2018)
(13) Morten Kamp Andersen (proacteur), Simon Svegaard (ISS) & Peter Ankerstjerne (ISS). (2015), Linking Customer Experience with Service Employee Engagement, ISS White Paper, pp 3-9.
(14) Morten Kamp Andersen (proacteur), Simon Svegaard (ISS) & Peter Ankerstjerne (ISS). (2015), Linking Customer Experience with Service Employee Engagement, ISS White Paper, pp 10-11.
(15) Lauren Jefferson (2017), Ethics Affects the Financial Results of a Company, https://strategiccfo.com/ethics-affects-the-financial-results-of-a-company/ (21 June 2019)
(16) Tanvir Haque (2018) Four Strategies To Boost Employee Engagement That Can Improve Your Bottom Line. https://www.entrepreneur.com/article/312697 (20 November 2018)
(17) Mohit Sharma, Talent Analytics: From Buzzword to Reality (2018), https://sightsinplus.com/2018/05/25/talent-analytics-from-buzzword-to-reality/
(24 September 2018)

(18) Erik van Vulpen (2018), How 11 Factors Influence Customer Service Performance. AIHR. https://www.analyticsinhr.com/blog/factors-influencing-customer-service-performance/ (15 November 2018)

(19) Michelangelo Vianello, Egidio Robusto, Pasquale Anselmi (2009), https://www.researchgate.net/publication/230559834_Implicit_conscientiousness_predicts_academic_performance *(13 November 2018)*

(20) Angelina R. Sutin, Paul T. Costa, Jr., Richard Miech, and William W. Eaton (2009), https://www.ncbi.nlm.nih.gov/pmc/articles/PMC2747784/ *(13 November 2018)*

(21) Tanya Pinto (2016). How the Big Five Personality Traits Can Predict Performance at Work. https://peakon.com/blog/people-management/big-five-personality-traits-in-the-workplace/ *(13 November 2018)*

(22) Joseph Walker, (2012) Meet the New Boss: Big Data. The Wall Street Journal. https://winintelligence.org/meet-the-new-boss-big-data/ *(24 September 2018)*

(23) John Boudreau (2014), Why Employee Turnover Hits the Bottom Line. http://ww2.cfo.com/workplace-issues/2014/03/employee-turnover-hits-bottom-line/ *(27 November 2018)*

(24) Josh Bersin (2018). What Emails Reveal About Your Performance At Work. https://joshbersin.com/2018/10/what-emails-reveal-about-your-performance-at-work/# *(12 November 2018)*

(25) Ryan Fuller (2015) What Makes Great Salespeople. Harvard Business Review. https://hbr.org/2015/07/what-makes-great-salespeople *(13 November 2018)*

(26) John Boudreau (2014), Why Employee Turnover Hits the Bottom Line. http://ww2.cfo.com/workplace-issues/2014/03/employee-turnover-hits-bottom-line/ *(27 November 2018)*

(27) Deloitte Australia (Deloitte) and the Victorian Equal Opportunity and Human Rights Commission (2013) A new recipe to improve business performance, https://www2.deloitte.com/content/dam/Deloitte/au/Documents/human-capital/deloitte-au-hc-diversity-inclusion-soup-0513.pdf *(9 November 2018)*

(28) Tanvir Haque (2018) Four Strategies To Boost Employee Engagement That Can Improve Your Bottom Line. https://www.entrepreneur.com/article/312697 *(20 November 2018)*

(29) Erik van Vulpen (2018), How 11 Factors Influence Customer Service Performance. AIHR. https://www.analyticsinhr.com/blog/factors-influencing-customer-service-performance/ *(15 November 2018)*

20.1) Case 1: Deloitte – Characteristics of High-Performing Salesperson In Financial Services

Characteristics of high-performing sales candidates

What didn't matter
College degree or reputation of college
Grade point average
Quality of references

What did matter
Lack of typos or misspellings in resumes
Successful experience selling related products (autos and real estate)
Successful experience completing some academic degree
(which one did not matter)

Graphic: Deloitte University Press | DUPress.com
Source: Josh Bersin (2014) Deloitte Review issue 14: The datafication of HR
https://www2.deloitte.com/insights/us/en/deloitte-review/issue-14/dr14-datafication-of-hr.html   (11 November 2018)

One of the greatest benefits of HR analytics is the debunking of typical management myths. In this example, by changing to a new method of assessing sales candidates, the financial services generated more than $4 million of new revenue in the first six months. A financial services company noticed that there were big variations in sales performance and retention among its salesforce, so they developed a model to predict sales representative performance and retention. Traditionally this company hired its salesforce from top universities and hired applicants with good grade point averages. [1]

After much study, the analytics team discovered that:
- High performers were not from the top schools.
- High performers did they have the best grades.
- References do not matter for High performers.

Instead, the analytics team discovered that the characteristics of high-performing sales candidates are:
- Lack of typos or misspellings in resumes
- Successful experience selling related products (Autos and real estate)
- Successful experience completing some academic degree (Which one did not matter)

References:
(1) Josh Bersin (2014) Deloitte Review issue 14: The datafication of HR
https://www2.deloitte.com/insights/us/en/deloitte-review/issue-14/dr14-datafication-of-hr.html   (11 November 2018)

20.2) Case 2: HBR –What Makes Great Salespeople

VoloMetrix found three metrics (time with customers; larger internal networks; time with managers and senior leadership) that highly correlated with top performing salespeople regardless of region, territory, or sales role. Building from the above findings, VoloMetrix developed a broader framework for those behaviors (two were combined), and an additional one. These key behavioral metrics are strong predictors of sales results and are highly actionable metrics to monitor, consider in territory design and to improve sales performance [1].

1. **Customer engagement.** This not only includes time with customers, but also the number of accounts touched; time spent with each account; frequency of interactions with each account; and breadth and depth of relationships with customers. Customer engagement does not merely imply spending time with more customers. Depth is more important than breadth - top performing salespeople concentrate on establishing deep relationships with fewer customers, instead of spending time establishing shallow relationship with many customers. However, the right balance varies by company depending on what is sold. For example, depth is better for consultative sales, while breadth is better for transactional sales.

2. **Internal networks.** Successful salespeople with a bigger relationship are exposed to more ideas, have easier access to expertise, and have a better understanding about what's happening. There are three sub-categories for internal network:
   - **General**: This is the overall number of relationships in the organization, time spent with colleagues, and influence in the network.
   - **Support resources**: These are the relationships that **salespeople** establish with sales support staff, including pre-sales staff, inside sales representatives, etc.
   - **Management**: These are the relationships between **salespeople** and their direct managers, and company leadership. Relationship with Management is an important area of internal networking. More exposure to senior leadership correlates with successful sales results.

3. **Energy:** This an additional perspective that is linked to customer engagement and internal networks. It measures the overall time and effort of the salespeople.

Based on **VoloMetrix** findings, recommendations to improve salespeople include:
- Use onboarding to help your employees grow their internal networks, because new hires should meet and mingle with many diverse colleagues to be effective in their work.
- Let your salespeople have access to consistent support staff. Having salespeople to start over with a different support staff for each account reduces their effectiveness.

References:
(1) Ryan Fuller (2015) *What Makes Great Salespeople. Harvard Business Review.* https://hbr.org/2015/07/what-makes-great-salespeople (13 November 2018)

20.3) Correlation Example: Predict how successful a candidate will be if hired

Research has shown that personality traits like conscientiousness and extraversion provide better customer service experience. Conscientiousness individuals will usually do what is expected of them, while extraverted people usually show high-energy when they interact with customers. [1]

In the data below, we want to predict a Service Staff's Customer Service performance rating using Excel correlation. The data represents events that have already happened.

	A	B	C	D	E
1	Service Staff	Customer Service performance	Conscientiousness Trait	Extraversion Trait	Agreeableness Trait
2	Anne	51%	55%	56%	60%
3	Bond	67%	70%	75%	81%
4	Charle	39%	40%	57%	51%
5	Dan	54%	55%	60%	55%
6	Eric	71%	70%	75%	50%
7	Fanny	53%	50%	56%	38%
8	Gaia	45%	50%	59%	59%
9	Han	70%	68%	75%	72%
10	Ian	69%	65%	73%	68%
11	John	63%	70%	72%	76%
12	Kate	56%	60%	55%	70%

1) Install "Analysis ToolPak", an Excel add-in

"Analysis ToolPak" is an add-in for Microsoft Excel that comes with Microsoft Excel. To be able to run regression using Excel, you need to first install "Analysis ToolPak", an Excel add-in program that provides data analysis tools. To load the Analysis ToolPak add-in, follow these steps:

- On the File tab, click Options.

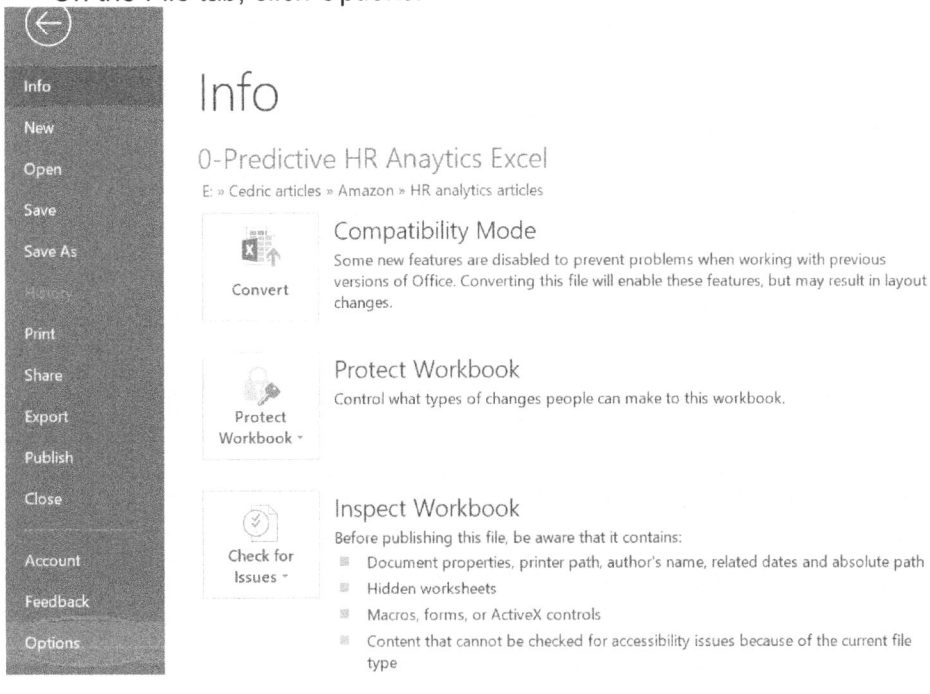

- Under Add-ins, click Analysis ToolPak and click the "Go" button.

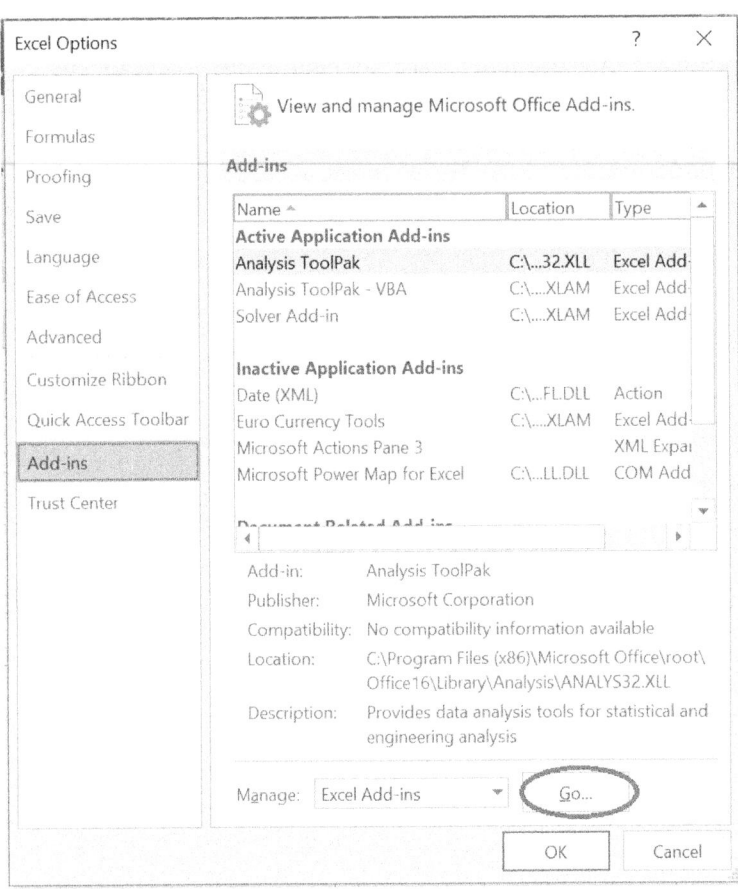

- Click "Analysis ToolPak" and click on OK.

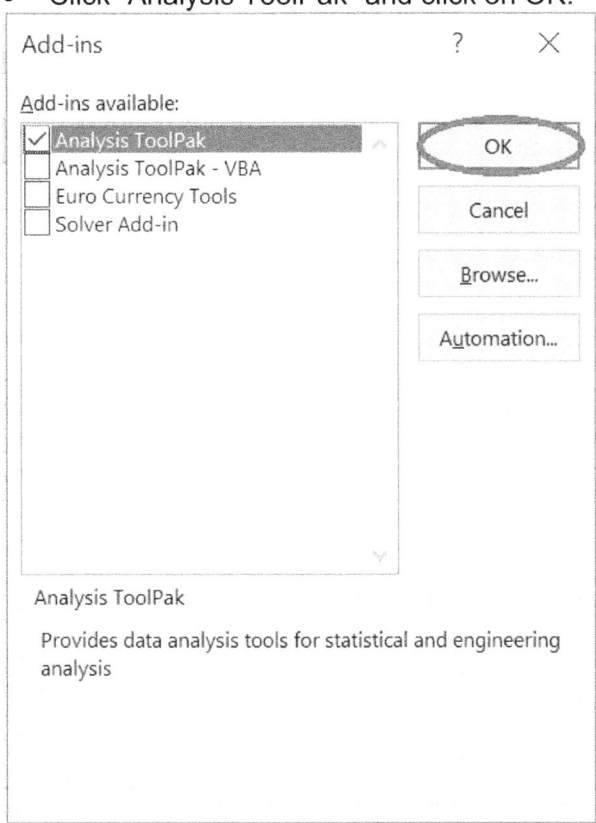

- On the Data tab, in the Analysis group, you are now able to click on "Data Analysis".

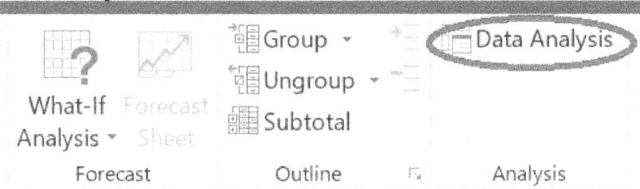

2) Copy the example data in the following table, and paste it in cell A1 of a new Excel worksheet.

	A	B	C	D	E
1	Service Staff	Customer Service performance	Conscientiousness Trait	Extraversion Trait	Agreeableness Trait
2	Anne	51%	55%	56%	60%
3	Bond	67%	70%	75%	81%
4	Charle	39%	40%	57%	51%
5	Dan	54%	55%	60%	55%
6	Eric	71%	70%	75%	50%
7	Fanny	53%	50%	56%	38%
8	Gaia	45%	50%	59%	59%
9	Han	70%	68%	75%	72%
10	Ian	69%	65%	73%	68%
11	John	63%	70%	72%	76%
12	Kate	56%	60%	55%	70%

3) Select "Correlation" and click "OK".

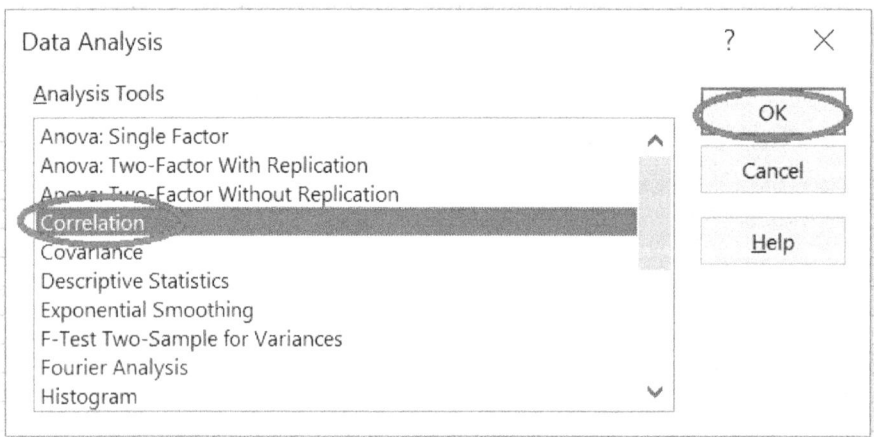

4) After you click OK in the "Data Analysis" dialog box, you will see a "Correlation" dialog box.
5) For "Input Range", select cells (B1:E12).
6) Check "Labels in first row"
7) For "Output Range", select cells (A14).
8) Click "OK"

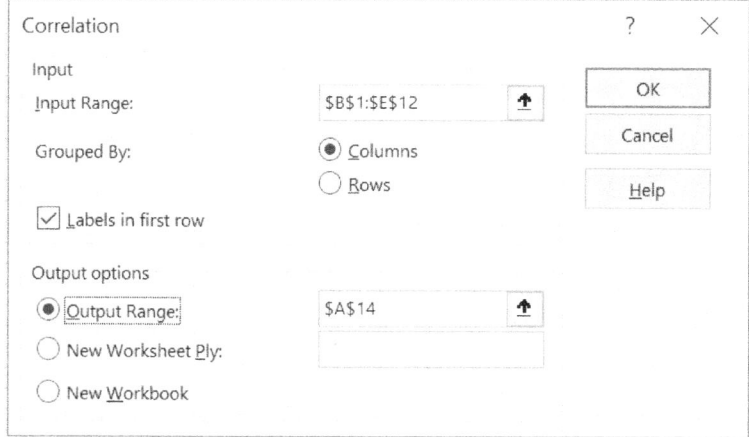

After you click "OK", Excel generates the following Correlation analysis.

	Customer Service performance	Conscientiousness Trait	Extraversion Trait	Agreeableness Trait
Customer Service performance	1			
Conscientiousness Trait	0.94	1		
Extraversion Trait	0.87	0.85	1	
Agreeableness Trait	0.48	0.65	0.53	1

A negative correlation coefficient means that an increase in X is associated with a decrease in Y. Similar to a positive correlation, a negative correlation shows a connection between two variables, and the relative strengths are the same. In other words, a correlation coefficient of 0.85 has the same strength as a correlation coefficient of -0.85. Correlation coefficients are always values between -1 and 1, where "-1" means that there is a perfect linear negative correlation, while "1" shows a perfect linear positive correlation. A correlation coefficient of zero, or near to zero, means that there is no meaningful relationship between variables. Correlation coefficient of 0.91 or -0.92 shows a very strong positive and negative correlation respectively. However, correlation does not mean causation. An example of negative correlation is the amount of snowfall and the temperature. As the temperature increases, the amount of snowfall decreases. An example of positive correlation is the relationship between temperature and ice cream sales. As temperature increases, so do ice cream sales.

9) Highlight those numbers that have strong positive correlation (above 0.85) and strong negative correlation (below -0.85) with Customer Service performance.

	Customer Service performance	Conscientiousness Trait	Extraversion Trait	Agreeableness Trait
Customer Service performance	1			
Conscientiousness Trait	0.94	1		
Extraversion Trait	0.87	0.85	1	
Agreeableness Trait	0.48	0.65	0.53	1

From the Excel Correlation analysis, these variables are good predictors of Customer Service performance as they have strong correlation of below -0.75 and above 0.75:
- Conscientiousness Trait (correlation of 0.94 with Customer Service performance)
- Extraversion Trait (correlation of 0.87 Customer Service performance)

From the Excel Correlation analysis, Agreeableness Trait (correlation of 0.48) has very little impact on Customer Service performance as they have very weak correlation of between -0.50 to 0.50.

References:
(1) Erik van Vulpen (2018), How 11 Factors Influence Customer Service Performance. AIHR. https://www.analyticsinhr.com/blog/factors-influencing-customer-service-performance/ (15 November 2018)

20.4) Multiple Regression Example: Predict an employee's performance rating based on their "Social Network", "Skillsets", & "Personality Traits".

Research have shown that Performance is affected by social network, time spend with senior management, and personality traits:
- Sackett and Walmsley discovered that conscientiousness (i.e. being dependable, persevering, and orderly) was most closely associated with overall job performance for a wide variety of jobs from construction to health care. But conscientiousness does not refer to employees who works long hours in the office. Conscientiousness refers to self-motivated individuals who sets and achieves their own ambitious work goals and always completes the assigned tasks on time. [1]
- VoloMetrix found that the size and amount of a salesperson's network inside their company is a more important leading indicator of sales, than the time salespeople spend with customers. Corporate buyers want a salesperson who is credible, who understands their needs, and who is able to address their concerns quickly and competently. Doing this well requires a salesperson is able to get access to management when needed, and have a comprehensive understanding of what their company can offer to the buyer above and beyond the current sales. Their research discovered that regardless of what you are sell, who you are sell it to, or your work location anywhere in the world, success in sales correlates highly with three actionable metrics. [2]
  ➢ Spending sufficient time with customers.
  ➢ Having a big network in your own organization.
  ➢ Spending time with your manager and other senior people in your organization.

In the data below, we want to predict a person's performance rating using Excel multiple regression. The data represents events that have already happened.

	A	B	C	D	E	F	G
1	Employee	Social Network Size	Social Network Diversity Index	Competency	Time with Management (hours)	Conscientiousness Trait	Performance Rating
2	Ann	30	3.2	9	150	10	10
3	Andy	24	3.3	9	145	9	9
4	Ben	21	3.3	8	135	9	8
5	Bond	20	3.3	7	120	7	6
6	Carl	18	3.3	6	130	7	6
7	Candy	15	3.4	6	120	6	5
8	Dave	8	2.5	5	70	5	5
9	Eddy	5	1.7	5	80	5	4
10	Faith	5	1.0	3	10	2	4
11	Gary	2	1.0	2	0	2	3

1) Install "Analysis ToolPak", an Excel add-in

"Analysis ToolPak" is an add-in for Microsoft Excel that comes with Microsoft Excel. To be able to run regression using Excel, you need to first install "Analysis ToolPak", an Excel add-in program that provides data analysis tools. To load the Analysis ToolPak add-in, follow these steps:

On the File tab, click Options.

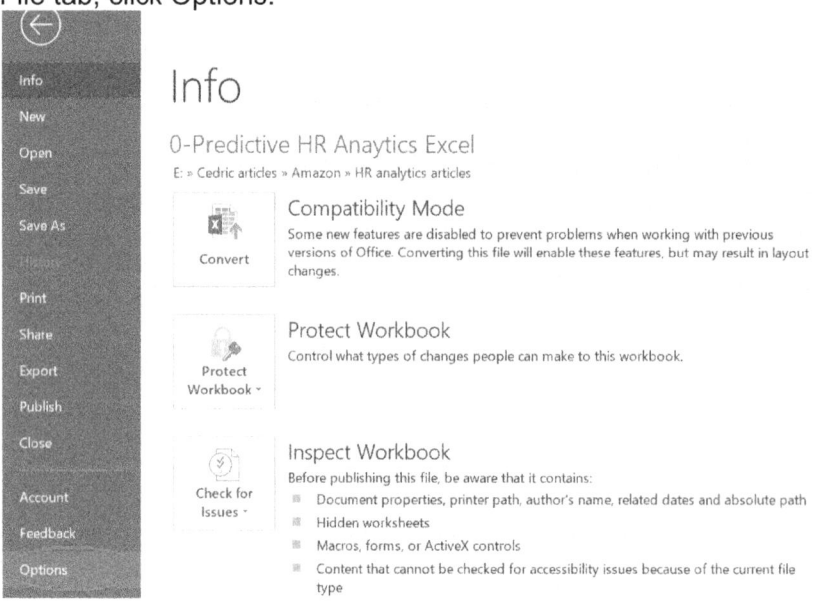

Under Add-ins, click Analysis ToolPak and click the "Go" button.

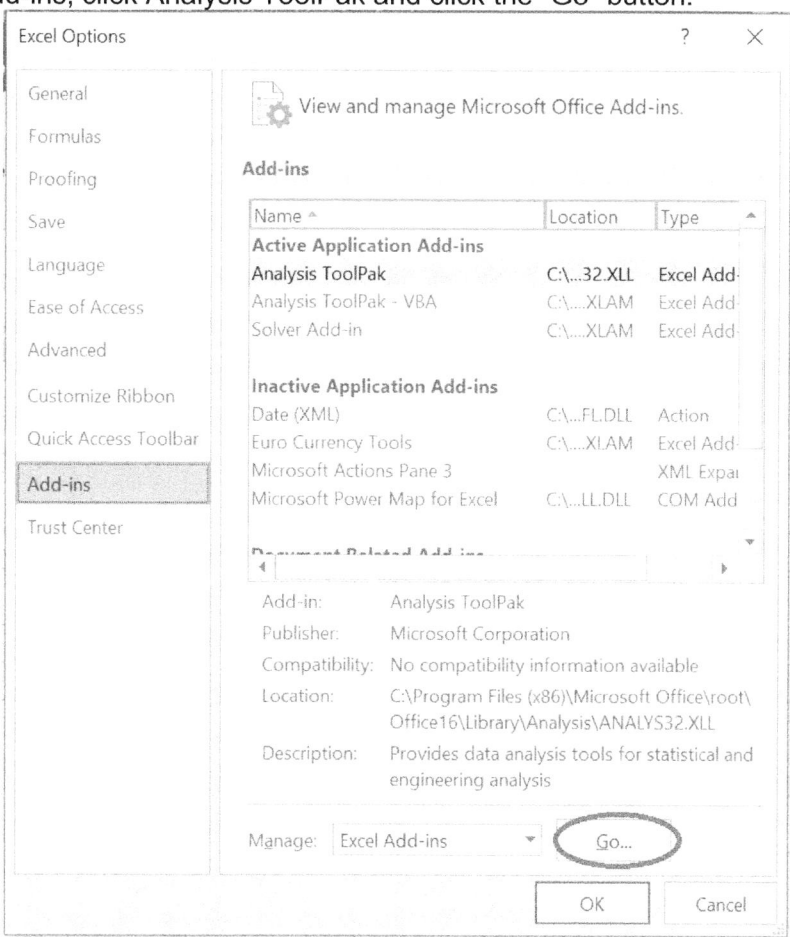

Click "Analysis ToolPak" and click on OK.

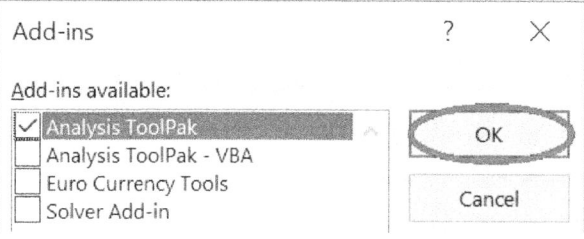

On the Data tab, in Analysis group, you are now able to click on "Data Analysis".

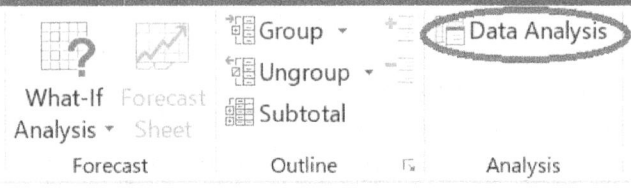

2) Copy the data below, and paste it in cell A1 of a new Excel worksheet.

	A	B	C	D	E	F	G
1	Employee	Social Network Size	Social Network Diversity Index	Competency	Time with Management (hours)	Conscientiousness Trait	Performance Rating
2	Ann	30	3.2	9	150	10	10
3	Andy	24	3.3	9	145	9	9
4	Ben	21	3.3	8	135	9	8
5	Bond	20	3.3	7	120	7	6
6	Carl	18	3.3	6	130	7	6
7	Candy	15	3.4	6	120	6	5
8	Dave	8	2.5	5	70	5	5
9	Eddy	5	1.7	5	80	5	4
10	Faith	5	1.0	3	10	2	4
11	Gary	2	1.0	2	0	2	3

3) On the Data tab, in the Analysis group, click on "Data Analysis".

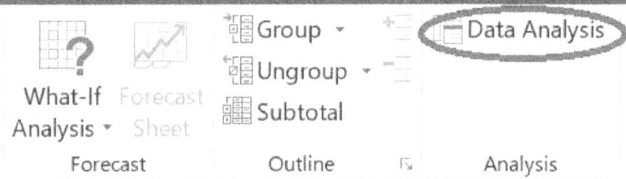

4) Select "Regression" and click "OK".

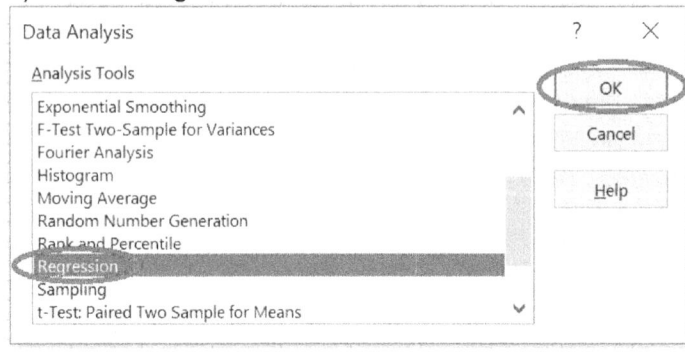

5) After you click OK in the "Data Analysis" dialog box, you will see a "Regression" dialog box.
6) For "Input Y Range", select cells (G1:G11). This is the predictor variable or dependent variable.
7) For "Input X Range", select cells (B1:F11). These are the explanatory variables or independent variables.
8) Check "Labels" box.
9) Click the "Output Range" box, and select cell A13.
10) Click "OK".

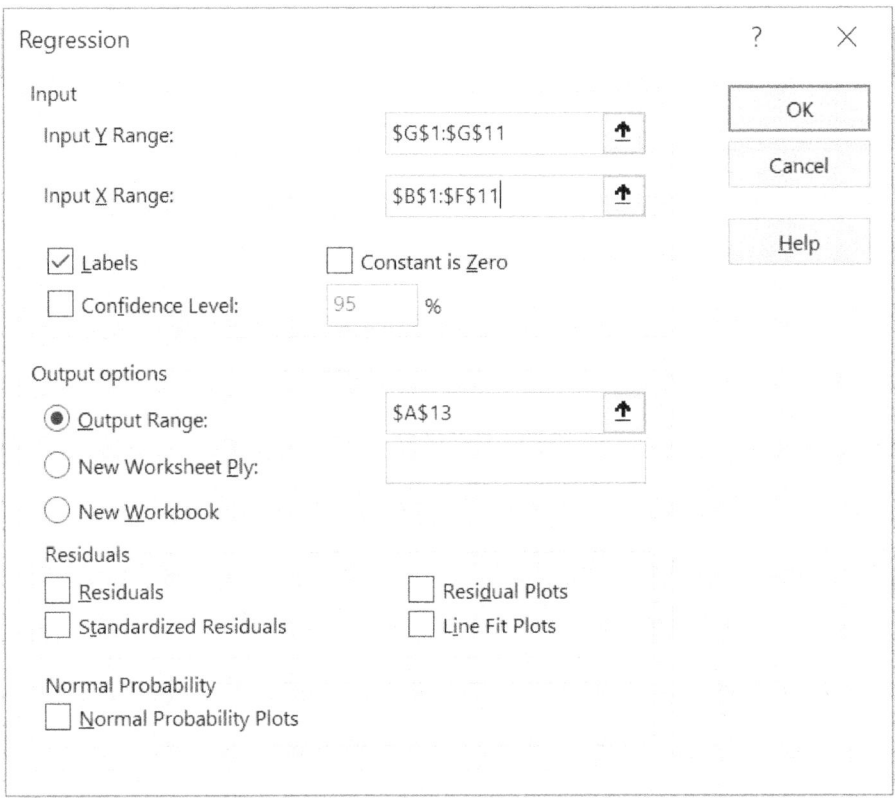

After you click "OK", Excel generates the following Summary Output. Round the numbers to 3 decimal places.

SUMMARY OUTPUT

Regression Statistics	
Multiple R	0.990
R Square	0.980
Adjusted R Square	0.954
Standard Error	0.496
Observations	10

ANOVA

	df	SS	MS	F	Significance
Regression	5	47.017	9.403	38.250	0.002
Residual	4	0.983	0.246		
Total	9	48			

	Coefficients	Standard Error	t Stat	P-value	Lower 95%	Upper 95%	Lower 95.0%	Upper 95.0%
Intercept	1.015	1.103	0.920	0.410	-2.048	4.079	-2.048	4.079
Social Network Size	0.133	0.060	2.228	0.090	-0.033	0.299	-0.033	0.299
Social Network Diversity Index	-0.287	0.602	-0.476	0.659	-1.957	1.384	-1.957	1.384
Competency	0.484	0.402	1.206	0.294	-0.631	1.600	-0.631	1.600
Time with Management	-0.030	0.019	-1.624	0.180	-0.083	0.022	-0.083	0.022
Conscientiousness Trait	0.610	0.404	1.511	0.205	-0.511	1.730	-0.511	1.730

R Square: In the output, R Square is 0.980, which means it is a very good fit. 98% of the variation in Performance Rating (Output) is explained by the independent variables (Input), Social Network Size, Social Network Diversity Index, Competency, Time with Management, and Conscientiousness Trait. The closer R Square is to "1", the better the regression line fits the data.

Significance F and P-values: To determine if your results are statistically significant (i.e. reliable), check "Significance F" (0.001). If the value of "Significance F" is less than 0.05, it is statistically significant (i.e. reliable). If "Significance F" is bigger than 0.05, don't use this set of independent variables. Delete those variables with "P-value" that is bigger than 0.05 and run the regression again until "Significance F" drops below 0.05. Most or all of your P-values should be lower than 0.05. In our example below "P-value" is 0.410, 0.090, 0.659, 0.294, 0.180 and 0.205, for Intercept, Social Network Size, Social Network Diversity Index, Competency, Time with Management, and Conscientiousness Trait, respectively.

Coefficients

From the Summary Output, the regression line is:

	Coefficients	Standard Error	t Stat	P-value	Lower 95%	Upper 95%	Lower 95.0%	Upper 95.0%
Intercept	1.015	1.103	0.920	0.410	-2.048	4.079	-2.048	4.079
Social Network Size	0.133	0.060	2.228	0.090	-0.033	0.299	-0.033	0.299
Social Network Diversity Index	-0.287	0.602	-0.476	0.659	-1.957	1.384	-1.957	1.384
Competency	0.484	0.402	1.206	0.294	-0.631	1.600	-0.631	1.600
Time with Management	-0.030	0.019	-1.624	0.180	-0.083	0.022	-0.083	0.022
Conscientiousness Trait	0.610	0.404	1.511	0.205	-0.511	1.730	-0.511	1.730

Performance Rating (Output)
= 1.015 + (0.133 * Social Network Size) - (0.287 * Social Network Diversity Index) + (0.484 * Competency) - (0.030 * Time with Management) + (0.610 * Conscientiousness Trait)

Based on the above regression formula, For each unit increase in Social Network Size, Performance Rating increase by 0.133. For each unit increase in Social Network Diversity Index, Performance Rating increase by 0.287, etc.

Coefficients can also be used for forecasting. For example, if "Social Network Size" is 21, "Social Network Diversity Index" is 3, "Competency" is 6, "Time with Management" is 110 hours, "Conscientiousness Trait" is 8, **predicted "Performance Rating"**
= 1.015 + (0.133 * Social Network Size) - (0.287 * Social Network Diversity Index) + (0.484 * Competency) - (0.030 * Time with Management) + (0.610 * Conscientiousness Trait)
= 1.015 + (0.133*21) - (0.287*3) + (0.484*6) - (0.030*110) + (0.610*8) = **7.4**
= *0.287 + (0.517*8) + (0.106*7) + (0.429*6)* = **7.7**

References:
(1) Tanya Pinto (2016). How the Big Five Personality Traits Can Predict Performance at Work. https://peakon.com/blog/people-management/big-five-personality-traits-in-the-workplace/ (13 November 2018)
(2) Ryan Fuller (2015) What Makes Great Salespeople. Harvard Business Review. https://hbr.org/2015/07/what-makes-great-salespeople (13 November 2018)

21) Compensation & Benefits Analytics

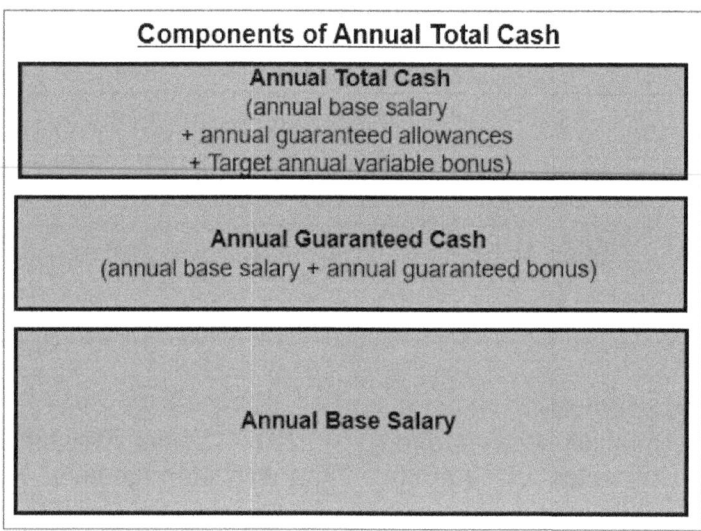

Research have shown that Compensation & Benefits can impact your organization's absenteeism, job satisfaction, performance, retention, sales and service:

Compensation & Benefit's impact on Absenteeism
- In 2005, Royal adopted an incentive scheme where workers with a clean attendance sheet were eligible for prize draws with cars, holidays and gift vouchers. Its sickness absence rates fell by 1%, after the cash incentives was introduced. [1]

Compensation & Benefit's impact on Job Satisfaction

Chidiebere Ogbonnaya, Kevin Daniels, and Karina Nielsen published a study on how incentive pay (Performance-related pay, Profit-related pay, and Share ownership) affects employees' experience of well-being, as measured by job satisfaction, organizational commitment, and trust in management. [2]

- **Performance-related pay** – The study showed that performance-related pay was positively associated with job satisfaction, organizational commitment, and trust in management.
- **Profit-related pay** - Positive effects of Profit-related pay depends on the extent that profit-related pay is given to a large proportion of the workforce. When employee participation in profit-related pay is high, and where organizational profits are perceived to be equitably distributed, more employees benefit and thus experience job satisfaction, organizational commitment, and trust in management. When the extend of employee participation in Profit-related pay is low to medium, the study found lower levels of job satisfaction, organizational commitment, and trust in management.
- **Share ownership** - the study found that share ownership has a negative relationship with job satisfaction and no relationships with employees' commitment and trust in management. High employee uptake of share ownership also revealed no relationships with employee well-being.

Compensation & Benefit's impact on Net Income Growth

- The bigger the pay spread, the better a company's performance. There is a correlation of 0.4 between net income growth and standard deviation of merit increase. [3]

Compensation & Benefit's impact on Performance

- A global technology company discovered that there is a direct correlation between the turnover of high performer and compensation. The issue was not just paying employees more, but paying them differently. The statisticians discovered that midlevel performers, who loved working at the company, would not resign even if their salary increases were as low as 90 percent of industry average. However high performers (i.e. top ten percent of employees) were much more sensitive, and will very likely resign when their annual raises were not at least 115 percent of the industry average. [4]

Compensation & Benefit's impact on Retention
- In 2011, two Hewlett-Packard (HP) scientists analyzed two years data and generated Flight Risk score. Flight Risk score predicts the probability of leaving of each of HP's 300,000 employees. Higher pay, promotions and better performance ratings, where negatively related to flight risk. However, there is a complicated relationship between these findings. For example, when an employee got a promotion but did not get a substantial pay raise, this employee would still be more likely to quit. [5]
- Google found that new salespeople, who do not get a promotion within 4 years, are more likely to leave. [6]
- TechnologyAdvice found that 56 percent of employees say they would trade a salary increase for certain on-the-job perks. [7]
- According to a study by Culture Amp on retention intention at Box, a worker's pay or relationship with his boss matters far less than how connected the worker feels to his team. [8]
- Research by American Institute of CPAs (AICPA) found that Americans 4 times as likely to choose a job with benefits over an identical job that offered 30 percent more salary but no benefits. [9]
- According to the 2018 Aflac study, more than half (55%) of employees surveyed would be at least somewhat likely to accept a job with lower compensation but a more robust benefits package. [10]
- According to Access Perks, 45% of employees said they consider a prospective company's work-life balance a crucial factor when researching a job. [11]

Compensation & Benefit's impact on Service
- In 2015, The New York Times reported that Walmart's sales fell for five straight quarters. Customers were unhappy with long queues and missing-in-action staff, and only 16 percent of stores met their customer service goals. Thus, Walmart announced that it will start paying its employees more - subsequently, sales started to rise and stores hitting their customer service targets increased back to 75 percent by year 2016. [12]
- In year 2015, Ivey Business School in London, found that raising the wages of low-skill workers (nursing assistants, orderlies, and food workers) in American hospitals, where the turnover and customer service was bad -> led to less turnover, fewer days taken off by staff, and more respect for customer and employer.[12]

References:
(1) Alison Coleman (2011) How incentives help reduce sickness absence. https://www.employeebenefits.co.uk/issues/june-2011/how-incentives-help-reduce-sickness-absence/ (20 November 2018)
(2) Chidiebere Ogbonnaya, Kevin Daniels, Karina Nielsen (2017) How Incentive Pay Affects Employee Engagement, Satisfaction, and Trust. Harvard Business Review. https://hbr.org/2017/03/research-how-incentive-pay-affects-employee-engagement-satisfaction-and-trust (18 Oct 2018)
(3) Jac FITZ-ENZ (2010), The New HR Analytics: Predicting the Economic Value of Your Company's Human Capital Investments, AMACOM
(4) Josh Bersin (2014) Deloitte Review issue 14: The datafication of HR https://www2.deloitte.com/insights/us/en/deloitte-review/issue-14/dr14-datafication-of-hr.html (11 November 2018)
(5) Erik van Vulpen, Predictive Analytics in Human Resources - Tutorial and 7 case studies, https://www.analyticsinhr.com/blog/predictive-analytics-human-resources/ (24 September 2018)
(6) Tom McKeown, Snapshots in Time (2017), https://www.trendata.com/snapshots-in-time/ (24 September 2018)
(7) Savita V Jayaram (2016), The Correlation between Benefits and Employee Retention http://www.hrinasia.com/employee-retention/the-correlation-between-benefits-and-employee-retention/ (27 September 2018)
(8) Rachel Emma Silverman and Nikki Waller (2015) The Algorithm That Tells the Boss Who Might Quit. https://www.visier.com/the-algorithm-that-tells-the-boss-who-might-quit/ (26 November 2018)
(9) AICPA (2018) Americans want employers to show them the benefits https://blog.aicpa.org/2018/11/americans-want-employers-to-show-them-the-benefits.html#sthash.5cfl96kY.dpbs (15 January 2020)
(10) HR Daily Advisor (2018) Research Shows Strong Job Satisfaction, Benefits https://hrdailyadvisor.blr.com/2018/10/12/research-shows-strong-job-satisfaction-benefits/?source=HAC&effort=44&utm_source=BLR&utm_medium=Email&utm_campaign=HRDAEmail&emailid=4332938&spMailingID=14465349&spUserID=MjU3NzE4NTEzNzYwS0&spJobID=1501512298&spReportId=MTUwMTUxMjI5OAS2 (19 June 2019)
(11) Access Perks (2018) 2018 Employee Benefits and Perks Statistics https://blog.accessperks.com/2018-employee-benefits-perks-statistics (19 June 2019)
(12) Lauren Pelley, Workopolis (2017) The Walmart Effect: Can Raising Salaries Increase Productivity? https://www.payscale.com/compensation-today/2017/05/walmart-effect-can-raising-salaries-increase-productivity (27 September 2018)

21.1) Market-Ratio Analytics

$$\text{Market-ratio} = \frac{\text{Salary (actual)}}{\text{Market pay midpoint}}$$

Market-ratios measures employee pay relative to the market pay midpoint.
- A market-ratio of 1.0 means that you are paying your employee exactly where you are targeting.
- A market-ratio of 0.8 means your employees are probably under-paid.
- A market-ratio of 1.2 means your employees are probably over-paid. In some cases, paying above 1.2 market-ratio is ok if your employee has many years of relevant experience, and unique skillsets.

Market pay midpoint depends on your pay strategy. Use 50th percentile of the market, if your pay strategy is 50th percentile of the market for the job. Use 75th percentile of the market, if your pay strategy is 75th percentile of the market for the job.

Critical jobs market-ratio analytics:
- **Market-ratio of a critical employee:** To get the market-ratio of a critical employee, divide that employee's pay by the market pay of that job.
- **Market-ratio for a critical job:** To get the average market-ratio for a critical job, compute each critical employee's market-ratio for a critical job, then get the average of all their market-ratios. Divide their average market-ratios by the Market Pay of the Job.

If you have difficulty keeping and hiring employees in location A, but you can't get rid of employees in location B. You can calculate the average salary market-ratio for location A and B to diagnose the problem. If the average market-ratio for employees in location A is 0.6 (below market), while the average market-ratio for employees in location B is 1.4 (percent above market) -> it shows that you're paying too high in location B and too low in location A - These market-ratios explain your challenges to recruit in location A, and maintain natural attrition in in location B.

21.2) Compa-Ratio Analytics

$$\text{Compa-ratio} = \frac{\text{Salary (actual)}}{\text{Salary Structure Midpoint}}$$

Compa-ratio tells you whether you are paying your people according to your salary ranges. Salary ranges are your company's pay guidelines that are developed based on your company's pay strategy. Thus, you should analyze to see how your employee's pay compares to your salary range midpoints.

A compa-ratio of 1.0 means you are paying your employee exactly at the midpoint. Take note that you don't want all your employees compa-ratio to be exactly 1.0. You want their pay to differ depending on their experience, skills, and performance.

Critical jobs compa-ratio analytics:
- To get the compa-ratio of a critical employee, divide that employee's pay by the midpoint of your salary range.
- To get the average compa-ratio for a critical job, compute each critical employee's compa-ratio for a critical job, then get the average of all their compa-ratios. Divide their average compa-ratios by the midpoint of your salary range.

21.3) Flight Risk Formula

Flight risk tells you if you are in danger of losing key employees.

In compensation analytics, flight risk are employees who are both high-performing and whose pay is low relative to the market (i.e. these employees have low salary market-ratio). Other than performance, it can also be used to track critical skillsets or experience that is critical to your business.

Flight risk is determined by plotting two factors (the "employee's performance" and their "salary market-ratio") against each other. Employees who fall into the "High Performance Low Pay" quadrant are considered flight risks. Employees who are in the "High Performance Low Pay" because of they are recently promotion or newly hired, are unlikely to be flight risks.

		Employee's performance		
		Below Expectation	Meets Expectation	Above Expectation
Employee's pay relative to market (salary market-ratio)	Below 80%			Flight risk
	80%-89%			Flight risk
	90%-109%			
	110%-119%			
	120% & above			

22) Training & Development Analytics

In an organization context, Training Analytics is the measurement, collection, analysis, and reporting of data related to the effectiveness and business impact of development programmes, to improve individual and organizational performance. Training analytics is the lynchpin for the training function to demonstrate how its programmes affect business results. Without measuring the contribution of learning, the training function is seen as a cost centre.

Research have shown that Training & Development can affect productivity, service, and retention.:

Training's impact on Productivity:
- Gallup found that improving opportunities to learn and grow from 40% to 80% can allow your business to achieve a 44% drop in absenteeism and a 16% jump in productivity. [1]

Training's impact on Service:
- Training improves employee's skills, and more skilled employees tend to provide better service (Ployhart et al., 2011). However, this does not apply for all jobs. Interestingly, a research by Ployhart et al. (2011) that focused on restaurants, showed that service training does not lead to higher service levels. Employees with less complex service jobs (e.g. waiting tables) seem to benefit less from training compared to employees with more complicated service jobs. This is probably because an employee who attended training does not necessarily mean that they will learn or apply their new skills to the job. It might also be that relatively easy service training such as serving food in a restaurant don't impact employees' customer service levels, as compared to more complicated and time-intensive service offering like software sales. [2]

Training's impact on Retention:
- Bridge survey data finds that offering career training and development would keep 86 percent of millennials from leaving their current position. [3]

References:
(1) Tanvir Haque (2018) Four Strategies To Boost Employee Engagement That Can Improve Your Bottom Line. https://www.entrepreneur.com/article/312697 (20 November 2018)
(2) Erik van Vulpen (2018), How 11 Factors Influence Customer Service Performance. AIHR. https://www.analyticsinhr.com/blog/factors-influencing-customer-service-performance/ (15 November 2018)
(3) Bride (2018) Millennials Are Most Likely to Stay Loyal to Jobs With Development Opportunities . https://www.instructure.com/bridge/news/press-releases/millennials-are-most-likely-stay-loyal-jobs-development-opportunities?newhome=bridge (19 June 2019)

22.1) Four Levels Of Training Evaluation - The Kirkpatrick Model

One of the models used to evaluate training programs is the Kirkpatrick's Four Levels of Evaluation, created by Donald Kirkpatrick. In the late 1950s, Donald Kirkpatrick developed an approach to evaluate the impact of training on performance with four levels (Reaction, Learning, Behavior and Results), while consulting with a client. It is today, considered a standard for education and training.

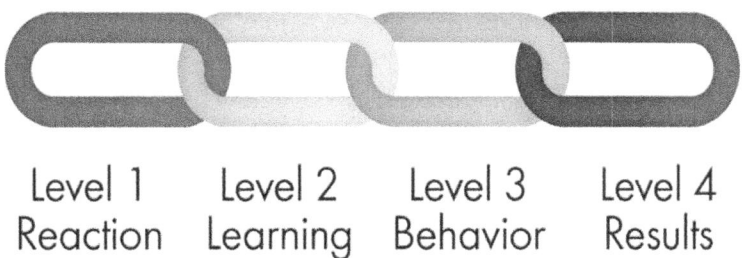

Source: Kirkpatrick Partners, The Kirkpatrick Model
https://www.kirkpatrickpartners.com/Our-Philosophy/The-Kirkpatrick-Model (27 October 2018)

- **Level 1 Reaction** – Reaction is the degree which participants find the training good, interesting and relevant to their jobs. In other words, Reaction measures satisfaction. Level 1, determines whether participants liked the training, in terms of the trainer, the content, the facilities, and the overall experience. The majority of organizations collect learning satisfaction information (i.e. Reaction) to ascertain learning impact. Although, Reaction is the most frequently gathered data, but it is the least valuable measure with for business impact. Methods used to evaluate level 1 include: feedback surveys, and Reaction sheets.

- **Level 2 Learning** – Learning is the degree which participants absorb the knowledge, attitude, and skills from the training. Did participants learn new knowledge and skills? Methods used to evaluate level 2 include: Pre-and post-training tests to assess the change in knowledge or skill, using control groups to compare participants' performance.

- **Level 3 Behaviour** – Behaviour is the degree which participants apply what they learned during training on the job. Methods used to evaluate level 3 include: interviewing the participant's immediate bosses to ascertain if there is any change in performance or behavior.

- **Level 4 Results** – Results measure the impact participants have on the business outcomes after finishing the training. Does training help improve important business metrics such as employee engagement, customer satisfaction, sales, productivity, profitability, etc.? Methods used to evaluate level 3 include: tracking improvements of Key Performance Indicators (KPI).

22.2) Fifth Level of Training Evaluation - ROI by Jack Philips

In the Kirkpatrick model, developed by Donald Kirkpatrick, there are four levels of training program evaluation.

In the Phillip Return on Investment (ROI) Methodology, developed by Jack Philips, there is an additional level 0 (Inputs) and level 5 (Return on Investment or ROI). Phillips added level 0 to include measures of training activity such as cost of training, number of people trained, the learning methodology used. The focus on level 5 (ROI) arises from Phillips' experience with business heads who wants to know the ROI of training. Phillip's levels 1 to 4 are similar in both name and function to Kirkpatrick's four levels of training evaluation. The ROI methodology developed by Jack Phillips, allows you to estimate the training program cost and predict whether running it will be cost-effective.

Level	Kirkpatrick Model	Phillip ROI Methodology
0		Inputs
1	Reaction	Reaction, Satisfaction, and Planned Application
2	Learning	Learning
3	Behavior	Application and implementation
4	Result	Business impact
5		Return on Investment (ROI)

The following describes the 5 levels of the Phillip ROI Methodology:
- **Level 0 - Inputs/ Indicators:** Level 0 measures inputs into training. Most companies report what is available and easy to gather, such as number of training programs, number of people trained, costs of training, and learning methodology used to deliver training. When combined, these data provides insight such as cost per course. But most companies wants to know more than just participation rates - they also want to measure quality of training, which is measured in the other levels.
- **Level 1 - Reaction, Satisfaction, and Planned Application:** Level 1 measures participants' satisfaction with a program. Most companies use "smile sheets" or post-training surveys to measure the satisfaction of the course participants. Example, Do the participants like the instructor? Did the participants understand the course material? However, participants who are satisfied with the training program, does not mean that they have acquired knowledge or new skills.
- **Level 2 – Learning:** Level 2 ascertain how much participants have learned, through tests, and other assessment. Note that participants may have learned new skills and knowledge. But it does not mean that participants will apply them on the job.
- **Level 3 - Application and Implementation:** Level 3 evaluation determine whether participants applied their newly acquired knowledge and skills in their job or whether there are any changes in their behavior. Note that participants applying their newly acquired knowledge and skills in their job, does not mean that there will be positive impact on business results
- **Level 4 - Business Impact:** Level 4 measures whether there is any improvement in business results after training. Examples of Level 4 measures are: quality, costs, output, and time. Take note that, even if the training program leads to improvement in business results, the program's costs might outweigh its business benefits.
- **Level 5 - Return on Investment (ROI):** ROI compares the monetary benefits of the program with the program costs. Usually the CFO would want to know if training programmes is worth the investment, and whether it is more cost effective to build or buy talent. The ROI coefficient is a percentage, showing the relationship between the estimated profit and the estimated costs of a training program, computed using to the following formula:

$$ROI = (Net\ Benefits\ /\ Costs) \times 100\%$$
$$OR$$
$$ROI = [(Benefits - Costs)\ /\ Costs] \times 100\%$$

Example,
A sales training programme results in an increase in sales of $135,000 per year, and the cost of the program was $90,000. The ROI of the program is:

ROI = [($135,000 -$90,000) / $90,000] x 100%
    = ($45,000 / $90,000) x 100%
    = 50%

This means that for every dollar spent on the sales training programme, $0.50 is returned as profit.

As adopting the fourth and fifth levels of evaluation is costly in terms of time and money, the ROI evaluation is usually done for less than ten percent of the training programmes. The fourth and fifth levels of evaluation are usually only done for training programs that are strategic of that require significant budget. While training programs of lesser importance or cost, typically just measure the first three levels of the evaluation.

Isolation techniques for Training programmes:

Another critical step in a training evaluation introduced by Phillips is the use of "isolation" strategies to separate the business results due to the training programme, from the influence of other factors (e.g., seasonal effects, employee bonus programs, marketing programmes, etc.). These are some of the isolation techniques [1]:

- **Control Groups**: Compare the results of the pilot group of training participants with the results of the control group of non-participants. This is the most effective isolation technique,
- **Trend lines**: Trend lines are typically used to forecast business results. Compare the forecasted business results, with the actual business results after the training initiative is completed to ascertain if training has an impact.
- **Participants estimates**: Ask the course participants to estimate the amount of improvement directly related to the training programme.

22.3) How to compute impact of Training on Earnings Per Share (EPS)

Reducing number of employees or reducing training spend is the easiest way for HR to use data to show it's impacts on business.

In the example below, the HR Director cuts training by $100 this year to increase earnings per share (EPS), so that stock price can increase, and the company can attract more investments to fund its regional expansion.

The approach to calculate the estimated change in Earnings Per Share (EPS) before and after the cut in training budget, is as follows,

Before company cuts training by $100	After company cuts training by $100
Earnings per share (EPS) = EBITDA/ number of shares = $5000/200 = $25	Earnings per share (EPS) = EBITDA/ number of shares = ($5000 +$100)/200 = $5100/200 = $25.5

ROI Results of real-world evaluation studies

Training evaluations can be done on various training programs. The table below shows some the training results from real-world evaluation studies. [1]

Study/Setting	Target Group	Program Description	Business Measures	ROI
Cracker Box	Managers, Manager trainees	Performance management training	Reduced turnover, absenteeism, & waste	298%
Healthcare	Managers, Supervisors, Later all employees	Anti-sexual harassment training	Reduced turnover & grievances	1,052%
Hewlett-Packard	Sales management team, Sales reps	Sales training for complex systems	Inside sales	195%
Verizon Communications	Training staff, Customer service	Customer service skills training	Reduced call escalations	(-85%)
Canadian Valve Company	New employees	Equipment operations training	Reduced time, scrap, & turnover, Improved safety	132%
Retail Merchandise Company	Sales associates	Retail sales skills	Increased sales revenues	118%
U.S. Department of Veterans Affairs	Managers, Supervisors	Leadership competencies	Cost, time savings, Reduced staff requirements	159%
Garrett Engine (Allied Signal)	Maintenance staff, Hourly employees	Team building	Reduced equipment downtime	125%
High Tech	Managers, Supervisors, Project leaders	Meeting skills	Time savings (reduced number & duration of meetings)	506%
Nortel Networks	Future leaders	Executive coaching	Output productivity, sales, employee retention	788%
Metro Transit Authority	Supervisors, Drivers	New hire screening, employee coaching	Reduced schedule delays & absenteeism, employee satisfaction	822%
US Federal Intelligence Agency	High value experts	Internal Masters degree program	Professional employee retention, turnover	153%

Source: Allan Bailey, CEO, Learning Designs Online, The Kirkpatrick/Phillips Model for Evaluating Human Resource Development and Training http://www.buscouncil.ca/busgurus/media/pdf/the-kirkpatrick-phillips-evaluation-model-en.pdf (29 October 2018)

References:
(1) Allan Bailey, CEO, Learning Designs Online, The Kirkpatrick/Phillips Model for Evaluating Human Resource Development and Training http://www.buscouncil.ca/busgurus/media/pdf/the-kirkpatrick-phillips-evaluation-model-en.pdf (29 October 2018)

22.4) Case 1: Hilton Worldwide University Business Impact

In this case study, Hilton Worldwide University (HWU) partnered with CEB to determine the financial impact of Hilton's "The Revenue Management at Work" training programme for those in hotel-level revenue management positions, regional revenue management positions, Revenue Management Consolidated Center (RMCC) positions and Revenue Analysis Teams. Analysis of pre- and post-Program revenue figures showed that participating hotels met revenue goals, achieving greater market-share growth. Of this growth, the study attributed $3.35 million to the HWU training [1].

Revenue Management

The School of Revenue Management is responsible for developing and delivering a comprehensive Revenue Management curriculum, as well as numerous Revenue Management training programs, for all Hilton Worldwide brands.

Revenue Management training is provided to those in hotel-level revenue management positions, regional revenue management positions, Revenue Management Consolidated Center (RMCC) positions and Revenue Analysis Teams.

Measurable Results: The Revenue Management at Work training program is designed to train Revenue Management professionals on the specific behaviors and actions proven to drive business results through an extensive curriculum that is measurable, research-based and best-practices oriented.

Improved ROI 28:1 | Market Share Growth | $20 MILLION more than non-participating hotels

A thorough analysis of pre- and post-Program revenue figures showed that participating hotels met revenue goals, achieving greater market-share growth. Of this growth, the metrics study directly attributed **$3.35 million to the HWU training** that had been given.

Source: Hilton Worldwide University (2018). Revenue Management, http://www.hwu-overview.com/team-members/revenue-management.html (1 November 2018)

To predict the extend that training improved performance, participants of Hilton's Revenue Management at Work" programme were asked three questions [2]:
Q1) Estimate performance improvement.
Q2) Isolate how much is due to the program.
Q3) Percentage work time spend on tasks (how much learning was actually applied)

The percentages of each question are multiplied together, then adjusted downward by 35 per cent (i.e. multiplied by 65 per cent) to factor for response bias and overestimation - a process called "estimation, isolation, isolation and adjustment".

Figure below shows the average participants evaluations for the revenue management programme. The estimated performance improvement of 18 per cent due to training, is higher than the travel and leisure industry benchmark of 9.0%. If the trained revenue management team achieved $1,000,000 in revenue last year, Hilton can predict that this same group would achieve $1,180,000 ($1,000,000 x 18% + $1,000,000 = $1,180,000) after attending training [4].

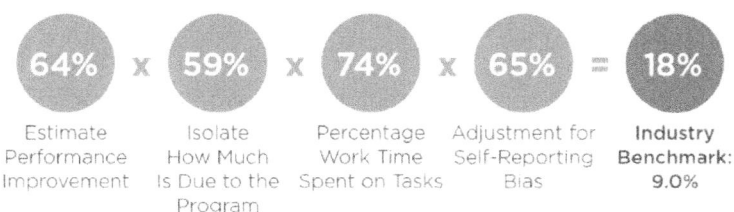

Source: John R. Mattox, II, Ph.D., Learning Analytics - Measurement Innovations to Support Talent Development, http://info.clomedia.com/hubfs/VL_SY_FA16_ResourceLibrary/CEB_Learning_Analytics_Measurement_Innovations_to_Support_Talent_Development.pdf (25 Oct 2018)

References:
(1) Hilton Worldwide University (2018). Revenue Management, http://www.hwu-overview.com/team-members/revenue-management.html (1 November 2018)
(2) John R. Mattox, II, Ph.D., Learning Analytics - Measurement Innovations to Support Talent Development, http://info.clomedia.com/hubfs/VL_SY_FA16_ResourceLibrary/CEB_Learning_Analytics_Measurement_Innovations_to_Support_Talent_Development.pdf (25 Oct 2018)

22.5) Case 2: JetBlue University Training Predictive Analytics

In 2012, the Assessment, Measurement and Evaluation (AME) group partnered with CEB to determine the key effectiveness drivers of its curricula. AME is a group formed by JetBlue University (JBU) to evaluate JBU's training programmes efficiency and effectiveness. Two surveys were used. A post-event evaluation survey, gathered participant's feedback immediately after they complete the training – it asked the participants to estimate how much learning they would apply on the job, and how it would improve their performance. A second follow-up evaluation survey, gathered participant's feedback sixty days after training, and asked participants to estimate how much learning they actually applied on the job, and how much it actually improved their job performance. CEB used one year's sample of JBU's training evaluation data, to develop three regression models, and predicted three key drivers of effective training [1]:

- **Learning effectiveness** – the extent participants gained new knowledge and skills from training.

- **Application on the job** – the extent participants applied new knowledge and skills on the job.

- **Job performance** – the extend training improved the participant's job performance.

The figure below shows the evaluation measures from the participant survey, that significantly predict job performance. The R-squared values show the variance of the predictor. The values on the lines between the predictors and the outcome are the weights. The figure shows that three factors contribute to expected improvement in job performance [2]:

1) Learning Effectiveness: I learned new knowledge and skills from this training.
2) Job Impact: I will be able to apply the knowledge and skills learned in this class to my job.
3) Instructor: The instructor was knowledgeable about the subject.

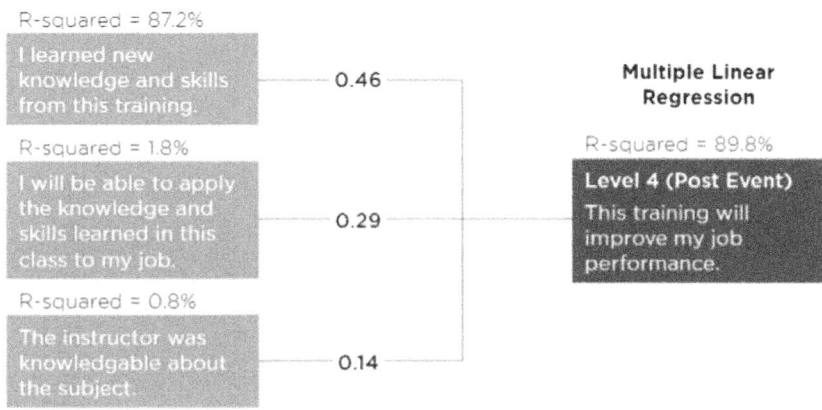

Source: John R. Mattox, II, Ph.D. (2016). Learning Analytics. http://info.clomedia.com/hubfs/VL_SY_FA16_ResourceLibrary/CEB_Learning_Analytics_Measurement_Innovations_to_Support_Talent_Development.pdf (31 October 2018)

This is the prediction equation from the JetBlue University regression analysis [2]:

Level 4 performance improvement =
0.66 + 0.46 (I learned new knowledge and skills from this training + 0.29 (I will be able to apply the knowledge and skills learned in this class to my job) + 0.14 (The instructor was knowledgeable about the subject).

For comparisons, CEB created a table for AME containing the predictors, benchmark values and course scores. The numbers shown in the table below are not real, and are for illustration only. The course score for job impact (4.88) is below benchmark value (5.57), while the others are above their benchmarks values. Based on CEB's table, improvement need to be made on "Job Impact", to help participants apply the knowledge and skills learned to their job [2].

Table 2.2 benchmark comparison table.

Comparison to Benchmarks

Category	Item	Benchmark Value	Course Scores
Learning Effectiveness	I learned new knowledge and skills from this training.	5.50	5.89
Job Impact	I will be able to apply the knowledge and skills learned in this class to my job.	5.57	**4.88**
Instructor	The Instructor was knowledgeable about the subject.	5.68	5.85

Source: John R. Mattox, II, Ph.D. (2016). Learning Analytics. http://info.clomedia.com/hubfs/VL_SY_FA16_ResourceLibrary/CEB_Learning_Analytics_Measurement_Innovations_to_Support_Talent_Development.pdf (31 October 2018)

CEB provided AME with recommendations for improving training as shown in the table below [2].

Key Driver	Suggestions to improve responses to this factor
Course Objectives	Align the materials to the course objectives; clearly describe the objectives; provide examples of course materials, topics and objectives to help learners determine if the course is what they need. Ensure truth in advertising. If the course description indicates there will be opportunities to practice and master skills in class, then live up to the promise; provide practice and feedback periods. Do not "bait and switch", where practice is advertised but lecture is the only learning method. Ensure opportunities to demonstrate and build knowledge and skills on the outlined course objectives—tying the activities to the objectives.
Pace of the Course	**Engage the Learner by Providing a Challenging Pace**—especially with web-based and other online learning, where they can review or ask questions. More often than not with training, the pace is somewhat to substantially too slow for learners. Generally, with a knowledgeable population, faster is going to be better. **Images:** Methodologies, process and graphs convey substantially more information in a fraction of the time than written or verbal descriptions. Emphasize the use of images rather than words. Information that is supplemented with images is also memorable for a longer period of time. **Self-Directed Learning:** Self-paced action and discovery (e.g., clicking various components of a methodology to learn more about it) is a better learning experience than listening to an audio description.
Instructor Effectiveness	**Prepare the instructor:** Preparedness for online instruction is just as important for online learning as for instructor led training. Get instructors familiar and comfortable teaching and trouble shooting in a virtual environment. Prepare the instructors to use various instructional strategies (lecture, discussion, visuals, etc.).

Source: John R. Mattox, II, Ph.D. (2016). Learning Analytics. http://info.clomedia.com/hubfs/VL_SY_FA16_ResourceLibrary/CEB_Learning_Analytics_Measurement_Innovations_to_Support_Talent_Development.pdf (31 October 2018)

<u>References:</u>
(1) John R Mattox II, Mark Van Buren, (2016). Learning Analytics, Kogan Page.

22.6) Multiple Regression Example: Predict training's Impact on Customer Service

Research have shown that customer service is affected by training:
- Training improves employee's skills, and more skilled employees tend to provide better service (Ployhart et al., 2011). However, this does not apply for all jobs. Interestingly, a research by Ployhart et al. (2011) that focused on restaurants, showed that service training does not lead to higher service levels. Employees with less complex service jobs (e.g. waiting tables) seem to benefit less from training compared to employees with more complicated service jobs. This is probably because an employee who attended training does not necessarily mean that they will learn or apply their new skills to the job. It might also be that relatively easy service training such as serving food in a restaurant don't impact employees' customer service levels, as compared to more complicated and time-intensive service offering like software sales. [1]

This section shows you step-by-step how to run an Excel multiple linear regression analysis to predict the effect of a training programme on Customer Service. You will be taught how to derive the multiple regression equation for the variables using Excel, and use this equation to evaluate a hypothesis.

These are the dependent X variables (three training evaluation questions) that we will be testing to determine its effect independent Y variable (Customer Service Rating):
Q1) Learning Effectiveness: I learned new knowledge and skills from this training.
Q2) Job Impact: I will be able to apply the knowledge and skills learned in this class to my job.
Q3) Instructor: The instructor was knowledgeable about the subject.

In the data below, we want to predict "Customer Service Rating" using Excel multiple regression. The data represents events that have already happened.

Employee	Trainer: The Trainer was knowledgeable about the subject.	Learning: I learned new knowledge and skills from this course	Application: I will be able to apply what I learned during training in my job	Customer Service Rating
Ann	9.80	8.00	9.00	10.0
Andy	9.50	10.00	8.00	9.5
Ben	9.50	9.00	7.60	9.5
Bond	9.30	9.00	7.20	9.0
Carl	9.30	8.55	7.20	9.0
Candy	8.70	8.55	7.20	9.0
Dave	8.50	8.08	6.80	8.5
Eddy	8.50	8.08	6.80	8.5
Faith	8.50	8.00	6.80	8.5
Gary	8.20	8.00	6.80	8.5
Han	8.00	7.60	6.40	8.0
Ian	8.00	7.60	6.40	8.0
Joe	7.50	7.13	6.00	7.5
Kathy	7.50	7.13	6.00	7.5
Larry	7.00	6.65	5.60	7.0
Molly	7.00	6.65	5.60	7.0
Maddy	6.00	5.70	4.80	6.0
Neil	6.00	5.70	4.80	6.0
Randy	6.00	5.00	4.80	6.0
Ted	5.50	5.00	4.40	5.5

1) Install "Analysis ToolPak", an Excel add-in

"Analysis ToolPak" is an add-in for Microsoft Excel that comes with Microsoft Excel. To be able to run regression using Excel, you need to first install "Analysis ToolPak", an Excel add-in program that provides data analysis tools. To load the Analysis ToolPak add-in, follow these steps:

On the File tab, click Options.

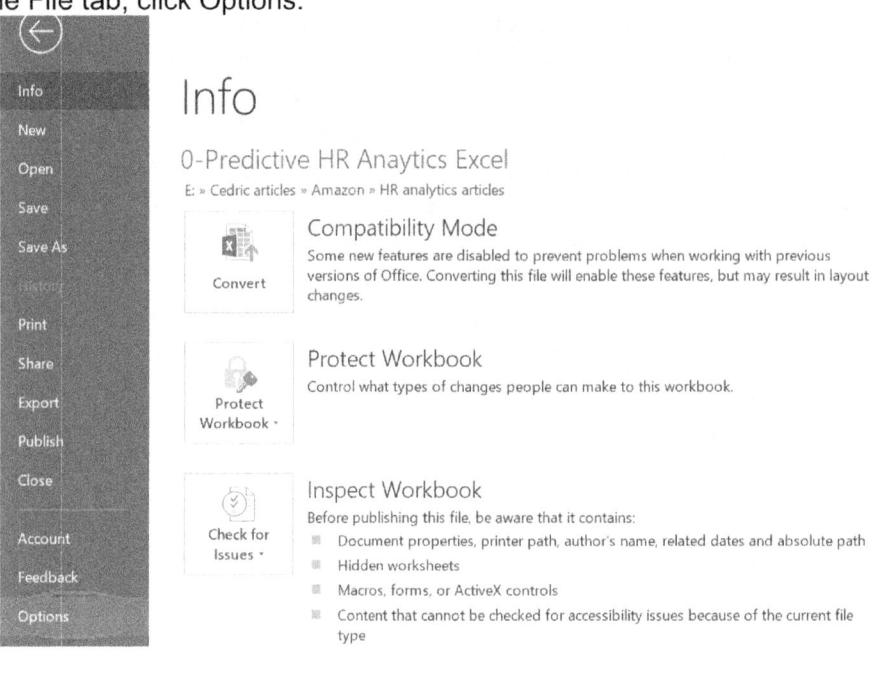

Under Add-ins, click Analysis ToolPak and click the "Go" button.

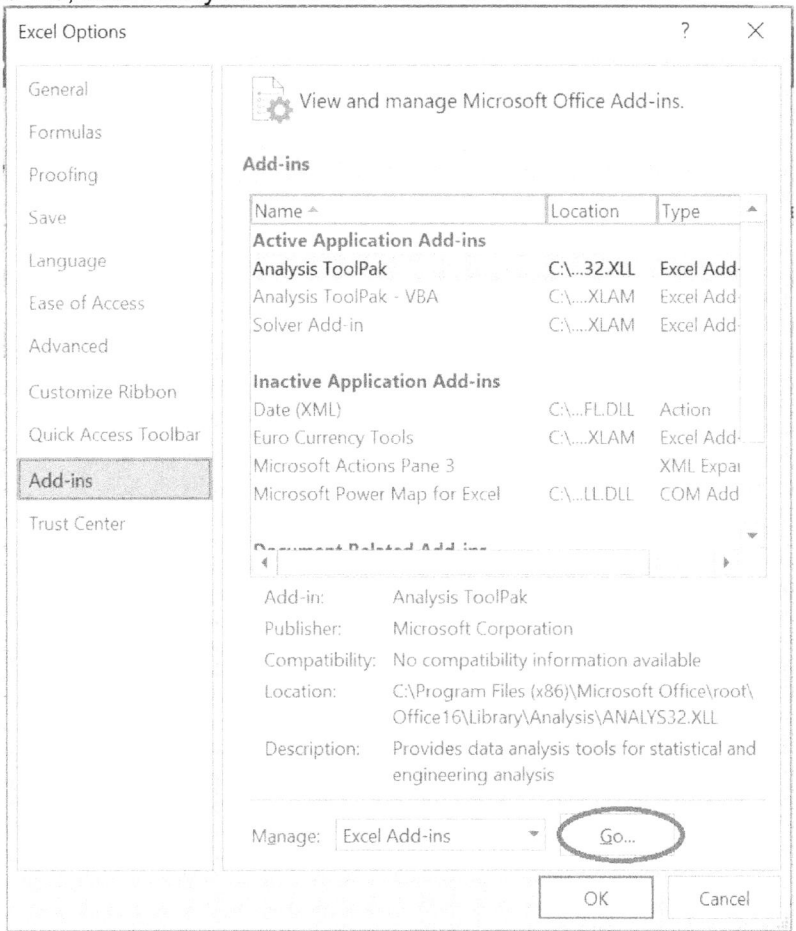

Click "Analysis ToolPak" and click on OK.

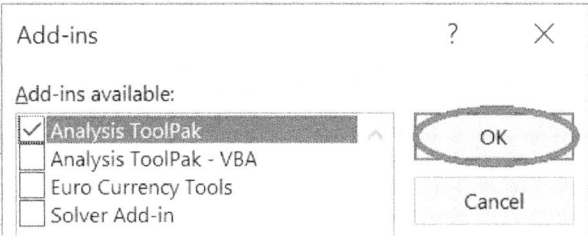

On the Data tab, in the Analysis group, you are now able to click on "Data Analysis".

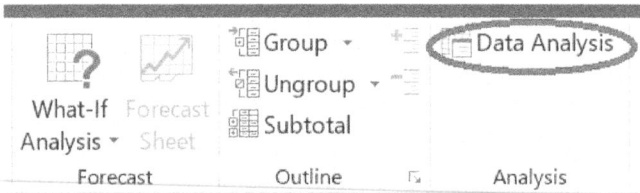

2) Copy the data below, and paste it in cell A1 of a new Excel worksheet.

	A	B	C	D	E
1	Employee	Trainer: The Trainer was knowledgeable about the subject.	Learning: I learned new knowledge and skills from this course	Application: I will be able to apply what I learned during training in my job	Customer Service Rating
2	Ann	9.80	8.00	9.00	10.0
3	Andy	9.50	10.00	8.00	9.5
4	Ben	9.50	9.00	7.60	9.5
5	Bond	9.30	9.00	7.20	9.0
6	Carl	9.30	8.55	7.20	9.0
7	Candy	8.70	8.55	7.20	9.0
8	Dave	8.50	8.08	6.80	8.5
9	Eddy	8.50	8.08	6.80	8.5
10	Faith	8.50	8.00	6.80	8.5
11	Gary	8.20	8.00	6.80	8.5
12	Han	8.00	7.60	6.40	8.0
13	Ian	8.00	7.60	6.40	8.0
14	Joe	7.50	7.13	6.00	7.5
15	Kathy	7.50	7.13	6.00	7.5
16	Larry	7.00	6.65	5.60	7.0
17	Molly	7.00	6.65	5.60	7.0
18	Maddy	6.00	5.70	4.80	6.0
19	Neil	6.00	5.70	4.80	6.0
20	Randy	6.00	5.00	4.80	6.0
21	Ted	5.50	5.00	4.40	5.5

3) On the Data tab, in the Analysis group, click on "Data Analysis".

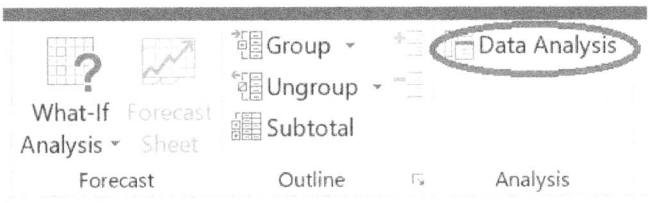

4) Select "Regression" and click "OK".

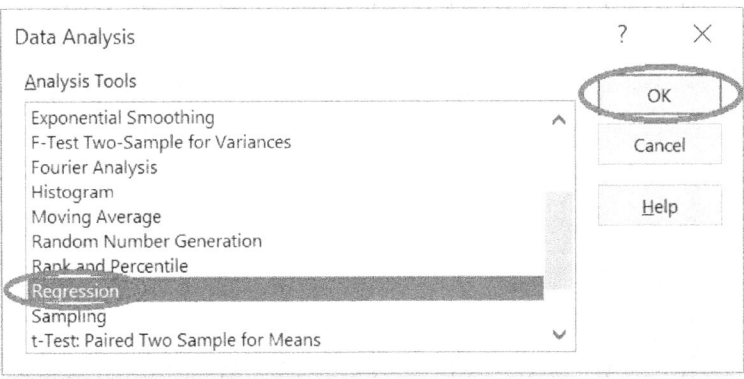

5) After you click OK in the "Data Analysis" dialog box, you will see a "Regression" dialog box.
6) For "Input Y Range", select cells (E1:E21). This is the predictor variable or dependent variable.
7) For "Input X Range", select cells (B1:D21). These are the explanatory variables or independent variables.
8) Check "Labels" box.
9) Click the "Output Range" box, and select cell A23.
10) Click "OK".

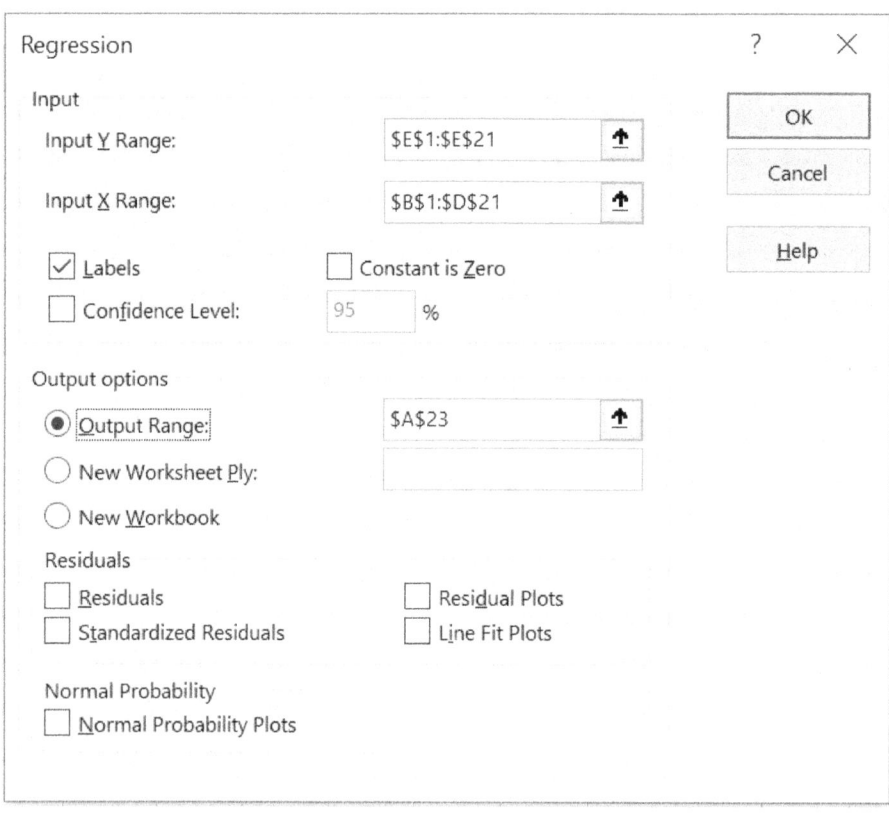

After you click "OK", Excel generates the following Summary Output. Round the numbers to 3 decimal places.

SUMMARY OUTPUT

Regression Statistics	
Multiple R	0.997
R Square	0.994
Adjusted R Square	0.993
Standard Error	0.107
Observations	20

ANOVA

	df	SS	MS	F	Significance F
Regression	3	32.954	10.985	958.275	0.000
Residual	16	0.183	0.011		
Total	19	33.138			

	Coefficients	Standard Error	t Stat	P-value	Lower 95%	Upper 95%	Lower 95.0%	Upper 95.0%
Intercept	0.287	0.161	1.782	0.094	-0.055	0.629	-0.055	0.629
Trainer: The Trainer was knowledgeable about the subject.	0.517	0.130	3.977	0.001	0.242	0.793	0.242	0.793
Learning: I learned new knowledge and skills from this course	0.106	0.065	1.624	0.124	-0.032	0.244	-0.032	0.244
Application: I will be able to apply what I learned during training in my job	0.429	0.100	4.309	0.001	0.218	0.641	0.218	0.641

R Square

In the output, R Square is 0.994, which means it is a very good fit. 99% of the variation in Sales (Output) is explained by the independent variables (Input), Trainer, Learning and Application. The closer R Square is to "1", the better the regression line fits the data.

Significance F and P-values

To determine if your results are statistically significant (i.e. reliable), check "Significance F" (0.001). If the value of "Significance F" is less than 0.05, it is statistically significant (i.e. reliable). If "Significance F" is bigger than 0.05, don't use this set of independent variables. Delete those variables with "P-value" that is bigger than 0.05 and run the regression again until "Significance F" drops below 0.05. Most or all of your P-values should be lower than 0.05. In our example below "P-value" is 0.094, 0.001, 0.124 and 0.001, for Intercept, Trainer, Learning, and Application, respectively.

Coefficients

From the Summary Output, the regression line is:

SUMMARY OUTPUT

Regression Statistics	
Multiple R	0.997
R Square	0.994
Adjusted R Square	0.993
Standard Error	0.107
Observations	20

ANOVA

	df	SS	MS	F	Significance F
Regression	3	32.954	10.985	958.275	0.000
Residual	16	0.183	0.011		
Total	19	33.138			

	Coefficients	Standard Error	t Stat	P-value	Lower 95%	Upper 95%	Lower 95.0%	Upper 95.0%
Intercept	0.287	0.161	1.782	0.094	-0.055	0.629	-0.055	0.629
Trainer: The Trainer was knowledgeable about the subject	0.517	0.130	3.977	0.001	0.242	0.793	0.242	0.793
Learning: I learned new knowledge and skills from this course	0.106	0.065	1.624	0.124	-0.032	0.244	-0.032	0.244
Application: I will be able to apply what I learned during training in my job	0.429	0.100	4.309	0.001	0.218	0.641	0.218	0.641

Y = Customer Service Rating
= 0.287 + (0.517 * Trainer) + (0.106 * Learning) + (0.429 * Application)

Based on the above regression formula,
- For each unit increase in Trainer evaluation score, Customer Service Rating increase by *0.517*.
- For each unit increase in Learning evaluation score, Customer Service Rating increase by *0.106*.
- For each unit increase in Application evaluation score, Customer Service Rating increase by *0.429*.

Coefficients can also be used for forecasting. For example, if "Trainer evaluation score" is 8, "Learning evaluation score" is 7, "Application evaluation score" is 6, **predicted "Customer Service Rating"**
= 0.287 + (0.517*8) + (0.106*7) + (0.429*6) = **7.7**

References:
(1) Erik van Vulpen (2018), How 11 Factors Influence Customer Service Performance. AIHR. https://www.analyticsinhr.com/blog/factors-influencing-customer-service-performance/
(15 November 2018)

23) Health, Safety & Environment (HSE) Analytics

Research have shown that Health, Safety and Environment (HSE) are affected by air quality, employee engagement, age, and tenure.

Air Quality's impact on Health, Safety and Environment:
- A study by Harvard and Syracuse Universities on green-certified buildings found employees had 30% fewer headaches and respiratory complaints in the green office environment. The study also showed employees performed nearly 27% better on cognitive tasks. Finally, workers in the study slept better at night, tracked by a wristband measuring sleep quality. [1]

Compensation & Benefits impact on Health, Safety and Environment:
- States with high workers' compensation premiums have 5% lower injury rate, compared with those in lower-premium states. In states where workplace injuries are more costly, managers are more diligent about their workers' safety and less willing to increase workloads and demands on employees. [2]
- In 2005, Royal adopted an incentive scheme where workers with a clean attendance sheet were eligible for prize draws with cars, holidays and gift vouchers. Its sickness absence rates fell by 1%, after the cash incentives was introduced. [3]

Demographics' impact on Health, Safety and Environment:

- AIHR found that 6% of the road traffic accidents involved people that are between the ages of 31 and 35 years. (4)

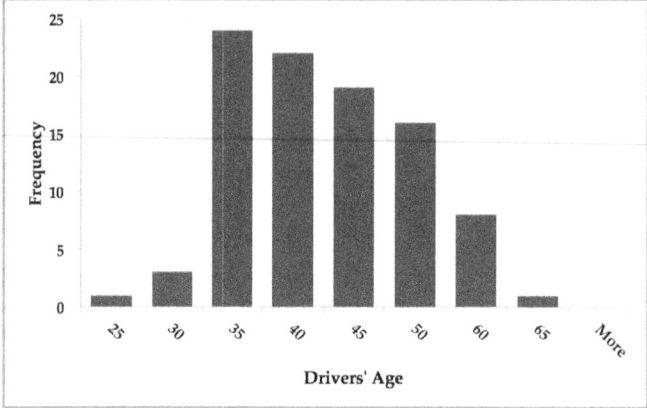

Source: Memory Nguwi (2018). Reducing Workplace Accident using People Analytics. AIHR. https://www.analyticsinhr.com/blog/reducing-workplace-accident-people-analytics/ (19 November 2018)

- AIHR found that 52% of the road traffic accidents involved drivers/riders with 5 years of service and below. (4)

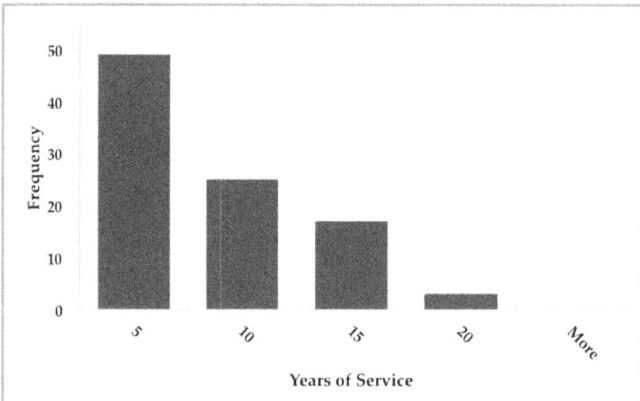

Source: Memory Nguwi (2018). Reducing Workplace Accident using People Analytics. AIHR. https://www.analyticsinhr.com/blog/reducing-workplace-accident-people-analytics/ (19 November 2018)

- IBM Kenexa Talent Insights, a data discovery application allows HR professionals and managers to explore, analyze, and navigate talent data so they can observe patterns or gain actionable insights. (5) In the diagram below, companies can analyze safety incidents by age group, and/or supervisor communication score, to see if there is a relationship.

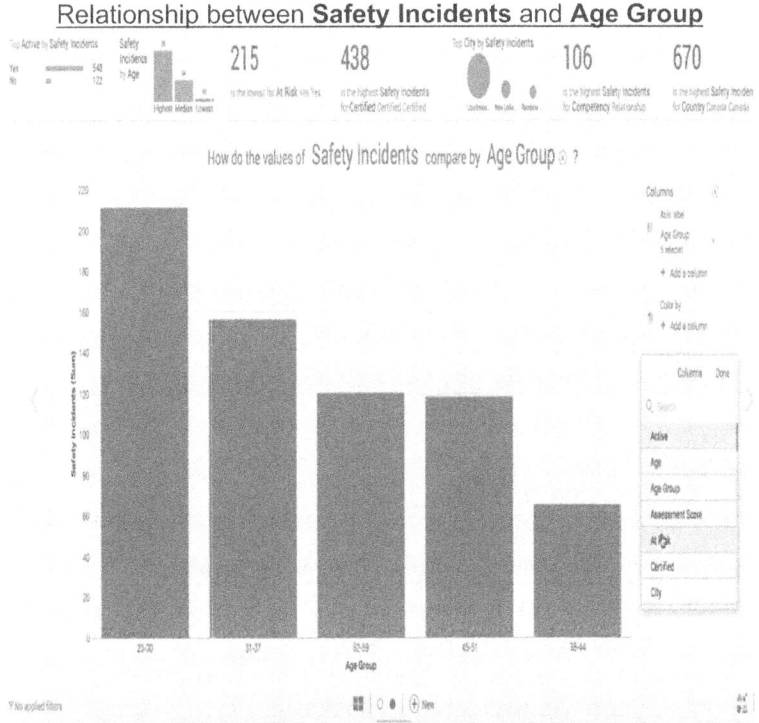

Source: IBM (2017) IBM Kenexa Talent Insights Powered by Watson Analytics
htttps://www.youtube.com/watch?v=AXoiyi2XlEs (1 March 2017)

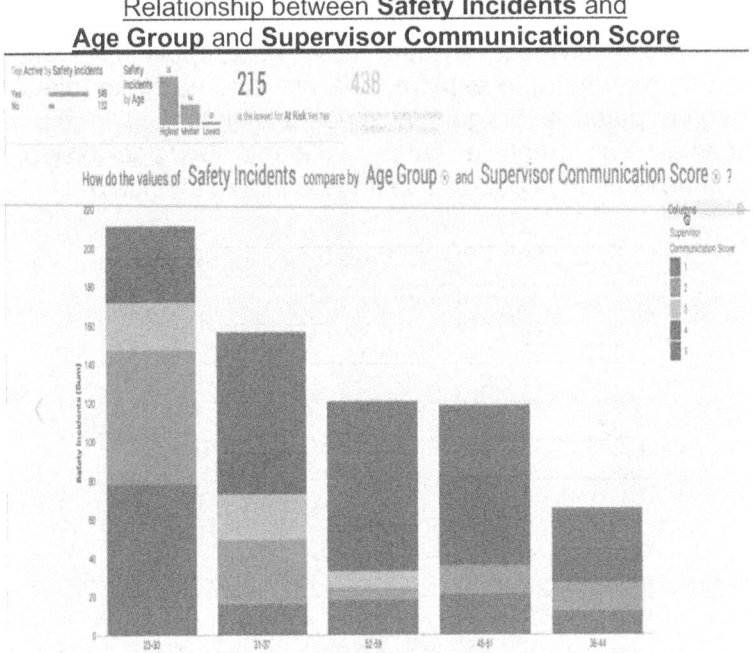

Source: IBM (2017) IBM Kenexa Talent Insights Powered by Watson Analytics
https://www.youtube.com/watch?v=AXoiyi2XlEs (1 March 2017)

Engagement's impact on Health, Safety and Environment:
- Engaged employees take an average of 2.7 sick days per year, whereas disengaged employees take 6.2 sick days per year. [6]
- MolsonCoors found that engaged employees were five times less likely than non-engaged employees to have a safety incident. [7]
- At Shell, a 1% increase in employee engagement resulted in a 4% drop of 'recordable case frequency', a key industry safety standard. [8]

Government's impact on Health, Safety and Environment:
- Companies doing business with the government have better workplace safety records. Governments typically require that companies submitting bids for contracts maintain adequate workplace safety. Companies that do not meet certain workplace safety benchmarks may be barred from competing for such work. The contract requirements cause managers to remain cognizant of workplace safety as they race to meet expectations. [9]

Revenue forecast's impact on Health, Safety and Environment:

- In a study published in the Journal of Accounting and Economics by Judson Caskey and N. Bugra Ozel, injury/illness rates are 5%–15% higher in periods where a firm meets or just beats analyst forecasts. They found that pressure to meet earnings forecasts can relate to workplace safety in at least two ways: high workload and cuts to safety-related expenditures. When managers believe their company may be close to missing earnings benchmarks, they may increase employees' workloads by pressuring them to work faster or for longer hours. In addition, employees may compromise their own safety by overexerting themselves or ignoring safety protocols that slow workflows. Managers may also circumvent or overlook explicit and implicit safety-related measures, such as maintenance spending on equipment and employee training. [10]
- According to the injury data from OSHA, one in every 24 employees is injured in firms that meet or just beat analyst earnings forecasts, compared with about one in 27 workers in firms that miss or comfortably beat forecasts. [10]

Union's impact on Health, Safety and Environment:

- Industries with high unionization report lower injury rates than those in industries with low unionization by about 6.4%. That's because unions typically serve as a proxy for employees' power to ensure safe work environments. They negotiate safety protocols and compliance into their contracts, and workers can report safety issues to their union representatives. [11]

References:
(1) Libby Sander (2017) Research shows if you improve the air quality at work, you improve productivity. (26 November 2018)
(2) Judson Caskey and N. Bugra Ozel (2017) https://hbr.org/2017/05/research-workplace-injuries-are-more-common-when-companies-face-earnings-pressure (22 June 2019)
(3) Alison Coleman (2011) How incentives help reduce sickness absence. https://www.employeebenefits.co.uk/issues/june-2011/how-incentives-help-reduce-sickness-absence/ (20 November 2018)
(4) Memory Nguwi (2018). Reducing Workplace Accident using People Analytics. AIHR. https://www.analyticsinhr.com/blog/reducing-workplace-accident-people-analytics/ (19 November 2018)
(5) IBM (2017) IBM Kenexa Talent Insights Powered by Watson Analytics
https://www.youtube.com/watch?v=AXoiyi2XlEs (1 March 2017)
(6) Harter, J.K., Schmidt, F. L., Kilham, E. A., Asplund, J.W., (2006), Gallup Q12 Meta-Analysis. Cited in MacLeod, 2009, p. 36.
(7) Kevin Kruse (2012) Why Employee Engagement? (These 28 Research Studies Prove the Benefits) https://www.forbes.com/sites/kevinkruse/2012/09/04/why-employee-engagement/#36bbedf63aab (20 November 2018)
(8) Erik van Vulpen (2018). 15 HR Analytics Case Studies with Business Impact. (26 November 2018)
(9) Judson Caskey and N. Bugra Ozel (2017) https://hbr.org/2017/05/research-workplace-injuries-are-more-common-when-companies-face-earnings-pressure
(10) Judson Caskey and N. Bugra Ozel (2017) https://hbr.org/2017/05/research-workplace-injuries-are-more-common-when-companies-face-earnings-pressure (22 June 2019)
(11) Judson Caskey and N. Bugra Ozel (2017) https://hbr.org/2017/05/research-workplace-injuries-are-more-common-when-companies-face-earnings-pressure (22 June 2019)

23.1) Correlation Example: Predict Workplace Accident

Research have shown that Health, Safety and Environment (HSE) is affected by employee's engagement, tenure, and forecasted revenue target achievement:
- At Shell, a 1% increase in employee engagement resulted in a 4% drop of 'recordable case frequency', a key industry safety standard. [1]
- According to the injury data from OSHA, one in every 24 employees is injured in firms that meet or just beat analyst earnings forecasts, compared with about one in 27 workers in firms that miss or comfortably beat forecasts. [2]
- AIHR found that 52% of the road traffic accidents involved drivers/riders with 5 years of service and below. [3]

In this example, we use correlation to find out whether average staff engagement score, average employee tenure, and forecasted revenue target achievement, affects number of accidents.

Year	Average staff engagement score	Average employee tenure	Forecasted revenue target achievement	Number of Accidents
2010	9	15	1.4	1
2011	8	16	1.5	1
2012	7	15	1.2	2
2013	8	14	1.1	2
2014	6	15	1.0	3
2015	7	14	0.9	3
2016	7	13	0.9	3
2017	6	13	0.7	4
2018	7	12	0.7	4
2019	6	9	0.5	5

1) Install "Analysis ToolPak", an Excel add-in

"Analysis ToolPak" is an add-in for Microsoft Excel that comes with Microsoft Excel. To be able to run regression using Excel, you need to first install "Analysis ToolPak", an Excel add-in program that provides data analysis tools. To load the Analysis ToolPak add-in, follow these steps:

- On the File tab, click Options.

- Under Add-ins, click Analysis ToolPak and click the "Go" button.

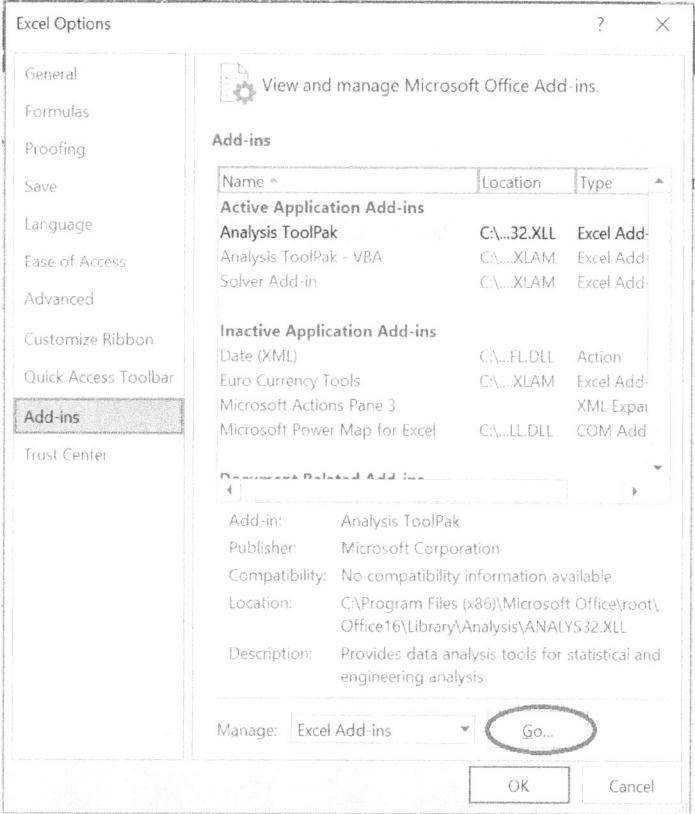

- Click "Analysis ToolPak" and click on OK.

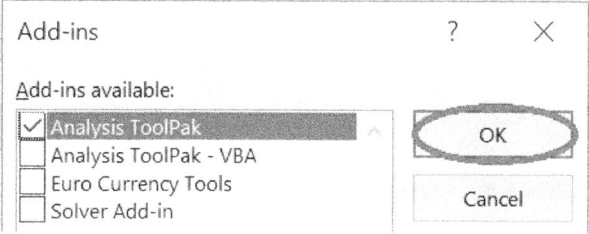

- On the Data tab, in the Analysis group, you are now able to click on "Data Analysis".

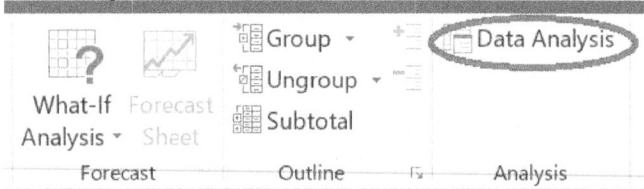

2) Copy the example data in the following table, and paste it in cell A1 of a new Excel worksheet.

	A	B	C	D	E
1	Year	Average staff engagement score	Average employee tenure	Forecasted revenue target achievement	Number of Accidents
2	2010	9	15	1.4	1
3	2011	8	16	1.5	1
4	2012	7	15	1.2	2
5	2013	8	14	1.1	2
6	2014	6	15	1.0	3
7	2015	7	14	0.9	3
8	2016	7	13	0.9	3
9	2017	6	13	0.7	4
10	2018	7	12	0.7	4
11	2019	6	9	0.5	5

3) Select "Correlation" and click "OK".

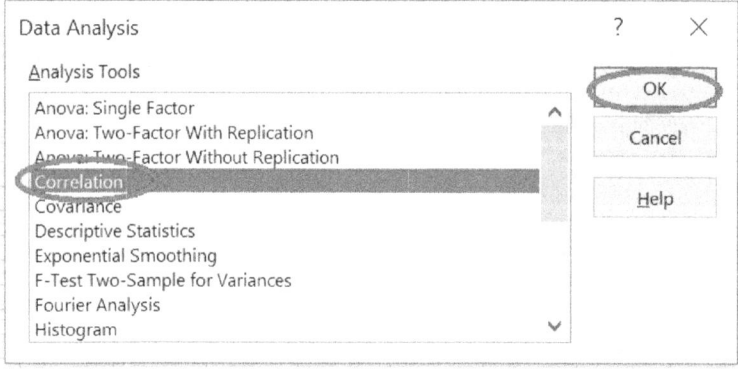

4) After you click OK in the "Data Analysis" dialog box, you will see a "Correlation" dialog box.
5) For "Input Range", select cells (B1:E11).
6) Check "Labels in first row".
7) For "Output Range", select cells (A13).
8) Click "OK".

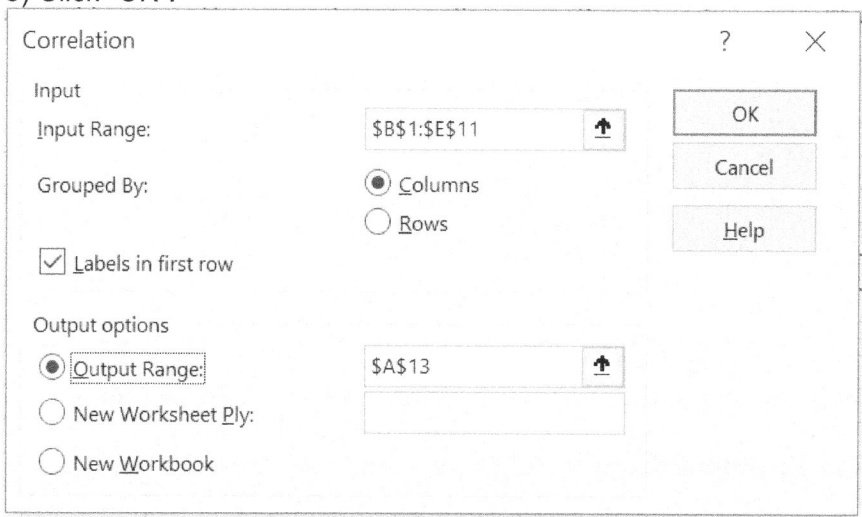

After you click "OK", Excel generates the following Correlation analysis.

	Average staff engagement score	Average employee tenure	Forecasted revenue target achievement	Number of Accidents
Average staff engagement score	1.00			
Average employee tenure	0.52	1.00		
Forecasted revenue target achievement	0.78	0.88	1.00	
Number of Accidents	-0.83	-0.87	-0.99	1.00

A negative correlation coefficient means that an increase in X is associated with a decrease in Y. Similar to a positive correlation, a negative correlation shows a connection between two variables, and the relative strengths are the same. In other words, a correlation coefficient of 0.85 has the same strength as a correlation coefficient of -0.85. Correlation coefficients are always values between -1 and 1, where "-1" means that there is a perfect linear negative correlation, while "1" shows a perfect linear positive correlation. A correlation coefficient of zero, or near to zero, means that there is no meaningful relationship between variables. Correlation coefficient of 0.91 or -0.92 shows a very strong positive and negative correlation respectively. However, correlation does not mean causation. An example of negative correlation is the amount of snowfall and the temperature. As the temperature increases, the amount of snowfall decreases. An example of positive correlation is the relationship between temperature and ice cream sales. As temperature increases, so do ice cream sales.

9) Observations from the above Excel Correlation analysis:

	Average staff engagement score	Average employee tenure	Forecasted revenue target achievement	Number of Accidents
Average staff engagement score	1.00			
Average employee tenure	0.52	1.00		
Forecasted revenue target achievement	0.78	0.88	1.00	
Number of Accidents	-0.83	-0.87	-0.99	1.00

From the Excel Correlation analysis, these variables are good predictors of "Number of Accidents" as they have strong correlation of below -0.75 and above 0.75:

- **Average staff engagement score:** -0.83 Correlation coefficient with Number of Accidents.
- **Average employee tenure:** -0.87 Correlation coefficient with Number of Accidents.
- **Forecasted revenue target achievement:** -0.99 Correlation coefficient with Number of Accidents.

References:
(1) Erik van Vulpen (2018). 15 HR Analytics Case Studies with Business Impact. (26 November 2018)
(2) Judson Caskey and N. Bugra Ozel (2017) https://hbr.org/2017/05/research-workplace-injuries-are-more-common-when-companies-face-earnings-pressure (22 June 2019)
(3) Memory Nguwi (2018). Reducing Workplace Accident using People Analytics. AIHR. https://www.analyticsinhr.com/blog/reducing-workplace-accident-people-analytics/ (19 November 2018)

23.2) Logistic Regression Example: Predict Workplace Accident

Research have shown that Health, Safety and Environment (HSE) is affected by employee's age, tenure, number of sick leave taken, and engagement:
- AIHR found that 6% of the road traffic accidents involved people that are between the ages of 31 and 35 years. [1]
- AIHR found that 52% of the road traffic accidents involved drivers/riders with 5 years of service and below. [1]
- Engaged employees take an average of 2.7 sick days per year, whereas disengaged employees take 6.2 sick days per year. [2]
- MolsonCoors found that engaged employees were five times less likely than non-engaged employees to have a safety incident. [3]

In the data below, we want to predict the non-numeric dependent variable, Employee Accident (i.e. whether the Employee "Had Accident" or "No Accident") using Excel Logistic regression. The data represents events that have already happened. As the variables can only take 1 or 0,
- For the "Accident" column: No Accident = 0, Had Accident = 1.

Accident	Age	Years of service	Sick leave taken in days
0	60	30	2
0	55	20	4
0	50	15	5
1	50	15	5
1	23	2	12
1	20	1	14

1) Our objective is to create an equation with coefficients b_0 to b_3 and then enter values for Age, Years of service, and Sick leave taken in days to predict Employee Accident ("Had Accident" or "No Accident"). We have 4 coefficients (Constant, Age, Years of service, and Sick leave taken in days). The logistic equation is:

$$\text{Logit(Accident)} = b_0 + b_1 \cdot \text{Age} + b_2 \cdot \text{Years of service} + b_3 \cdot \text{Sick leave taken in days}$$

2) Logit is a function that takes a probability of an event as input and returns the logarithm of the odds of that event as output. We need to first assign an arbitrary value (e.g. 0.000) for these coefficients (b0, b1, b2, b3) - Later you will be shown how to use the Excel Solver to replace these starting arbitrary coefficients (e.g. 0.000) with optimized coefficients to create an equation to predict probability.

	A	B	C	D	E	F	G	H
1							b_0 (Intercept)=	0.000
2							b_1=	0.000
3							b_2=	0.000
4							b_3=	0.000
5								
6								
7	Accident	Age	Years of service	Sick leave taken in days				
8	0	60	30	2				
9	0	55	20	4				
10	0	50	15	5				
11	1	50	15	5				
12	1	23	2	12				
13	1	20	1	14				

3) Here, you need to calculate a Logit for each record. Enter the Logit formula below for all the data records, in your Excel spreadsheet:

Logit(Accident)
= b0 + b1*Age + b2*Years of service + b3* Sick leave taken in days

E8 f_x =$H$1+$H$2*B8+$H$3*C8+$H$4*D8

	A	B	C	D	E	F	G	H
1							b_0 (Intercept)=	0.000
2							b_1=	0.000
3							b_2=	0.000
4							b_3=	0.000
5								
6								
7	Accident	Age	Years of service	Sick leave taken in days	Logit (L)			
8	0	60	30	2	0.000			
9	0	55	20	4	0.000			
10	0	50	15	5	0.000			
11	1	50	15	5	0.000			
12	1	23	2	12	0.000			
13	1	20	1	14	0.000			

4) Here, you need to calculate e^L for each record. The number e is the base of the natural logarithm. It is approximately equal to 2.71828163. e^L must be calculated for each record. Enter the Exponential formula below, in your Excel spreadsheet:

F8   fx   =EXP(E8)

	A	B	C	D	E	F	G	H
1							b_0 (Intercept)=	0.000
2							b_1=	0.000
3							b_2=	0.000
4							b_3=	0.000
5								
6								
7	Accident	Age	Years of service	Sick leave taken in days	Logit (L)	e^L		
8	0	60	30	2	0.000	1.000		
9	0	55	20	4	0.000	1.000		
10	0	50	15	5	0.000	1.000		
11	1	50	15	5	0.000	1.000		
12	1	23	2	12	0.000	1.000		
13	1	20	1	14	0.000	1.000		

5) Here, you need to calculate P(X) for each record. P(X) is the probability of event X occurring. Enter the formula for probability of the event (i.e. Had Accident) in your Excel spreadsheet, using the formula below:

$$P(X) = e^L / (1 + e^L)$$

G8   fx   =F8/(1+F8)

	A	B	C	D	E	F	G	H
1							b_0 (Intercept)=	0.000
2							b_1=	0.000
3							b_2=	0.000
4							b_3=	0.000
5								
6								
7	Accident	Age	Years of service	Sick leave taken in days	Logit (L)	e^L	P (X)	
8	0	60	30	2	0.000	1.000	0.500	
9	0	55	20	4	0.000	1.000	0.500	
10	0	50	15	5	0.000	1.000	0.500	
11	1	50	15	5	0.000	1.000	0.500	
12	1	23	2	12	0.000	1.000	0.500	
13	1	20	1	14	0.000	1.000	0.500	

6) Here, you need to calculate LL, the Log-Likelihood Function. The log-likelihood function computes a probability based on the input variables values. Enter the log-likelihood formula below, in your Excel spreadsheet:

H8 f_x =A8*LN(G8)+(1-A8)*(LN(1-G8))

	A	B	C	D	E	F	G	H
1							b_0 (Intercept)=	0.000
2							b_1=	0.000
3							b_2=	0.000
4							b_3=	0.000
5								
6								
7	Accident	Age	Years of service	Sick leave taken in days	Logit (L)	e^L	P (X)	Log-Likelihood (LL)
8	0	60	30	2	0.000	1.000	0.500	-0.693
9	0	55	20	4	0.000	1.000	0.500	-0.693
10	0	50	15	5	0.000	1.000	0.500	-0.693
11	1	50	15	5	0.000	1.000	0.500	-0.693
12	1	23	2	12	0.000	1.000	0.500	-0.693
13	1	20	1	14	0.000	1.000	0.500	-0.693

7) The sum that we wish to maximize is the total of log-likelihood (LL):

H14 f_x =SUM(H8:H13)

	A	B	C	D	E	F	G	H
1							b_0 (Intercept)=	0.000
2							b_1=	0.000
3							b_2=	0.000
4							b_3=	0.000
5								
6								
7	Accident	Age	Years of service	Sick leave taken in days	Logit (L)	e^L	P (X)	Log-Likelihood (LL)
8	0	60	30	2	0.000	1.000	0.500	-0.693
9	0	55	20	4	0.000	1.000	0.500	-0.693
10	0	50	15	5	0.000	1.000	0.500	-0.693
11	1	50	15	5	0.000	1.000	0.500	-0.693
12	1	23	2	12	0.000	1.000	0.500	-0.693
13	1	20	1	14	0.000	1.000	0.500	-0.693
14							sum of LL:	-4.159

The objective of Logistic Regression is find the coefficients of the Logit (b_0, b_1, b_2 +...+ b_k) that maximize LL, the Log-Likelihood Function in cell H14, to produce the Maximum Log-Likelihood (MLL) Function. The only values we can change are the guesses for the coefficient b_0 through b_3, which we have assigned an arbitrary value of 0.000. We don't have to optimize them ourselves, as we can use Solver, an Excel add-in, that adjusts the coefficient to maximize or minimize the value in the cell.

8) The Excel Solver is an add-in that is included in Excel. But it must be manually activated by you before it can be utilized for the first time. To use Solver, on the File tab, click Options.

9) Under Add-ins, click Excel Add-ins and click the "Go" button.

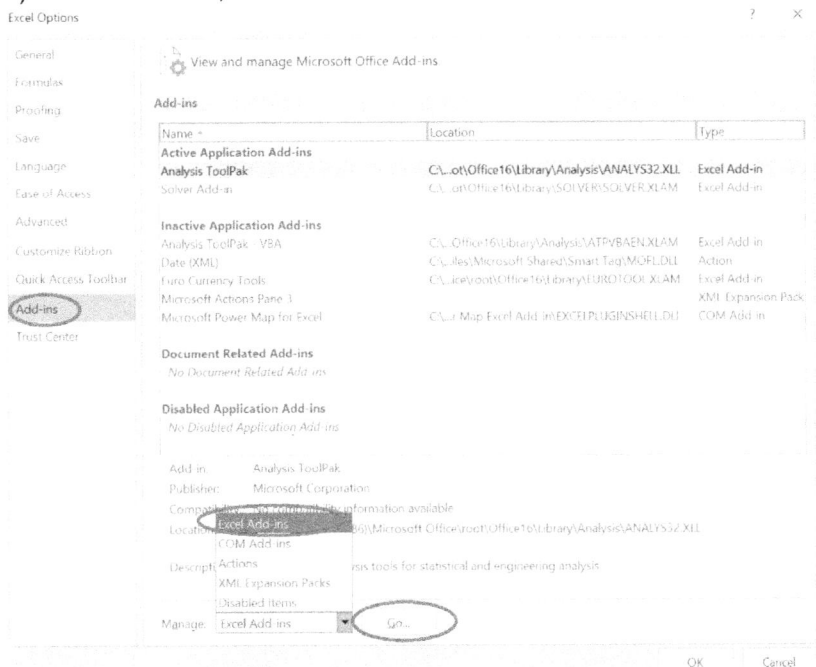

10) Click "Solver Add-In" and click on OK.

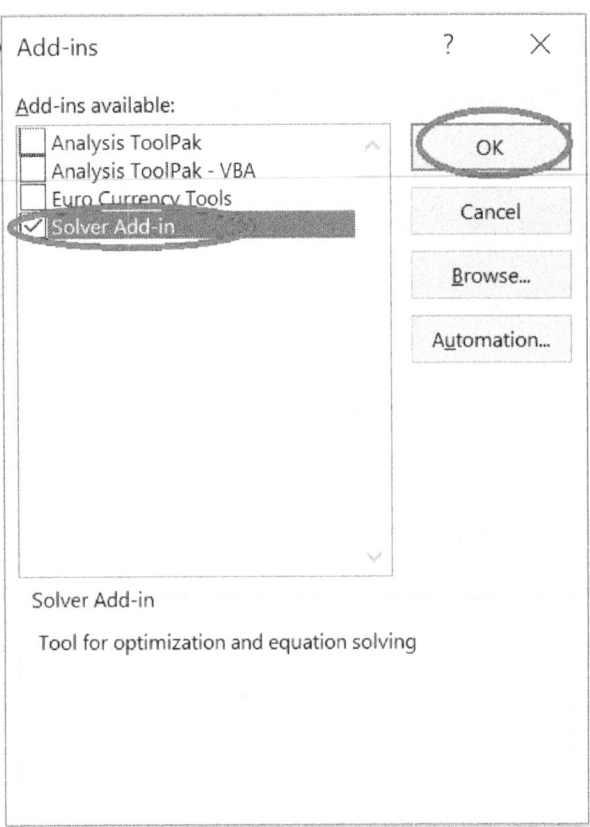

11) On the Data tab, in the Analysis group, you are now able to click on "Solver". Click the Solver button.

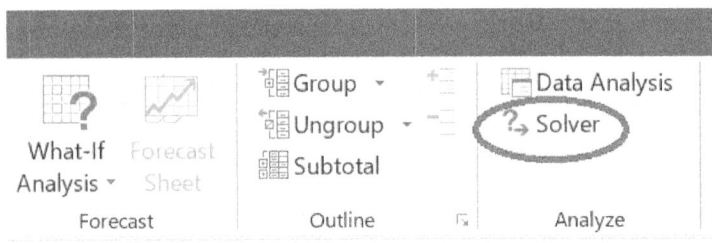

12) The objective is to maximize the sum of the log-likelihood column (LL), by changing the values in H1:H4, representing coefficients b_0-b_3.
- For "Set Objective", select cell (H14), the sum of the log-likelihood column (LL).
- Uncheck the box labeled "Make Unconstrained Variables Non-Negative".
- For "Select a Solving Method", select "GRG Nonlinear" because we are not performing a linear optimization.
- Click "Solve"

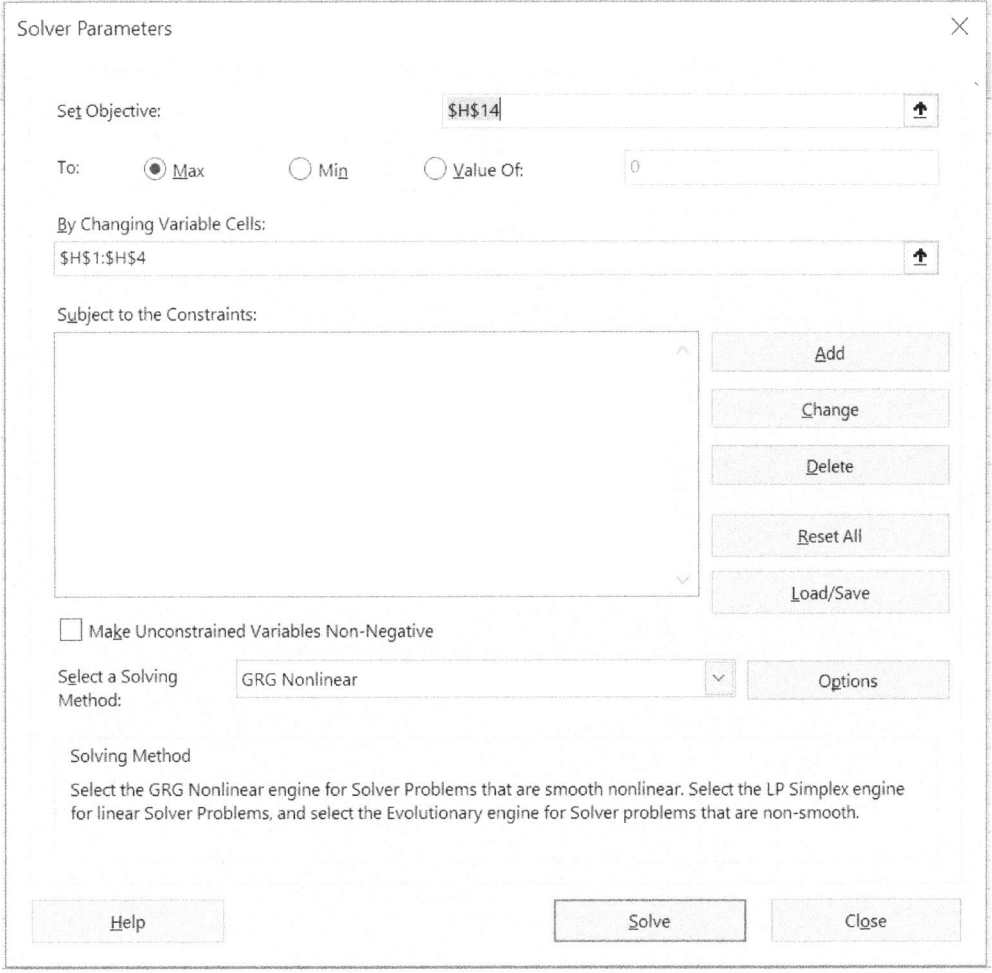

13) After clicking "Solve", you will see the screen "Solver Results". Check "Keep Solver Solution".

14) After clicking "Solve" you get new values in H1:H4, representing coefficients b_0-b_3.

	A	B	C	D	E	F	G	H
1							b_0 (Intercept)=	0.126
2							b_1=	0.719
3							b_2=	-2.681
4							b_3=	0.831
5								
6								
7	Accident	Age	Years of service	Sick leave taken in days	Logit (L)	e^L	P (X)	Log-Likelihood (LL)
8	0	60	30	2	-35.524	0.000	0.000	0.000
9	0	55	20	4	-10.646	0.000	0.000	0.000
10	0	50	15	5	-0.003	0.997	0.499	-0.692
11	1	50	15	5	-0.003	0.997	0.499	-0.695
12	1	23	2	12	21.265	1718373617.5	1.000	0.000
13	1	20	1	14	23.452	15315131982.4	1.000	0.000
14							sum of LL:	-1.386

15) To predict whether an employee whose Age is 30, Years of Service is 10, and Sick leave taken in days is 6 may have Accident, copy those figures that are circled in your spreadsheet.

	A	B	C	D	E	F	G	H
							b_0 (Intercept)=	0.126
1								
2							$b_1=$	0.719
3							$b_2=$	-2.681
4							$b_3=$	0.831
5								
6								
7	Accident	Age	Years of service	Sick leave taken in days	Logit (L)	e^L	P (X)	Log-Likelihood
8	0	60	30	2	-35.524	0.000	0.000	0.000
9	0	55	20	4	-10.646	0.000	0.000	0.000
10	0	50	15	5	-0.003	0.997	0.499	-0.692
11	1	50	15	5	-0.003	0.997	0.499	-0.695
12	1	23	2	12	21.265	1718373617.5	1.000	0.000
13	1	20	1	14	23.452	15315131982.4	1.000	0.000
14							sum of LL:	-1.386
15								
16								
17		Age	Years of service	Sick leave taken in days	Logit (L)	e^L	P (X)	
18		30	10	6	-0.139			

Cell E18 formula: =$H$1+$H$2*B18+$H$3*C18+$H$4*D18

You will notice that the formula is cell E18 is:

Logit(L)
= $H$1+$H$2*B18+$H$3*C18+$H$4*D18
= b0 + b1*Age + b2*Years of service + b3* Sick leave taken in days
= 0.126 + (0.719*Age) – (2.681*Years of service) + (0.831*Sick leave taken in days)

Just like a multiple linear regression, you need to enter the new b_0, b_1, b_2 and b_3 coefficient values into your logistic regression equation to predict a value. But unlike a linear regression that predicts values like sales amount, the logistic regression equation predicts probabilities.

16) Next, you need to calculate e^L. The number e is the base of the natural logarithm. It is approximately equal to 2.71828163. Enter the Exponential formula below, in your Excel spreadsheet:

F18				f_x	=EXP(E18)			
	A	B	C	D	E	F	G	H
1							b_0 (Intercept)=	0.126
2							b_1=	0.719
3							b_2=	-2.681
4							b_3=	0.831
5								
6								
7	Accident	Age	Years of service	Sick leave taken in days	Logit (L)	e^L	P (X)	Log-Likelihood
8	0	60	30	2	-35.524	0.000	0.000	0.000
9	0	55	20	4	-10.646	0.000	0.000	0.000
10	0	50	15	5	-0.003	0.997	0.499	-0.692
11	1	50	15	5	-0.003	0.997	0.499	-0.695
12	1	23	2	12	21.265	1718373617.5	1.000	0.000
13	1	20	1	14	23.452	15315131982.4	1.000	0.000
14							sum of LL:	-1.386
15								
16								
17		Age	Years of service	Sick leave taken in days	Logit (L)	e^L	P (X)	
18		30	10	6	-0.139	0.870		

17) Next, you need to calculate P(X). P(X) is the probability of event X occurring. Enter the formula for probability of the event (i.e. Accident) in your Excel spreadsheet, using the formula below:

$$P(X) = e^L / (1 + e^L)$$

G18 f_x =F18/(1+F18)

	A	B	C	D	E	F	G	H
1							b_0 (Intercept)=	0.126
2							b_1=	0.719
3							b_2=	-2.681
4							b_3=	0.831
5								
6								
7	Accident	Age	Years of service	Sick leave taken in days	Logit (L)	e^L	P (X)	Log-Likelihood
8	0	60	30	2	-35.524	0.000	0.000	0.000
9	0	55	20	4	-10.646	0.000	0.000	0.000
10	0	50	15	5	-0.003	0.997	0.499	-0.692
11	1	50	15	5	-0.003	0.997	0.499	-0.695
12	1	23	2	12	21.265	1718373617.5	1.000	0.000
13	1	20	1	14	23.452	15315131982.4	1.000	0.000
14							sum of LL:	-1.386
15								
16								
17		Age	Years of	Sick leave	Logit (L)	e^L	P (X)	
18		30	10	6	-0.139	0.870	0.465	

From the spreadsheet, if an employee whose Age is 30, Years of Service is 10, and Sick leave taken in days is 6, the probably that he may have accident resign is 0.4655, which is closer to 0 than to 1. Closer to 1 means it is probably "Had Accident", while closer to 0 means it is probably "No Accident".

References:
(1) Memory Nguwi (2018). Reducing Workplace Accident using People Analytics. AIHR. https://www.analyticsinhr.com/blog/reducing-workplace-accident-people-analytics/ (19 November 2018)
(2) Harter, J.K., Schmidt, F. L., Kilham, E. A., Asplund, J.W., (2006), Gallup Q12 Meta-Analysis. Cited in MacLeod, 2009, p. 36.
(3) Kevin Kruse (2012) Why Employee Engagement? (These 28 Research Studies Prove the Benefits) https://www.forbes.com/sites/kevinkruse/2012/09/04/why-employee-engagement/#36bbedf63aab (20 November 2018)

24) Data Visualization with Excel

Data visualization is the presentation of qualitative and quantitative data in a graphical format. Excel charts are commonly used for data visualization, but selecting the right excel chart may be a challenge.

Data visualization is a crucial part of statistics and analytics. Graphs allows us to identify trends and relationships that we might overlook when we only look at numbers. Graphics enables the data scientist to explain data patterns to non-technical people.

Chart elements

In order to read and use Excel charts, you need to first understand the various components of the chart.

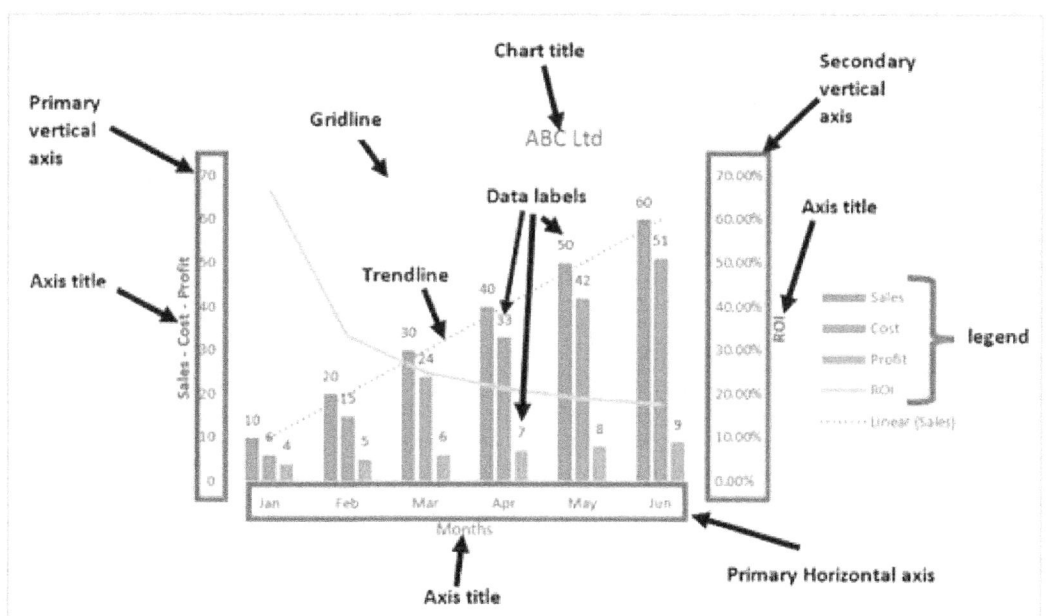

Source: Optimize Smart (2019) Best Excel Charts Types for Data Analysis, Presentation and Reporting   https://www.optimizesmart.com/how-to-select-best-excel-charts-for-your-data-analysis-reporting/#a13 *(17 May 2019)*

24.1) Clustered Column Chart

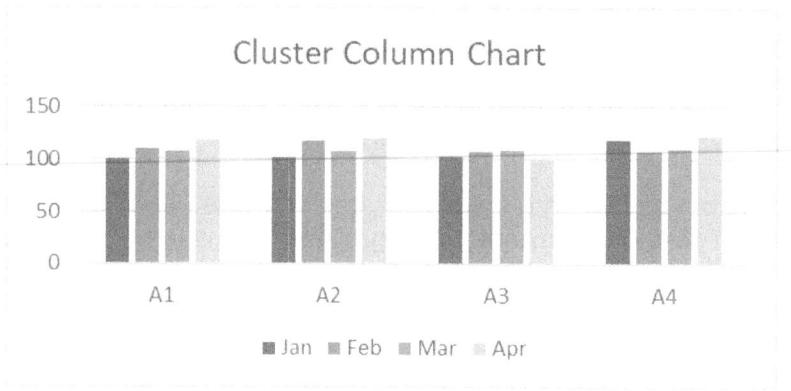

When to use a clustered column chart:
- When you want to compare 2 to 4 data series.
- When the data series you want to compare have the same measurement unit, and are of similar sizes, so that the values of one data series doesn't dwarf the values of the other data series.
- When you want to show the maximum and minimum values of the data series that you want to compare.
- When you want to study the short-term trend.

Use this,

Use a clustered column chart when you want to compare 2 to 4 data series.

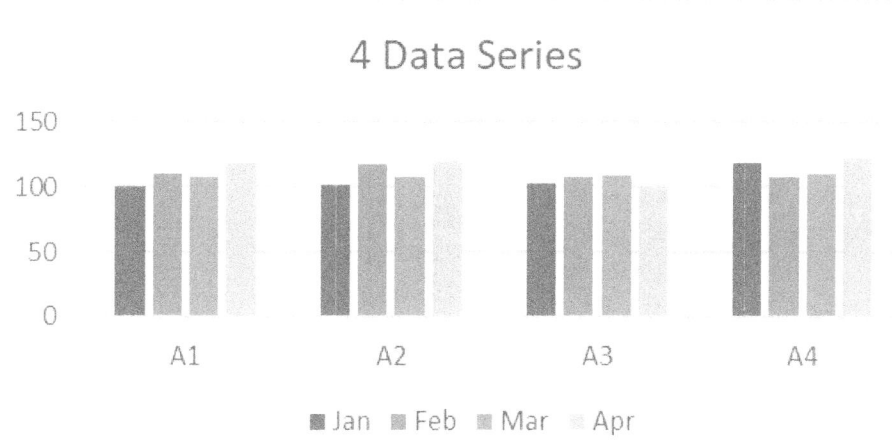

Avoid this,

A chart that has 10 data series (more than 4 data series) is difficult to read and understand. Avoid presenting too much data in one chart. Break it into smaller cluster column charts.

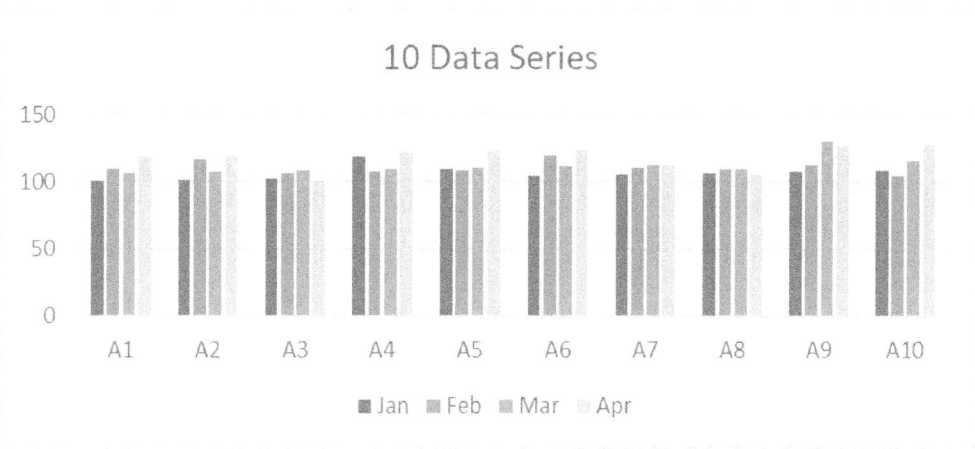

Use a clustered column chart when the data series you want to compare have the same measurement unit, and are of similar sizes, so that the values of one data series doesn't dwarf the values of the other data series.

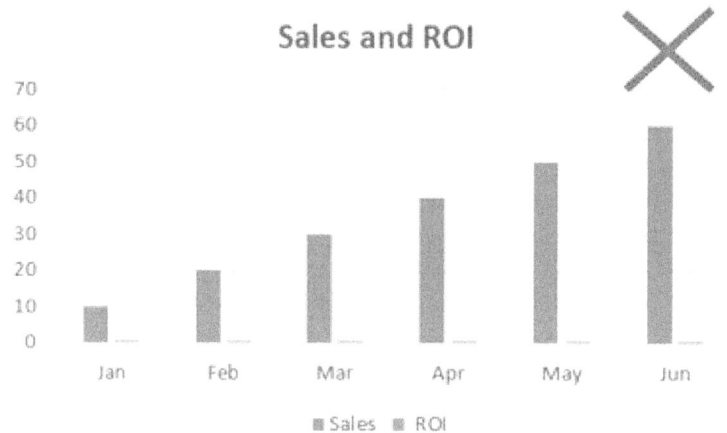

Source: Optimize Smart (2019) Best Excel Charts Types for Data Analysis, Presentation and Reporting   https://www.optimizesmart.com/how-to-select-best-excel-charts-for-your-data-analysis-reporting/#a13 *(17 May 2019)*

Steps to create a clustered column plot in Excel

1) Table below shows the median tenure of workers by age brackets. To better visualise this data, create a clustered column plot. Select two columns with numeric data, including the column headers. In our case, it is the range A1:B7.

	A	B
1	Age	Tenure with company
2	20 to 24 years	1.2
3	25 to 34 years	2.8
4	35 to 44 years	4.9
5	45 to 54 years	7.6
6	55 to 64 years	10.1
7	65 years and over	10.2

2) Go to the Inset tab > Charts group, click the "Clustered Column" chart icon.

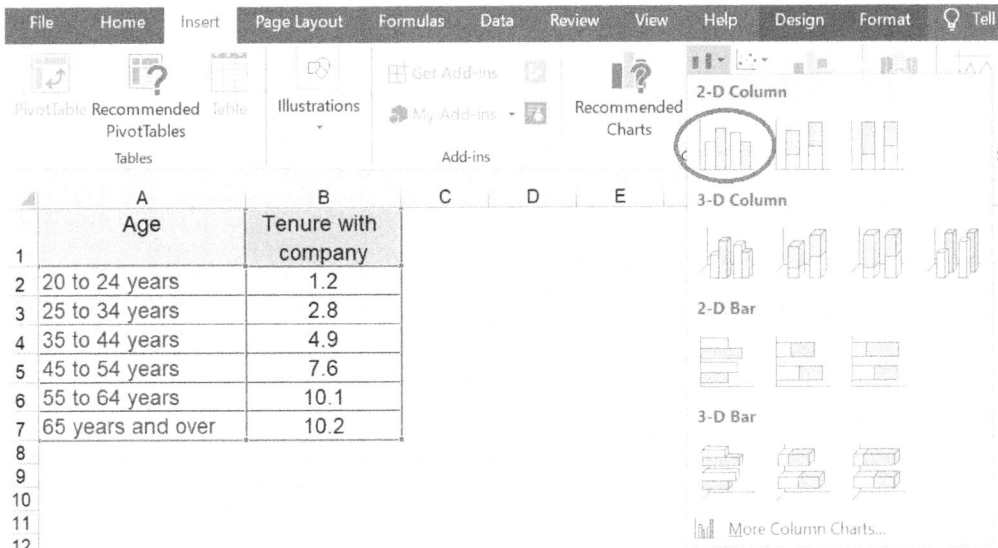

3) The cluster column chart will be immediately inserted in your worksheet:

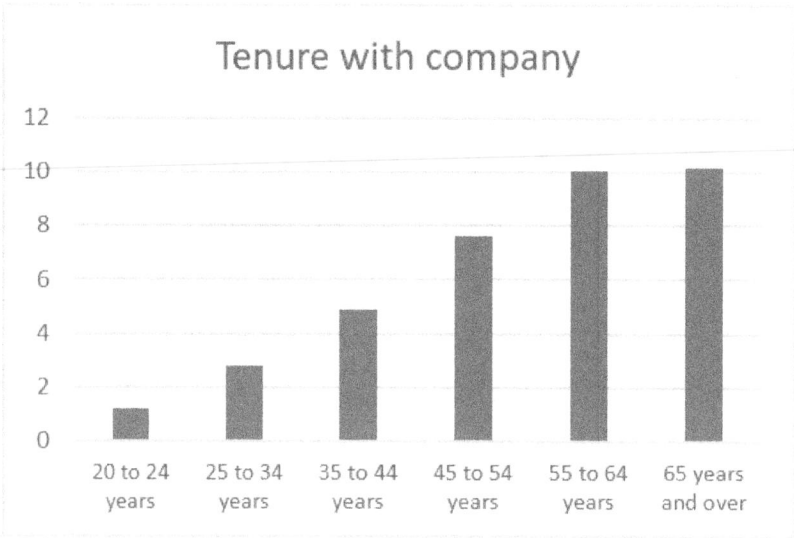

24.2) Combination Charts

A combination chart is a combination of two or more chart types. Use a combination chart:
- when you need to compare two or more data series that have different measurement units.
- when you need to compare two or more data series that have the same measurement units, but vastly different sizes.

Different Measurements

Source: Optimize Smart (2019) Best Excel Charts Types for Data Analysis, Presentation and Reporting https://www.optimizesmart.com/how-to-select-best-excel-charts-for-your-data-analysis-reporting/#a13 (17 May 2019)

Different Sizes

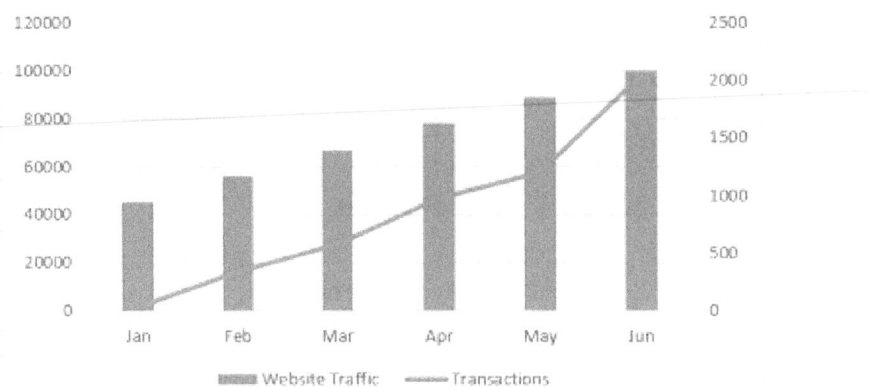

Source: Optimize Smart (2019) Best Excel Charts Types for Data Analysis, Presentation and Reporting   https://www.optimizesmart.com/how-to-select-best-excel-charts-for-your-data-analysis-reporting/#a13 (17 May 2019)

Steps to create a combination chart in Excel

1) Table below shows the month, days rained, and sales. To better visualise this data, create a combination chart. Select the range A1:C13.

	A	B	C
1	Month	Days rained	Sales
2	Jan	11	3500
3	Feb	10	4700
4	Mar	10	5300
5	Apr	9	6700
6	May	8	8800
7	Jun	6	12000
8	Jul	4	12500
9	Aug	5	11000
10	Sep	7	9700
11	Oct	8	9800
12	Nov	10	6600
13	Dec	11	5000

2) Go to the Inset tab > Charts group, click the "Combo" chart icon, and then click "Create Custom Combo Chart".

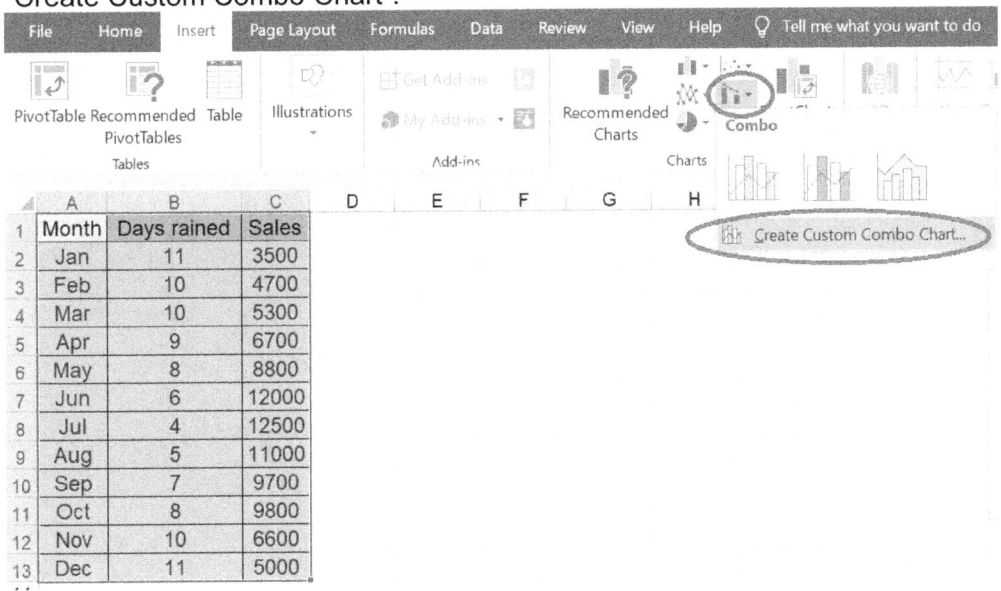

3) The "Insert Chart" dialog box appears.
- For the "Days rained" series, choose "Clustered Column" chart type.
- For the "Sales" series, choose "Line" chart type.
- Check the "secondary axis" for "Sales" Series.
- Click "OK"

4) The combination chart will be immediately inserted in your worksheet:

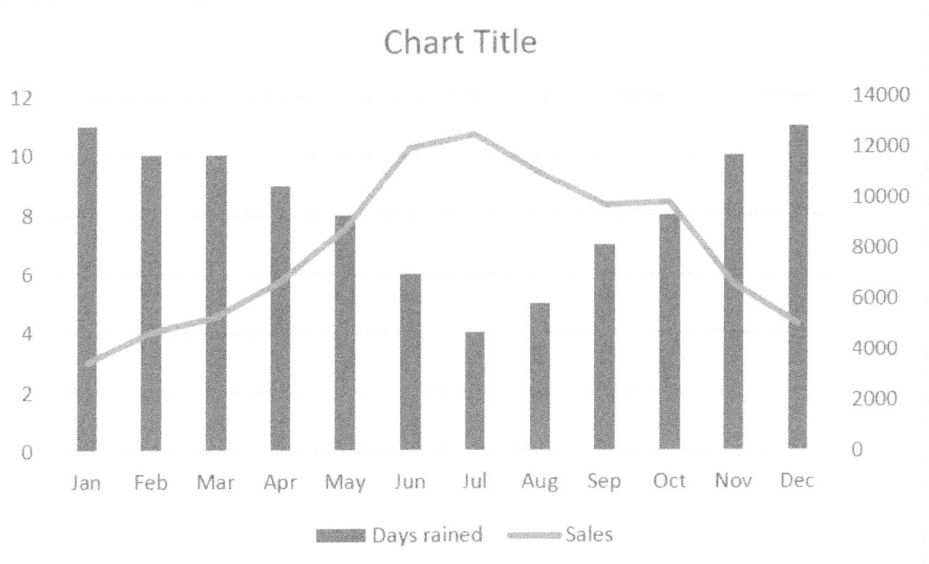

24.3) Pie Charts

- Use a pie chart to show composition of data when you have a data series with less than five categories.
- Use a pie chart to show data composition only when the pie slices are of comparable sizes.
- Order your pie slices in such a way that the biggest slice comes first, followed by the second biggest pie slice, etc. To create such pie chart, sort the data from largest to smallest.

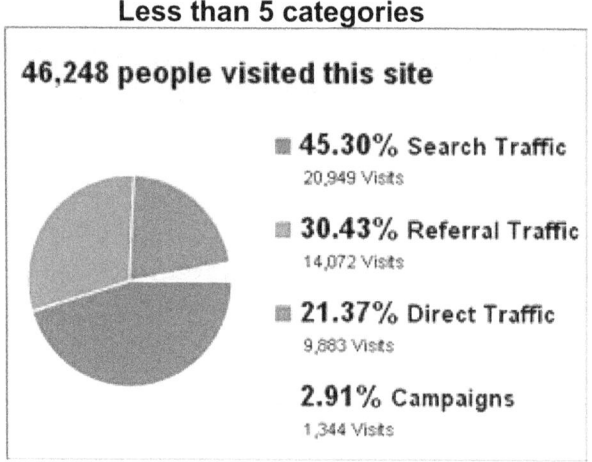

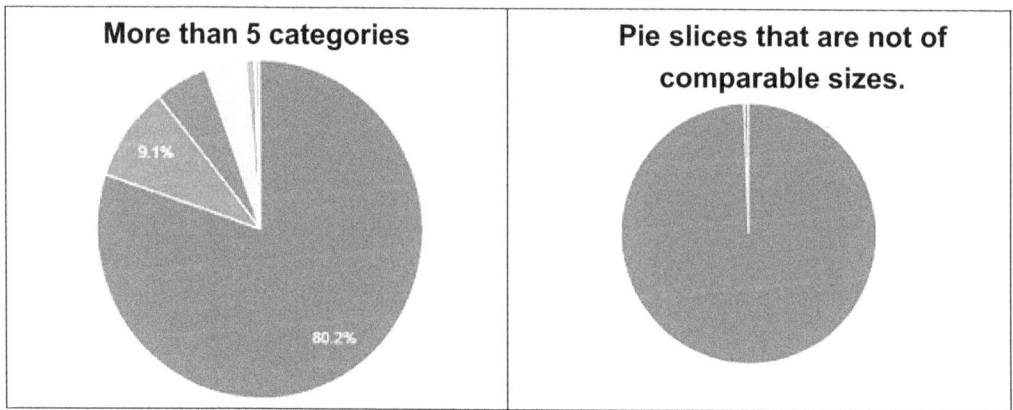

Source: Optimize Smart (2019) Best Excel Charts Types for Data Analysis, Presentation and Reporting   https://www.optimizesmart.com/how-to-select-best-excel-charts-for-your-data-analysis-reporting/#a13 *(17 May 2019)*

Steps to create a pie chart in Excel

1) Table below shows the "percentage of employees who resigned" by "Distance between home and work". To better visualise this data, create a pie chart. Select the range A1:B5.

	A	B
1	Distance between home and work	Percentage of employees who resigned
2	less than 15 km	5%
3	15 to 30 km	15%
4	31 to 45 km	20%
5	above 45 km	60%

2) Go to the Inset tab > Charts group, click the "pie" chart icon.

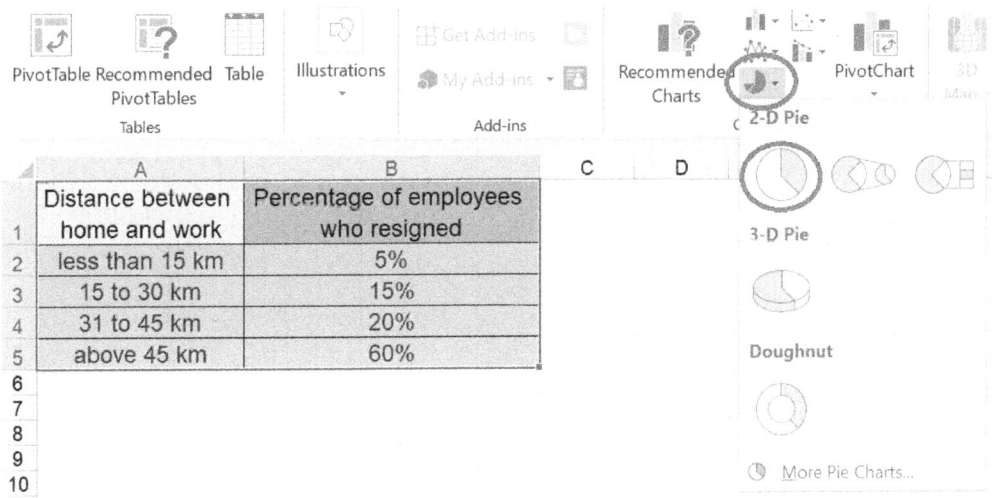

3) The pie chart will be immediately inserted in your worksheet:

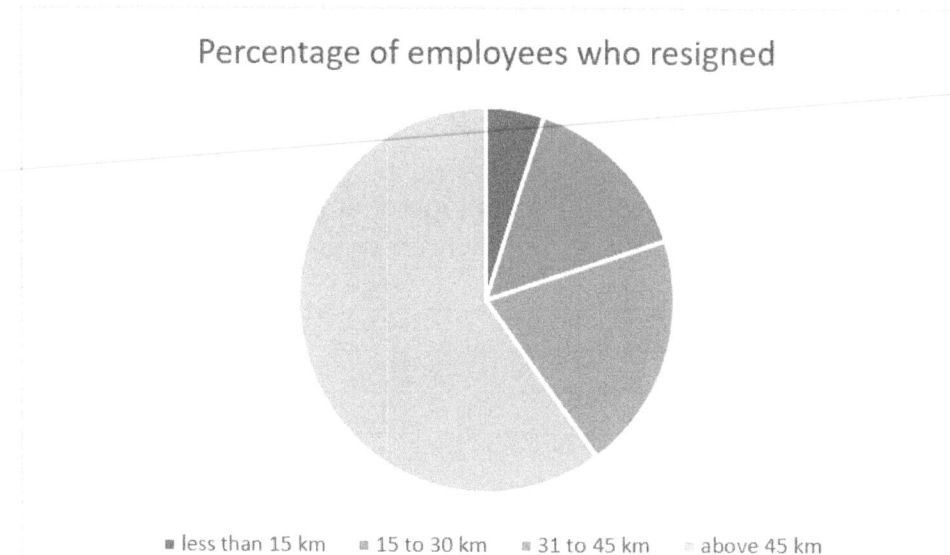

24.4) Scatter Plot

A scatter plot (also called XY graph) is a two-dimensional chart that shows the relationship between two variables. Usually, the independent variable is on the x-axis, and the dependent variable on the y-axis. The values are displayed at data points at the x and y axis intersection. The scatter plot shows how strength of correlation between two variables. The closer data points are along a straight line, the stronger the correlation.

From the chart below, we conclude that the relationship between the two variables are linear. i.e. as the value of x variable increases, there is a corresponding increase in the value of the y variable.

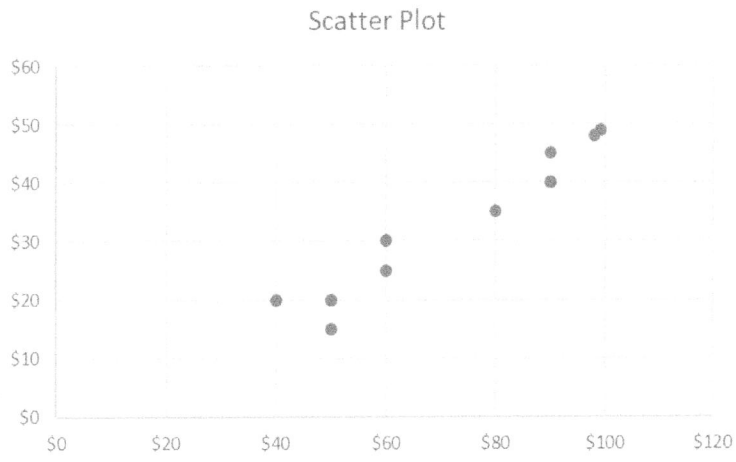

When to use a scatter plot:
- When you want to analyze and report the relationship/correlation between two variables.
- When there are at least 10 data points. The more the data points the better.

In this section, you will learn how to do a scatter plot in Excel to create a graphical representation of two correlated data sets. If you have to analyse the data below to identify patterns or trends, it would be a tedious task. However, a scatter plot can help you to visualize how columns of data are related to each other.

Month	Advertisement cost	Burgers sold
Jan	$50	$15
Feb	$60	$25
Mar	$50	$20
Apr	$80	$35
May	$90	$40
Jun	$99	$49
Jul	$98	$48
Aug	$90	$45
Sep	$90	$40
Oct	$50	$20
Nov	$40	$20
Dec	$60	$30

Steps to create a scatter plot in Excel

1) First, arrange your data. The independent variable should be in the left column as it will be plotted on the x axis. The dependent variable should be in the right column, as it will be plotted on the y axis. In our example, we are going to visualize the relationship between the Advertisement cost for a certain month (independent variable) and the Burgers sold (dependent variable):

	A	B	C
1	Month	Advertisement cost	Burgers sold
2	Jan	$50	$15
3	Feb	$60	$25
4	Mar	$50	$20
5	Apr	$80	$35
6	May	$90	$40
7	Jun	$99	$49
8	Jul	$98	$48
9	Aug	$90	$45
10	Sep	$90	$40
11	Oct	$50	$20
12	Nov	$40	$20
13	Dec	$60	$30

2) Select two columns with numeric data, including the column headers. In our case, it is the range B1:C13.

3) Go to the Inset tab > Charts group, click the "scatter" chart icon.

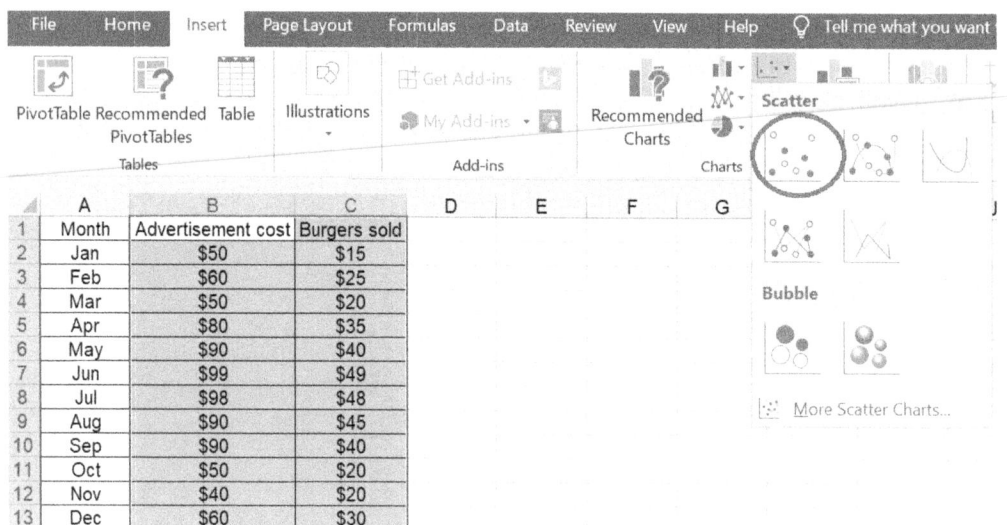

4) The scatter diagram will be immediately inserted in your worksheet:

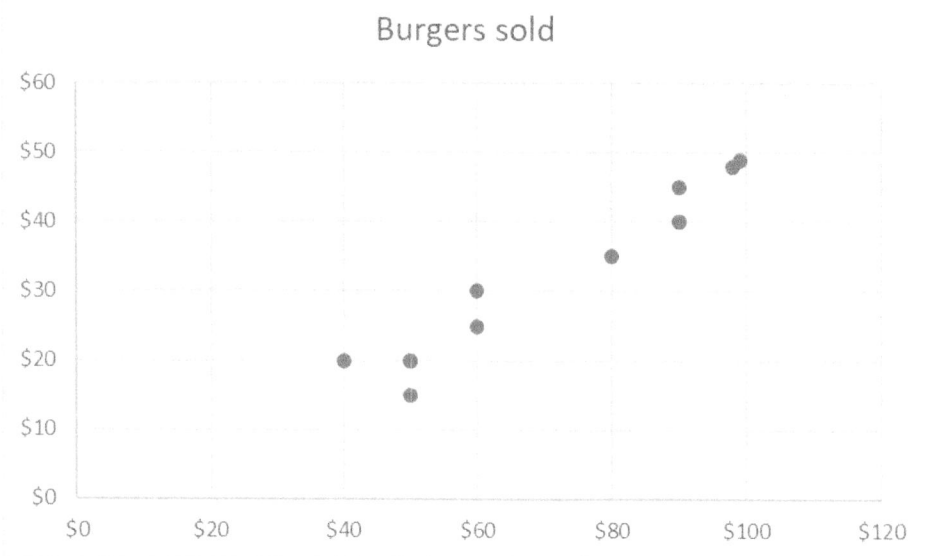

5) To better visualize the relationship between the two variables, you can draw a trendline in your scatter graph (also called a line of best fit). Right click on any data point and choose "Add Trendline" from the menu.

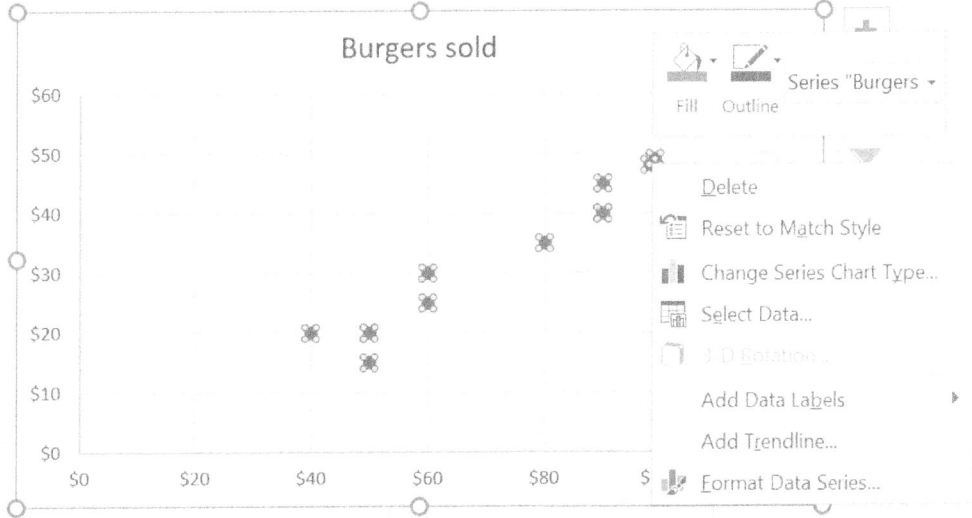

6) Check the linear trendline option. Excel will draw a line as close as possible to all data points so that there are as many points above the line as below.

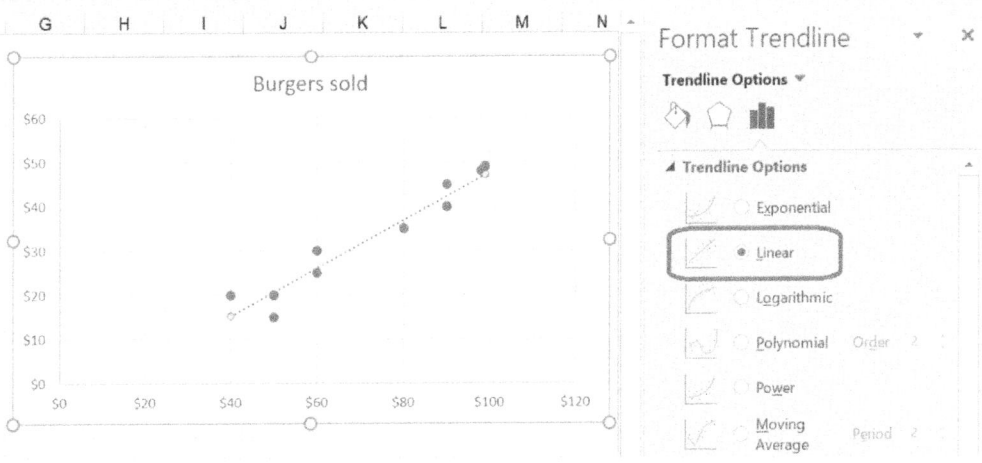

7) Check the "Display Equation on chart" the box - to show the equation for the trendline that mathematically describes the relationship between the two variables. Check the "Display R-squared value on chart" – to show the fit (near to 1 means it is a very good fit).

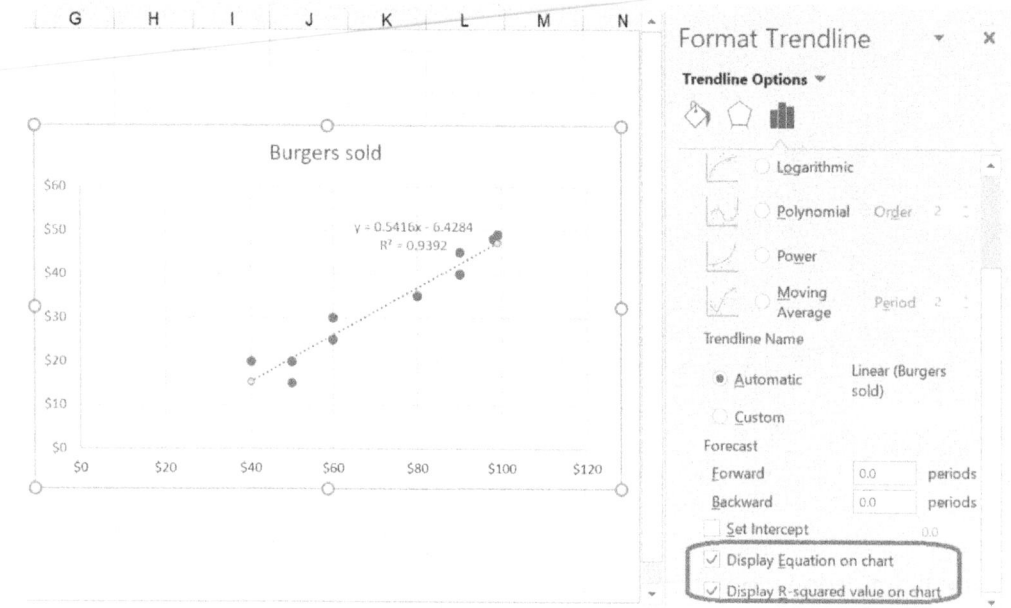

Annex A) Other Publications by the Author

https://www.amazon.com/s?k=ng+mong+shen
https://www.facebook.com/thehrdiary/

Index

A

Action Priority Matrix, 89
Active ONA, 260
Age, 348, 357
Analytics Maturity Model, 23
Apache Spark, 20
Azure Machine Learning, 196

B

Behavioral signs, 348
Best Buy, 241, 395
best performing engineers, 391, 400
best performing leaders, 391
Betweenness centrality, 257
Big Five model, 399

C

call center worker, 353, 399
Central node, 258
clear expectations, 353
Closeness centrality, 257
Clustered Column Chart, 478
cNPS, 396
colleague left, 348
Colors combinations, 62
Combination Charts, 483
commute, 348
Commute time, 348
Compa-Ratio Analytics, 427
Compensation & Benefits Analytics, 422
Compensation Metrics, 31
Corporate Culture, 349
Correlation, 108, 207, 233, 368
Correlations, 292
Court cases, 94
Court Rulings on Analytics Cases, 93
Co-worker satisfaction, 358
creative people, 353, 399

D

Data Gathering, 46
Data storytelling, 51
Data visualization, 477
Data Visualization, 477
Decision & Probability Trees, 100
decision tree, 100, 366
Degree centrality, 257
Demographics, 348
descriptive analytics, 25
diagnostic analysis, 25
diversity and inclusion, 352
Diversity and Inclusion, 323
divorced, 354

E

Eigenvector centrality, 257
Employee Net Promoter Score, 228
Engagement, 221
eNPS, 396
Excel, 21

F

Financial Services, 404
financial services company, 404
Flexible Work, 230
Flight Risk Formula, 428
Flight Risk scores, 349
Flight Risk Scoring, 363

G

Gender, 348
Glassdoor, 194, 207
Google, 349

H

having kids, 354

Health, Safety & Environment, 453
Heroes, 55
Hewlett-Packard, 349, 424
High performers, 405

I

Inquisitive people, 353, 399
ISS, 223, 397

K

Kenexa, 348
Kirkpatrick Model, 431
Kirkpatrick's Four Levels of Evaluation, 431
Knowledge broker, 258

L

Labour Relations Metrics, 36
Learning & Development Metrics, 37
Logistic regression, 145, 300, 315, 376, 408, 416, 445, 465
Logistic Regression, 145, 161, 376
Logistics Regression, 299

M

Machine learning, 16
Machine Learning, 16
major lifestyle change, 354
manager leaves, 354
Marital status, 348
Market-Ratio Analytics, 426
married, 354
Median employee tenure, 351
merit increase, 391
Minitab, 21
misspellings, 405
multiple linear regression, 444

N

net income growth, 391
Net Promoter Score, 228
NodeXL, 278

O

Operating margin, 395
Organizational Network Analysis, 258

P

Passive ONA, 260
Pay satisfaction, 358
pay spread, 391
Performance rating, 348
Peripheral, 258
Personality trait, 348
Personality traits, 398
Pie Charts, 488
Power Interest Matrix, 86
Predict Employee Attrition & Absenteeism, 348
Predict Performance, 391
Predictive analytics, 26
Predictive HR Analytics Framework, 39
Prescriptive analytics, 26
Prescriptive Analytics, 27
Pro Word Cloud, 180
Probability Tree, 366
productivity, 353
Productivity Metrics, 30
promotion, 349
Python, 21

Q

Quality, 399

R

R, 21
Recruitment Metrics, 32
Regression, 118, 214, 241, 315, 329, 338, 339
Reinforcement, 19
Reinforcement learning, 19
relationship, 406
relationships, 391
Retention Metrics, 36
ROI by Jack Philips, 433
ROI of training, 433

S

Salary market-ratio, 348
SAS, 21
Scatter Plot, 491
Sentiment Analysis, 189, 196
Sentiment Word Cloud, 192
Simpson's Diversity Index, 327
Social Capital, 261
Social media, 180
Social Media Analytics, 249
Social Network Analysis, 254
Sociogram, 254
SPSS, 22
SQL, 22
Stakeholder Analysis, 81
Stakeholder engagement strategy, 88
Standard Chartered Bank, 395
Stata, 22
success in sales, 400, 415
Supervised learning, 17
Supervised Learning, 17

T

Tableau, 22
tenure, 350
Tenure, 348, 357
Text Analytics, 166
text cloud, 180
Text mining, 180
Text Mining, 166
Ties, 258
Training, 401, 429, 444
Training & Development Analytics, 429
Triggering events, 348, 354
Turnover Predictors, 356

U

unsupervised learning, 18
Unsupervised Learning, 18

V

VBA, 168
Victims, 55
Villains, 55

W

Word cloud, 180
Word Cloud, 180
Workforce Demographics Metrics, 38
Workforce Efficiency Metrics, 37

X

Xerox, 353, 399

Made in the USA
Las Vegas, NV
29 November 2021